PROBABILITIES, HYPOTHETICALS, AND COUNTERFACTUALS IN ANCIENT GREEK THOUGHT

This volume explores the conceptual terrain defined by the Greek word *eikos*: the probable, likely, or reasonable. A term of art in Greek rhetoric, a defining feature of literary fiction, a seminal mode of historical, scientific, and philosophical inquiry, *eikos* was a way of thinking about the probable and improbable, the factual and counterfactual, the hypothetical and the real. These thirteen original and provocative studies examine the plausible arguments of courtroom speakers and the "likely stories" of philosophers, verisimilitude in art and literature, the likelihood of resemblance in human reproduction, the limits of human knowledge, and the possibilities of ethical and political agency. The first synthetic study of probabilistic thinking in ancient Greece, the volume illuminates a fascinating chapter in the history of Western thought.

VICTORIA WOHL is Professor of Classics at the University of Toronto. She works on the literature and culture of classical Athens. Her previously published work includes *Law's Cosmos: Juridical Discourse in Athenian Forensic Oratory* (2010), *Love among the Ruins: The Erotics of Democracy in Classical Athens* (2002), and *Intimate Commerce: Exchange, Gender, and Subjectivity in Greek Tragedy* (1998).

PROBABILITIES, HYPOTHETICALS, AND COUNTERFACTUALS IN ANCIENT GREEK THOUGHT

EDITED BY
VICTORIA WOHL

CAMBRIDGE
UNIVERSITY PRESS

CAMBRIDGE
UNIVERSITY PRESS

University Printing House, Cambridge CB2 8BS, United Kingdom

Cambridge University Press is part of the University of Cambridge.

It furthers the University's mission by disseminating knowledge in the pursuit of education, learning and research at the highest international levels of excellence.

www.cambridge.org
Information on this title: www.cambridge.org/9781107050495

First published 2014

Printed in the United Kingdom by Clays, St Ives plc

A catalogue record for this publication is available from the British Library

Library of Congress Cataloguing in Publication data
Probabilities, hypotheticals, and counterfactuals in ancient Greek thought / edited by Victoria Wohl.
pages cm
Includes bibliographical references and index.
ISBN 978-1-107-05049-5 (hardback)
1. Philosophy, Ancient. 2. Reasoning. 3. Probabilities. 4. Logic, Ancient.
5. Greece – Civilization – To 146 B.C. I. Wohl, Victoria, 1966– editor of compilation.
B187.R35P76 2014
180 – dc23 2014002500

ISBN 978-1-107-05049-5 Hardback

Contents

Illustrations

Notes on contributors

JAMES ALLEN is Professor of Philosophy at the University of Pittsburgh. He is the author of *Inference from Signs: Ancient Debates about the Nature of Evidence* (2001) and articles about aspects of Stoicism, Epicureanism, ancient scepticism, ancient medicine, and Aristotelian logic and dialectic, among other things. He is currently working on a book about the Aristotelian disciplines of argument (logic, rhetoric, and dialectic).

RYAN K. BALOT is Professor of Political Science and Classics at the University of Toronto. The author of *Greed and Injustice in Classical Athens* (2001), *Greek Political Thought* (2006), and *Courage and Its Critics in Democratic Athens* (2014), and editor of *A Companion to Greek and Roman Political Thought* (2009), Balot specializes in American, early modern, and classical political thought. His essays and reviews have appeared in *Political Theory*, *Ancient Philosophy*, *Social Research*, *Review of Politics*, *Arion*, *American Journal of Philology*, *Classical Quarterly*, *The Journal of Hellenic Studies*, and *Rhetorica*. His current projects include work on Machiavelli's republicanism.

KAREN BASSI is Professor of Classics and Literature at the University of California at Santa Cruz. In addition to articles on Greek literature and historiography, she is the author of *Acting Like Men. Gender, Drama and Nostalgia in Ancient Greece* (1998) and co-editor, with Peter Euben, of *When Worlds Elide: Classics, Politics, Culture* (2010). She is currently working on a book on the relationship between visual perception and the passing of time in ancient Greek narrative.

JENNY BRYAN is Lecturer in Classical Philosophy at University College London. She is the author of *Likeness and Likelihood in the Presocratics and Plato* (Cambridge, 2012).

VINCENT FARENGA is Professor of Classics and Comparative Literature at the University of Southern California. In addition to articles exploring poststructural theory and contemporary reinterpretations of archaic and classical Greek poetry, rhetoric, and politics (tyranny), he is the author of *Citizen and Self in Ancient Greece: Individuals Performing Justice and the Law* (Cambridge, 2006), an investigation into the productive tension between citizen identity and individualism in Greek culture from Homer to the end of the Athenian democracy. His current research project, *Authoring Justice*, contrasts contemporary narratives about the experience of injustice in multiple cultures to the efforts of political philosophers to elaborate a universal theory of justice.

MICHAEL GAGARIN is James R. Dougherty, Jr. Centennial Professor of Classics Emeritus at the University of Texas at Austin. He has published widely, primarily on ancient Greek law and rhetoric, most recently *Writing Greek Law* (Cambridge, 2008) and *Speeches from Athenian Law* (2011). He was Editor-in-Chief of the *Oxford Encyclopedia of Ancient Greece and Rome* (2010) in seven volumes. He is currently preparing, with a colleague, an edition of *The Laws of Ancient Crete*.

CATHERINE GALLAGHER is the emeritus Eggers Professor of English Literature at the University of California at Berkeley. Her books include *The Industrial Reformation of English Fiction: Social Discourse and Narrative Form, 1832–67* (1985), *Nobody's Story: The Vanishing Acts of Women Writers in the Literary Marketplace* (1994), *Practicing New Historicism* (2000, with Stephen Greenblatt), and *The Body Economic: Life, Death, and Sensation in Political Economy and the Victorian Novel* (2006). She is currently writing a history of modern counterfactual historical narratives.

SEAN GURD teaches at the University of Missouri-Columbia. He has written two books, *Iphigenias at Aulis* (2005) and *Work in Progress* (2012), edited one volume of essays, *Philology and its Histories* (2010), and translated one Greek tragedy, Euripides' *Hippolytus* (2012).

BROOKE HOLMES is an Associate Professor and the Elias Boudinot Bicentennial Preceptor in the Department of Classics at Princeton University. She is the author of *The Symptom and the Subject: The Emergence of the Physical Body in Ancient Greece* (2010) and *Gender: Antiquity and Its Legacy* (2012), and two co-edited volumes, *Aelius Aristides between Greece, Rome, and the Gods* (2008) and *Dynamic Reading: Studies in the Reception of Epicureanism* (2012), as well as numerous articles on Greek literature, the history of medicine and the body, and ancient philosophy.

DARYN LEHOUX is Professor of Classics at Queen's University. He is the author of *What Did the Romans Know? An Inquiry into Science and Worldmaking* (2012) and *Astronomy, Weather, and Calendars in the Ancient World* (Cambridge, 2007), as well as the co-editor (with A. D. Morrison and A. Sharrock) of *Lucretius: Poetry, Philosophy, Science* (2013).

VERITY PLATT is Associate Professor of Classics and the History of Art at Cornell University. She is the author of *Facing the Gods: Epiphany and Representation in Graeco-Roman Art, Literature and Religion* (Cambridge, 2011) and co-editor, with M. Squire, of *The Art of Art History in Graeco-Roman Antiquity* (2010). In particular, she focuses on the relationship between image and text in antiquity, and is currently working on the lives of ancient Greek artists.

ROBERT TORDOFF teaches Greek and Roman Literature in the Department of Humanities at York University in Toronto. His research and publications focus on Greek literature, Greek social history, and the reception of classical literature in the long eighteenth century. He has co-edited, with Ben Akrigg, *Slaves and Slavery in Ancient Greek Comic Drama* (Cambridge, 2013) and is finishing a book on Aristophanes' late plays, *Assembly Women* and *Wealth.*

VICTORIA WOHL is Professor of Classics at the University of Toronto. Her research focuses on the literature and culture of classical Athens, spanning a variety of genres, poetic and prosaic. She is the author of *Intimate Commerce: Exchange, Gender, and Subjectivity in Greek Tragedy* (1998), *Love among the Ruins: The Erotics of Democracy in Classical Athens* (2002), and *Law's Cosmos: Juridical Discourse in Athenian Forensic Oratory* (Cambridge, 2010). She is currently working on the politics of form in Euripides.

Preface

This volume began its rather unlikely existence as a workshop entitled "*Eikos*: Probabilities, Hypotheticals, and Counterfactuals in Ancient Greek Thought," held at the University of Toronto in May 2010. The purpose of the workshop was twofold. The first aim was to use the notion of *eikos* – the probable or likely – as a means of traversing the sub-disciplines of Classics; interdisciplinarity was not just a methodological principle of the workshop but also an object of study, as we explicitly examined the way in which the concept mutated from one discourse to another. Second, we explored whether, given these discursive variations, *eikos* in fact denoted a coherent set of issues for the Greeks and whether it thus constitutes a coherent object of scholarly analysis. We determined that it did and it does. As our discussions revealed the fruitful intersections and (sometimes surprising) connections between our different sub-areas, we concluded that *eikos* was valuable both as a vehicle for analyzing our field's interdisciplinarity and as a central category of ancient Greek thought. Hence this volume.

The workshop was sponsored by the Department of Classics at the University of Toronto and a workshop grant from the Social Sciences and Humanities Research Council of Canada. I would like to thank my then-chair, Alison Keith, for her support, and the graduate students, Adriana Brook Carter, Marie-Pierre Krück, Miranda Robinson, and Lee Sawchuk, who participated in the workshop as moderators. I am particularly grateful to Marie-Pierre Krück for her invaluable aid in organizing the event, as well as for her eagle-eyed assistance in copy-editing the volume. She was also responsible for finding the cover image, Marcel Duchamp's *Network of Stoppages*. Vincent Farenga deserves the credit (or blame) for encouraging me to publish the results of our workshop. The process has been full of contingencies, and at various points it looked as if this volume would itself remain purely hypothetical, if not counterfactual. My thanks to Michael Sharp for his help in making it actual, as well as to all the contributors for their patience and their superb contributions. Finally, I am grateful to Erik

Gunderson for sharing with me the fruits of his own editorial experience, for his advice and good judgment throughout the process, and for his unflagging support, both technical and moral.

Abbreviations of ancient authors and titles can be found in *The Oxford Classical Dictionary*, ed. S. Hornblower and A. Spawforth, 3rd edn., 1996.

Introduction
Eikos *in ancient Greek thought*

Victoria Wohl

This volume explores the cluster of concepts linked by the Greek word *eikos*. *Eikos* (εἰκός, τὸ εἰκός) in Greek refers to what is probable, likely, or reasonable. Derived from the verb *eoika* ("to be like, look like, seem"), it encompasses both the seemly (as opposed to the improper) and the merely seeming (as opposed to the real). It is a term of art in Greek legal thought, where it denotes a certain type of logical argument, but it also occurs in contemporary historiography, literature, political theory, philosophy, and science. At once a logical operation, a rhetorical trope, and a literary device, *eikos* is a way of thinking about the probable and the improbable, the factual and the counterfactual, the hypothetical and the real.[1]

Eikos was a seminal mode of argumentation in Greek law: in the absence of hard evidence, forensic orators argued from probabilities, tracing through a rhetoric of likelihood the "tracks of suspicion" (as Antiphon puts it, 3.10) leading from the courtroom back to the scene of the crime. Presented as a means of uncovering the truth, *eikos* was in fact a rhetorical device for producing a likely account of the truth and thus a means of reaching the equity (*to epieikes*) of a just verdict. The paradigmatic case is attributed to Corax and Tisias, the quasi-mythical inventors of oratory: in the absence of witnesses, who is more likely to have started a fight, the stronger man or the weaker? (Plato, *Phdr.* 273a–c.) The argument relies on strategic deployment of collective belief and accepted ethical typologies: "How would a man like me assault a man like him?" (*Phdr.* 273c1–2).[2] Aristotle notes a twist on this "classic" *eikos* argument: the stronger man should argue that he was unlikely to have started the fight for the very

[1] Important cognates include *eikōn* (image) and *to epieikes* (equity). *Eikēi* ("at random") is not etymologically related, although later folk etymology may have associated the two: Plut. *Quaest. conv.* 740c; Epictetus 2.12.22. Hoffman 2008: 1 n. 1 provides a thorough bibliography of scholarship on *eikos*.

[2] Plato's Tisias glosses *to eikos* as "what the majority believe," *to tōi plēthei dokoun*, *Phdr.* 273b1. Arist. *An. Pr.* 70a2–10 defines *eikos* as a "generally admitted proposition," based on "what people know for the most part"; cf. Anaximenes, *Rhetorica ad Alexandrum* 7.4.

reason that he would seem the more likely to have done so (*Rh.* 1402a17–23). This reverse *eikos* argument has not only the orator but the criminal himself figuring the probabilities, calculating before he throws the first punch what will seem likely to a jury of his peers. *Eikos* arguments thus draw on – and provide an invaluable window onto – the Greeks' shared assumptions about human psychology, social norms, and narrative verisimilitude.

While orators used *eikos* arguments to construct plausible accounts of an unwitnessed crime, historians used probability to recreate an unwitnessed past: lacking written records, Thucydides remarks that he composes the speeches in his *History* "as it seems to me each of the speakers would have spoken given the demands of the circumstances" (1.22.1). This practice situates the project of Thucydides' historiography within the realm of *eikos*: probable motives affect and effect real outcomes.[3] At the same time, Thucydides' counterfactual statements project *eikos* into the past in order to imagine alternative possible presents and to meditate upon the necessity or contingency of historical outcomes. Such probabilistic thinking also dominated ancient political debate, for, as Aristotle says, "we deliberate about those things that seem capable of turning out in two different ways since there is no point in deliberating about things that cannot be different in past, present, or future" (*Rh.* 1357a4–7). Situated between *anankē* (necessity) and *tukhē* (contingency), a regularity that never hardens into a rule, *eikos* delineates the field of collective political deliberation and individual ethical action.

Eikos operates widely across the various domains of ancient Greek thought. The *Epidemics* of Hippocrates show early medicine as a science of the probable, as doctors diagnosed hypothetical internal states from patients' external signs – complexion, sweat, excretions – and extrapolated from individual cases to general likelihoods useful for prognosis. Here *eikos* denotes a pattern – what happens, as Aristotle says, "for the most part" (*hōs epi to polu*) – that links universal and particular, genus and species, in a relation of likeness (*Rh.* 1357a34–b1, 1402a3–28). This mimetic relation between the general rule and its individual instance is the theoretical foundation of scientific thought, logical argumentation, and jurisprudence.[4] It is also a defining feature of *poēsis* (fiction) which universalizes the particular by eschewing the random quiddity of "what has happened" for

[3] On *eikos* in Thucydides see Westlake 1958, Woodruff 1994; and on Thucydides' counterfactuals, Flory 1988 and Tordoff in this volume.

[4] Anaximenes, *Rhet. ad Alex.* 7.4: *eikos* is when the audience have in mind an example (*paradeigma*) of what has been said. On jurisprudence see Eden 1986: 43–54: equity (*to epieikes*) extends general laws to particular cases through probable accounts of intention.

the patterned plot-lines of "what might happen according to probability" (*kata to eikos*, Arist. *Poet.* 1451a36–38). *Eikos* is the governing modality of art and literature inasmuch as they deal in *eikones* (images): *eikos* grants verisimilitude to art's mimesis, simultaneously likening it to and distancing it from reality.[5] This same ambiguous verisimilitude makes *eikos* an object of intense ambivalence in the discourse that aims at truth itself, not its likenesses. Plato, for one, associates *eikos* with the deceptive rhetoric of the sophists, for whom plausible fictions trump real facts (*Phdr.* 267a6–b1, 272d1–273a1).[6] Yet for us mortals such semblances may be the best we can do and in lieu of certain knowledge, philosophy builds whole theories – even whole cosmologies – on "likely stories" (*eikōs logos, eikōs muthos*).[7]

Eikos thus opens onto a broad array of questions, logical and psychological, physical and metaphysical, epistemological and ontological. It allows us to see how the ancient Greeks conceptualized historical truth and literary fiction, the laws of nature and human behavior, social propriety and logical necessity. A broad but coherent concept that served different purposes in science, jurisprudence, historiography, philosophy, politics, and literature, *eikos* also shows how these different discourses intersected and interacted at a time when they were only just beginning to emerge as discrete modes of thought and expression.

This volume looks at *eikos* in ancient Greek thought from a variety of disciplinary perspectives. Its chronological scope is broad, stretching from Homer to Galen, and its conceptual field capacious. The chapters examine *eikos* in all of its temporalities – probability in the present, hypotheticality in the future, counterfactuality in the past – and across its full semantic range: seeming and seemliness, likeness and likelihood, the regularly expected and the scarcely possible.[8] That said, the presence of the word *eikos* and its cognates is not our sole criterion: the volume does not offer a series of word searches.[9] Instead, *eikos* functions here as a lexical shorthand for a specific conceptual field. The historical emergence of a word to name a concept is not a matter of indifference, of course, but a narrow focus on the lexical entry would exclude the many other ways in which the probable was expressed in ancient Greek thought: embedded in the syntax of the Greek conditional sentence and potential optative, for example, or embodied in

5 On *eikos* and verisimilitude in ancient literature, see Lanza and Longo 1989, Gill and Wiseman 1993, Scodel 1999.

6 Cf. Pl. *Tht.* 162d3–163a1.

7 Pl. *Ti.* esp. 29b1–d3, on which see Johansen 2004: 48–68; Burnyeat 2005; Bryan 2012: 114–91.

8 On the semantic range of *eikos* see the detailed taxonomy of Synodinou 1981; cf. Schmitz 2000: 68–72 and Hoffman 2008.

9 For the limitations of such an approach, see Dalfen's review of Synodinou: Dalfen 1985: 3.

objects (like Odysseus' bed) or individuals (like Helen of Troy) which functioned as switch-points between alternative versions of reality, casting the shadow of the counterfactual over the real.

Finally, the volume does not aim for comprehensive coverage of the topic; with a topic such as this, *ouk eikos esti*: it is neither reasonable nor fitting. *Eikos* could be tracked in countless authors and areas beyond those discussed here and a counterfactual alternative volume might have contained numerous other hypothetical chapters. This actual volume presents, in place of a comprehensive survey, a series of case studies in ancient Greek probabilistic thinking, each examining the way in which *eikos*, broadly defined, operates – whether as a modality of analysis, its object, or both – in a particular Greek text, genre, or discourse. Our hope is that this necessarily selective study will serve as a provocation to further work on the topic and perhaps even other potential volumes.

The chapters are united by a recurring set of themes and issues: the relation between contingency and necessity; the possibility of certain knowledge and the constraints on ethical action; the unclear distinction between truth and its various semblances. But these issues take on different contours in different discourses. The theme of likeness, for instance, recurs in philosophy, art and literature, and science, but the relation between original and copy, and the significance of that relation, varies markedly. So too the question of contingency and necessity is differently articulated and bears different ramifications in speculative philosophy or retrospective historiography than in the urgent present of political debate. Certain texts feature centrally in a number of papers, especially Plato's *Phaedrus* and *Timaeus* and Aristotle's *Poetics* and *Rhetoric*. These paradigmatic philosophical treatments often influence but by no means exhaust the probabilistic thinking of Greek poets, artists, historians, and scientists: philosophy is not the only discourse with a stake in and take on questions of likeness and likelihood. By adopting a pointedly interdisciplinary approach to the topic, this volume aims to show both the coherence of *eikos* as a concept and its modulation across the diverse topographies of ancient Greek thought.

The history of modern probability has been written from two principal perspectives, both placing its origin in the early modern period. Ian Hacking examines the emergence of the modern mathematical and philosophical understanding of probability in the 1660s, when developments in inductive reasoning and in the concept of evidence effected a combination of the "aleatory" understanding of probability (as a tendency of chance to develop stable frequencies) with the "epistemological" sense of the term (as the degree of belief merited by the evidence).[10] Others have analyzed the role of probability in the "invention of fiction," in particular the rise of the novel in the eighteenth century. Robert Newsom, for instance, argues that the emergence of the novel was made possible by an "antinomy of fictional probability" by which the reader occupies real and fictional worlds simultaneously. "Of course Aristotle does not know about the antinomy of fictional probability," he remarks.[11]

But that claim is perhaps too hasty. Stephen Halliwell has shown how *eikos* functions in Aristotle to mediate between the "world-reflecting" and "world-creating" facets of literary mimesis and thus to situate fiction in its characteristically equivocal relation to reality.[12] Aristotle already had a well-developed theory of fictionality in which probability played a central role. Likewise, his association of *eikos* with what happens "for the most part" (*hōs epi to polu*) and his distinction between this regularity and the certainty of "signs" (*sēmeia*) anticipates both the aleatory and the epistemological aspects of probability that Hacking identifies,[13] although, to be sure, Aristotle understands both regularity and certainty and the relation between them in distinctly different ways from Leibniz. And it is not just Aristotle: historians, scientists, and forensic orators speculated about the epistemological and evidentiary reliability of *eikos*, while tragedy traced its "aleatory" oscillation between the pure randomness of *tukhē* and the iron regularity of *anankē*. This volume thus hopes to demonstrate that the Greeks were thinking in sophisticated ways about probability many centuries before its supposed emergence in the modern era.

This "prehistory" of probability itself has a history, of course. Plato's attribution of the *eikos* argument to Corax and Tisias locates its origin as a formal type of rhetorical argument in the early fifth century.[14] Needless to say, probabilistic thinking existed long before this "invention." Characters in Homer imagine both hypothetical futures (Hector and Andromache in *Iliad* 6, for instance) and unrealized counterfactuals that would have

[10] Hacking 1975: 17: "there was no probability until about 1660 . . . It is true that we may find in Aristotle sentences translated as, 'the probable is what usually happens', but that was too long ago for us." See further Sambursky 1956, B. J. Shapiro 1983, Daston 1988.

[11] Newsom 1988: 169. He also, however, stresses that "the logical space occupied by probability remains very much the same for us as it was for Aristotle" (13). See further Patey 1984, Pavel 1986, Gallagher 1994, and especially Gallagher 2006, who sees the suspension of disbelief required by the novel's plausible stories as one facet of the "expedient fictionality" that underwrites modern subjectivity.

[12] Halliwell 2002: 151–76. On the ancient Greek "discovery" of fictionality, see Rösler 1980, Bowie 1993, Lowe 2000.

[13] Arist. *An. Pr.* 70a2–10, *Rh.* 1357a34–b21.

[14] See Kennedy 1963: 30–32, 58–61; Goebel 1989; Cole 1991a: 23–27, 82–83, and 1991b; Gagarin 1994. For a concise history of the concept, see Schmitz 2000: 48–51.

produced a different present ("would that I had been borne off by a whirlwind on the day my mother bore me," says Helen at *Iliad* 6.345, hypothetically undoing the entire war and poem). In Homer one sometimes finds contradictory possibilities paratactically juxtaposed. In the paired stories of Helen and Menelaus in *Odyssey* 4, for instance, Helen is depicted first as aiding, then as betraying the Greeks. Helen introduces her speech with the promise (or warning) that she will speak *eoikota* (4.239) but it is not clear whether in this context the word means "what is fitting," "what is likely," or "what seems like truth" (i.e. lies), and the poet does not adjudicate which account is true. This oscillation between two different realities recurs in Stesichorus' palinode, where Helen both went to Troy and did not, and in the programmatic statement of Hesiod's Muses, who know how to speak both the truth (*alēthea*) and "lies similar to truth" (*pseudea . . . etumoisin homoia, Theog.* 27–28) but see no need to distinguish clearly between the two and offer no guidance as to how one might do so.

The need for such a distinction and *eikos* as a means of making it come to the fore only in the early fifth century, to judge both from Plato and from the first attested example of the "classic" *eikos* argument. In the *Homeric Hymn to Hermes* (dated to the early fifth century), the baby Hermes defends himself against the charge of having stolen the cattle of Apollo by noting the improbability that a one-day-old infant like himself could have committed such a crime (265–73). The ambiguous parataxis of alternate realities (Hermes stole the cattle; Hermes did not steal the cattle) will be resolved through appeal to *eikos*: the more likely story is the truth, Hermes proposes – disingenuously, since he did in fact steal the cattle.[15]

Such probabilistic reasoning permeated all modes of ancient Greek thought and both distinguished and united them. On the one hand, the role of *eikos* within a discourse or genre could serve to mark its boundaries and to identify it as a discrete intellectual domain: the philosophers' rejection of the likenesses and (mere) likelihoods of poetry or rhetoric defined philosophy, in contrast to these other fields, as a discourse of truth.[16] Meanwhile on the side of fiction, the attitude toward *eikos* – the degree of

[15] The argument is marked as deceptive (*muthoisin . . . kerdaleoisin*, 260) and is consonant with the trickery and theft that characterize the god (cf. 13–16). Contrast Gorgias' *Palamedes*, where the (truly innocent) speaker deploys multiple arguments from probability (ethical, logical, and practical) on his own behalf: see further Gagarin 1994: 54–55 and in this volume.

[16] Pl. *Phdr.* 273d2–274a5, *Soph.* 235d–236d. Bryan 2012 shows how the Presocratic philosophers engage with and distance themselves from poetry (Xenophanes) and rhetoric (Parmenides) on the nature of *eikos* and the epistemology it implies. Gagarin 1994 suggests that, despite Plato's opposition between philosophy and rhetoric on the question of *eikos*, early Greek orators and rhetorical theorists used *eikos* to examine serious philosophical issues, such as causality, agency, language, and truth.

verisimilitude, the place of the miraculous – was one distinguishing feature of literary genre. One can think of the difference between Old Comedy, with its ostentatious mélange of the probable and the improbable, the actual and the impossible, and New Comedy, where generic and social probabilities drive the plot with a force approaching inevitability.[17] Bound to its grounding logic and fundamental ontological presuppositions, *eikos* lets us see how a genre or discourse frames its own realities and imagines possibilities that fall beyond it.[18]

On the other hand, *eikos* cuts across these different realms of thought and points toward a shared mental substructure beneath them. Thus Kathy Eden shows how *eikos* joins Aristotle's thinking on law, ethics, rhetoric, and poetry. The same hypothetical reasoning allows for probable motive in the courts and plausible characters on the tragic stage: equity (*to epieikes*) in law and ethics is thus closely akin to poetic fiction.[19] Reviel Netz links the hypothetical projections of early Greek mathematicians and their "multilayered ontology" to the suspension of disbelief in contemporary literature, to the competing realities debated in the courts and political assembly and, more broadly, to a society "already willing to embrace the radical thought experiment, the counterintuitive, the merely hypothetical."[20] In making this argument he draws on the claim of G. E. R. Lloyd (1990) that the exceptional methodological self-awareness and explicitness about grounding principles that characterize Greek science had their roots in the open and agonistic political culture of the polis, especially the democratic polis, where every truth was subject to scrutiny, challenge, and revision, and each position had to justify itself in its contest against others. *Eikos* would feature in this cultural ethos both as a strategy of self-justification and as a mode of radical inquiry, both an expedient of eristic rhetoric and an imagination of alternatives to the status quo. Literature's (more or less) plausible fictions and art's (more or less) faithful likenesses, historiography's diagnostic counterfactuals and science's hypothetical prognostications, the "likely stories" of philosophy and the utopian projections of politics: all were instances of a probabilistic thinking – thinking through and about *eikos* – that characterized the mental world of the ancient Greeks. Studying

[17] The Tractatus Coislinianus observes that humor comes from the impossible, the illogical, the unexpected. O'Sullivan 1995: 61–62: "Not only does Old Comedy have no need of probability, but . . . it is quite hostile to it."

[18] I argue this point in regard to legal discourse in Wohl 2010: 133–45, 153–54.

[19] Eden 1986: 25–64. Schmitz 2000 likewise argues that the forensic use of *eikos* arguments must be understood in the context of contemporary Greek thought about the pseudo-referentiality of fictional mimesis.

[20] Netz 2009: 48.

eikos in all its various guises, this volume offers a unique perspective onto that world.

The volume begins with *eikos* in its most fully developed and familiar form. Michael Gagarin traces the development of the juridical *eikos* argument and its ubiquitous opposition to *erga*, facts or reality. This opposition is enshrined in Aristotle's distinction between "artistic proofs" (rhetorical arguments) and "nonartistic proofs" (such as witness testimony or written documents). Forensic speakers uphold this dichotomy in principle, giving evidentiary priority to documents and facts over arguments from probability. Yet while prioritizing *erga* in theory, in practice, Gagarin shows, forensic orators also acknowledged the inextricability of the two. Wills, for example, are introduced in inheritance cases as incontrovertible evidence, but when the validity of a will is contested, its authenticity is proved or disproved through *eikos* arguments. Likewise, laws are cited as an objective yardstick by which to measure the case, but laws must be interpreted and their relevance shown through argumentation, including *eikos* arguments. The very facticity of the fact – including the law itself – is the product of argumentation. In this sense, forensic practice both gives ammunition to Plato's charge in the *Phaedrus* that in the law courts no one cares about truth, only about what is persuasive or likely, and is one step ahead of it: the realities of the case are tacitly accepted to be artifacts of probability.

The ontological implications of such a hypothetical reality might make a philosopher uncomfortable, but the next two chapters show that the philosophical repudiation of *eikos* is not as clear-cut as one might expect. Both Jenny Bryan and James Allen take aim at the opposition, in Plato and Aristotle respectively, between *eikos* and philosophical truth. Jenny Bryan looks directly at the seminal passage in Plato's *Phaedrus* that contains the first explicit philosophical discussion of *eikos*. There we hear that "likelihood comes about for people because of a similarity to the truth (*alētheia*)" (*Phdr.* 273d3–4). While this passage is usually read as setting rhetoric's mere plausibilities in antithesis to philosophical truth, Bryan argues that its "similarity to truth" puts *eikos* in an intrinsic but ambiguous relation to *alētheia*. As an *eikōn* or likeness of truth, *eikos* can be redeemed as an instrument of persuasion by the dialectician who knows the truth and can recognize the nature of its similarities. Philosophy and rhetoric are thus distinguished less by *eikos per se* than by their understanding of it and the use to which they put it. In this way, Socrates rehabilitates *eikos* "as something perfectly compatible (indeed, only really compatible) with the project of philosophical psychagogic rhetoric informed by truth set out in the *Phaedrus*."

James Allen makes a similar case for *eikos* as a tool for philosophy in Aristotle. He too starts from Plato's distinction in the *Gorgias* and *Phaedrus* between rhetoric as the art of plausible likelihoods and dialectic as a practice aiming at truth. For Plato this antithesis entails, at best, the subordination of rhetoric to philosophy. Aristotle, by contrast, is more optimistic about the positive potential of rhetoric as an argumentative method that, like dialectic, leads toward the truth. Allen homes in on Aristotle's claim in the *Rhetoric* that "it belongs to the same faculty to see the true and what is like the true" (1355a14–18). For Allen, as for Bryan, this "likeness" is ambiguously situated in relation to truth, neither identical with nor fully divorced from it; this ambiguity allows rhetoric to draw out our natural affinity for the truth in a way not unlike the practice of dialectic. But whereas dialectic aims primarily at the understanding of fixed and immutable truths, "rhetoric is especially useful in spheres where immutable truths are few and prone to exceptions, where constantly renewed deliberation on the basis of shifting and inconclusive evidence is required to make the decisions on which action is based and where there is no such thing as a fixed and settled condition of the intellect in which all the truths proper to the domain are grasped." Argument by way of plausible likelihoods is a fitting vehicle for this pragmatic, fluid, partial – and quintessentially human – knowledge.

This latter is the realm of knowledge addressed in the next two chapters, the realm of politics. Like Allen, Ryan Balot argues that *eikos* has a positive role to play in this sphere, and he proposes that this is one thing that differentiates politics as a practice from philosophical dialectic. Balot examines the role of "likely stories" in Plato's *Laws* and the consequences of founding a city upon such stories. Are the Athenian Stranger's likely stories intended to lead the citizens toward greater wisdom, or are they simply a tool of social control? The status of these stories is insistently questioned in the text, and their relation to truth and justice, on the one hand, and political expediency, on the other, is left troublingly indeterminate. But this indeterminacy is, in Balot's view, precisely the point. By showing the Athenian's uncertainty about many of the most basic elements of his own political theory, the *Laws* reveals a fundamental problem of political life: our simultaneous inability, as human beings, to know "the whole" and our need to make political decisions even in the absence of such knowledge. The lawmaker's likely stories draw attention to this problem without resolving it, exposing "the discomforts involved in actualizing political theories in real legislation." *Eikos*, on this reading, characterizes the political sphere as such, in both its provisionality and its pragmatism, and recognizing our likely stories for what they are – merely likely, for all that we might wish to endow them with the certainty of truth and enshrine them in immutable law – can help to prevent politics from congealing into dogmatism.

Vincent Farenga's chapter excavates the cognitive underpinnings of this contingent political practice and proposes that *eikos* forms a cognitive bridge between the mind of the individual citizen and the collective decisions of the polis. One of the key challenges of democratic theory, both ancient and modern, is to understand the process by which the partial and subjective worldviews of individuals, realms of *doxa* (opinion) and *eikos*, are transformed into communal knowledge and collective agency. Farenga shows how the archaic thinkers Solon and Xenophanes addressed this problem and the place of *eikos*, as a limitation or a resource, in their thought. For the former, *eikos* – the subjective situatedness of the individual that makes him able to grasp only what seems true to his own mind – places a limit on the possibility of collective agency, a possibility embodied in an enlightened leader like Solon himself. For the latter, by contrast, it is precisely this epistemological uncertainty at the individual level that serves as an inducement toward collective knowledge: only through collaborative thinking can we transcend the mere seeming of individual intelligence and reach toward the holistic truth represented for Xenophanes by the mind of Zeus. As a practical instantiation of this theoretical process, Farenga points to the Athenian practice of *parrhēsia*, frank speech in the public realm: through *parrhēsia*, what seems plausible to the mind of the individual speaker becomes what seems best to the community, the communal *doxa* that enables political deliberation and action.

Politics is the sphere of probabilities in part because it deliberates about the unknowable future. But the past is no less hypothetical, as we see in the next two chapters. Robert Tordoff's contribution on Thucydides' counterfactual history shows how evaluation of the probabilities of the past enables the sort of contingent political deliberation discussed by Balot and Farenga. Tordoff observes that counterfactual thinking became prominent in Athens after the disastrous end of the Peloponnesian War: in forensic and epideictic oratory and even in the deliberative processes of the democratic Assembly, the Athenians deployed counterfactual logic to imagine how things might have turned out better. Thucydides, he argues, engaged with this contemporary debate, but, in contrast to his contemporaries, his counterfactuals pose the possibility that things might in fact have turned out worse. The Athenians' optimistic counterfactual thinking in the wake of the war allowed them to write off the recent calamities as mere contingencies. Thucydides' pessimistic counterfactual speculation is an antidote to this blithe interpretation of historical probabilities and a defense against its implications for the future: the rebuilding of the Athenian Empire. Far from an idle academic pastime, counterfactual history, Tordoff shows, has serious ethical and political consequences: the way a society reconstructs past potentialities affects its view of future possibilities.

Like Tordoff, Karen Bassi examines the temporal dimension of *eikos* and the hypothetical nature of the past. Her chapter deals with the prehistory, as it were, of *eikos* in Greece: long before the advent of the "classic" *eikos* argument discussed by Gagarin, probabilistic thought in Homer was embodied in an object, the Achaean wall. The Achaean wall occupies a strange ontological and temporal position in the *Iliad*. Simultaneously indestructible and utterly destroyed, it functions as a switch-point between two possible futures, only one of which will ultimately become the past. As the trace, simultaneously, of an actual and a merely potential past, the Achaean wall stands on the boundary between historical fact and poetic fiction, blurring the line Aristotle will later draw between history as "things that have happened" (*ta genomena*) and fiction as "the sort of things that might happen" (*hoia an genoito*, Arist. *Poet.* 1451a36–38). For Bassi, as for Tordoff, the hypotheticality of the past has ethical consequences for, she argues, the uncertain future the wall projects also reveals the limits of heroic glory, *kleos*. Like the future endurance of the wall, the future *kleos* won through heroic endeavor is a mere potentiality, set against the only future that is not hypothetical or contingent, death. The invisibility of the past motivates the heroic drive for *kleos*. It also motivates our own relation to history, our longing for a living and visible past. The archaeological record concretizes this longing and compensates for the impossibility of its satisfaction: the past remains purely virtual, a product of our own desire.

Aristotle's distinction between *historia* and *poiēsis* is also the starting-point for my own chapter, the first of two addressing the role of *eikos* in the creation and definition of literature. For Aristotle, *eikos* distinguishes fiction as the domain of the probable, as opposed to history's realities, but also of the regular and ordered, as opposed to the random contingencies of everyday life. But this distinction, I argue, turns out to be less clear-cut than Aristotle at first suggests: *eikos* comes to signify both the probable and the improbable and, in this double aspect, characterizes fiction and history alike. Euripides' *Helen* both illustrates and further complicates this tangled relation. Against the backdrop of the "history" of the Trojan War, this play stages an ostentatiously counterfactual history, Helen in Egypt, and represents this unlikely fiction as reality. Amidst its relentless play on appearance and reality, truth and fiction, the drama warns its audience to be skeptical of its improbabilities, even as it demands, and works hard to produce, credence in them. The play's self-conscious convolutions of probability breach the theoretical barrier, shaky already in Aristotle, between *poiēsis* and *historia*. I suggest that they also breach the wall between the fictional world of the play and the real world of Athens in 412 BCE, where the improbabilities of real life (including the unexpected Athenian defeat in Sicily) play out as a historical imitation of *Helen*'s unlikely action. The ontological instability that Bassi identifies in relation to the past here affects the present as well, as Euripides projects fiction's improbabilities as the logic governing reality itself.

Sean Gurd's chapter also investigates *eikos* as the defining condition of literature, not just in theory but in practice. Examining the process of revision in literary papyri, he shows how literary narrative emerges out of a pre-conceptual space of contingency – an open field of potentiality – and its emergence (*pace* Aristotle) is radically non-teleological: the text is a probability that becomes an actuality only after the fact and according to rules that are impossible to predetermine. Through detailed analysis of authorial revision in three literary papyri Gurd demonstrates that while one may find local teleologies, as new versions of the text replace prior ones, authors can also go back to earlier versions. Alternative versions of the text, its erasures and interlinear additions, are thus "counterfactual" only from a posterior perspective. From the author's perspective, not only the language and the plot but the story itself – including the causal logic and narrative coherence that Aristotle theorizes under the heading of *eikos* – is radically open and all versions are equally hypothetical. Literature thus emerges – literally and at the most fundamental level – within the realm of *eikos*: as Gurd writes, "counterfactuals, possibility, and the hypothetical are part of the genesis of what there is, and what there is is what it is because it might have been otherwise. In the case of literature, a text is what it is because it *was* otherwise."

While Gurd traces the working of *eikos* in the material practice of artistic production, Verity Platt examines it as a resource for the artist's theoretical reflection on that practice. With her chapter we move from the literary to the visual. As the realm of the *eikōn* visual art has an inherent investment in questions of likeness and likelihood, and Platt shows that Greek artists addressed these questions in sophisticated ways. Her chapter considers two different visual explorations of *eikos* in Greek art. The first involves *eikos* as a plausible "likeness to truth," as we find it in forensic oratory. She argues that classical naturalism, instead of aiming for an exact relation between image and model, develops a "visual rhetoric of likeness" as artists combine realistic and unrealistic elements to produce images that are more plausible for maintaining some distance from the literal truth. The second engages with Plato's ontology of the image in the *Timaeus* and his suggestion that an *eikōs muthos*, a likely story, is the most accurate way to represent a cosmos that is itself a likeness of an ideal Form. Plato's interest in the self-reflexive qualities of *eikos* finds a parallel in contemporary Greek depictions of images within images. Foregrounding the representational status of their own enterprise, they offer a theorization of the relation between the visual image and its ineffable prototype. The images Platt studies vividly illustrate the widespread interest in fifth- and fourth-century Greece in questions of verisimilitude and plausibility, representation and being, and the conceptual utility of *eikos* as a means of thinking and talking about these questions.

The final two chapters take up the question of likeness and likelihood, probability and necessity in ancient Greek scientific thought. Daryn Lehoux starts with the fascinating question of why, according to ancient medical thinkers, organisms produce offspring like themselves. The ancient understanding of genetic inheritance was, he shows, fundamentally probabilistic, a study of likely outcomes. *Eikos*, both as likelihood and as likeness, governs thinking on reproduction: the child becomes an *eikōn* of the father, following a template of models and copies that goes back to Plato's *Timaeus*. From this perspective, unlikeness becomes improbable, even monstrous, and requires scientific explanation. The explanation offered by Aristotle reaffirms likeness: where there is no resemblance to the specific parent, Aristotle affirms one to the larger human *genus*. Just as in the *Rhetoric*, where *eikos* ties the individual to a generic ethical type, and in the *Nicomachean Ethics*, where *eikos* connects individual actions to the general *hōs epi to polu*, in natural science, too, Aristotle sees probability as a link between the universal model and the particular instance. The limit case for this understanding of reproductive likeness is children born with different skin color than their parents. Here unlikeness is not only unlikely but also unseemly, another sense of *eikos*. But again, the defense against this monstrous difference is the reimposition of the rule of *eikos* – likeness, likeliness, seemliness – even through the most unlikely of means.

Previous chapters have shown *eikos* to be the governing modality of our human lives as social and political beings, as creatures of limited knowledge, as ethical agents bound by time. Brooke Holmes argues that *eikos* also governs our very existence *qua* living organisms. She shows that for Galen, human life – our survival as organisms – is likely but not necessary; it is a "terrain where the flourishing of life is a probable outcome . . . a space of dynamic open-endedness." Galen, she argues, eschews both mechanistic determinism and randomness in favor of a theory of natural intelligence that figures Nature as a skilled and deliberate demiurge (again, as in Platt's and Lehoux's chapters, Plato's *Timaeus* is a conspicuous reference point). Linked by sympathetic bonds to the Demiurge and his design, organisms have an innate but nonconscious tendency toward life, the probable success of which is shaped by unruly matter, on the one hand, and our own conscious choices, on the other. It is thus precisely because our existence is governed by probability, rather than by chance or necessity, that the practice of life becomes conscious and rational. Both as *politikon zōon* and as mere *zōon*, the human being becomes an ethical agent only in the realm of *eikos*, and *eikos* offers a way of thinking about the role and limits of ethical agency in our human existence.

Together the studies in this volume demonstrate the complexity and diversity of ancient thought on probability and argue for the centrality of *eikos* to Greek ethics, epistemology, and ontology. Is *eikos* just a curious historical byway, a cognitive counterfactual in the history of Western thought, or does it have a lasting legacy in modern discourses of probability? This is the question Catherine Gallagher addresses in her Afterword. Historians generally posit a rift between pre-modern and modern thinking on probability with the emergence of a mathematical theory of probabilities in the seventeenth century. Gallagher shows that this rift is profound: when probability escaped its role as "classical philosophy's rejected other" and "ascended to the very status of 'knowledge'," the result was a radical transformation not only in our theoretical discourses but even in our basic conception of the individual. But alongside this fundamental discontinuity Gallagher traces certain continuities, notably in the novel and counterfactual histories, where, despite significant shifts of emphasis, the enduring categories of type and individual, general rule and contingent particularity, suggest the persistence of classical *eikos* within early modern thought. In this way, she shows, an understanding of Greek *eikos* exposes the fissures within our own modern discourses of probability and reveals both the diversity of our classical Greek inheritance and the distinctiveness of modern thinking about the probable, hypothetical, and counterfactual.

CHAPTER I

Eikos arguments in Athenian forensic oratory

Michael Gagarin

Although *eikos* arguments – arguments that rely on likelihood or probability – are probably inherent in human discourse and certainly existed in Greece from the earliest times, the first explicit study of *eikos* arguments is traditionally attributed to two Sicilians, Corax and Tisias, in the middle of the fifth century BCE.[1] There is considerable dispute about just what (if anything) Corax and Tisias contributed to the study of what we may call "rhetoric," but there is little doubt that formal study of *eikos* argument began about this time and it is likely that Tisias, at least, was among those who contributed to the study.[2]

The earliest source for Tisias' work is Plato's *Phaedrus*, near the end of which the discussion turns to the question whether someone who wants to be "competent in rhetoric" (*hikanōs rhētorikos*, literally "sufficiently rhetorical") needs to be concerned with truth (272d–e). In language that recalls the condemnation of rhetoric in *Gorgias*, Socrates asserts that in the law courts no one cares at all about truth, only about what is persuasive (*pithanos*), and this is equivalent to what is likely (*eikos*). This in turn leads to a discussion of Tisias: "does he say that the likely (*to eikos*) is anything other than what most people think is the case?" After Phaedrus assents to this, Socrates presents an example of Tisias' argument from likelihood (273b4–c4):

> ἔγραψεν ὡς ἐάν τις ἀσθενὴς καὶ ἀνδρικὸς ἰσχυρὸν καὶ δειλὸν συγκόψας, ἱμάτιον ἤ τι ἄλλο ἀφελόμενος, εἰς δικαστήριον ἄγηται, δεῖ δὴ τἀληθὲς μηδέτερον λέγειν, ἀλλὰ τὸν μὲν δειλὸν μὴ ὑπὸ μόνου φάναι τοῦ ἀνδρικοῦ

[1] Hoffman 2008 surveys the historical development of the verb *eoika* from its basic meaning "I am like" to the sense of "likely" or "probable" that becomes common in the fifth century, especially in the participial forms *eikos* and *eikotōs*; he includes a good selection of recent studies of *eikos* (Hoffman 2008: 1–2 n. 1).

[2] For a traditional account, see Kennedy 1963: 52–124. Cole 1991a and Schiappa 1999 both challenge this account and argue that we should not even speak of rhetoric as the formal study of speech before the fourth century. Cole 1991b makes the case that Corax never existed.

> συγκεκόφθαι, τὸν δὲ τοῦτο μὲν ἐλέγχειν ὡς μόνω ἤστην, ἐκείνῳ δὲ καταχρήσασθαι τῷ πῶς δ' ἂν ἐγὼ τοιόσδε τοιῷδε ἐπεχείρησα; ὁ δ' οὐκ ἐρεῖ δὴ τὴν ἑαυτοῦ κάκην, ἀλλά τι ἄλλο ψεύδεσθαι ἐπιχειρῶν τάχ' ἂν ἔλεγχόν πῃ παραδοίη τῷ ἀντιδίκῳ.
>
> He [Tisias] wrote that if a weak but brave man beats a strong but cowardly man and steals his cloak or something else, and the man brings him to court, neither man must speak the truth. The coward must deny that he was beaten by a single brave man, whereas the other must contend that they were alone and must use that well-known argument, "How could someone like me [i.e. a weak man] have assaulted someone like him [i.e. a strong man]?" The other will certainly not admit his own cowardice but will try some other lie, perhaps giving his opponent an opportunity to refute him.

Only the second argument here is an *eikos* argument, and it closely resembles *eikos* arguments we find in earlier contexts. For example, the one-day-old Hermes in the *Homeric Hymn to Hermes* makes a nearly identical argument: after stealing Apollo's cattle, he returns home, wraps himself in swaddling clothes in his cradle, and when Apollo comes and accuses him of theft, he says, "I am not like a strong cattle thief" (οὐδὲ βοῶν ἐλατῆρι κραταιῷ φωτὶ ἔοικα, 265). Even though the poet uses a finite verb here, *eoika* ("I am like") rather than the participle *eikos*, the argument is the same: I am not (I do not resemble) the kind of person who commits this sort of wrong (stealing cattle, assaulting a strong man). I designate them both explicit *eikos* arguments – arguments where the word *eikos* or a close cognate occurs. When a similar argument is made without using an *eikos* word, I will call it an implicit *eikos* argument.[3]

Judging solely from Plato's account, Tisias taught pupils to lie in court and to use a well-known traditional type of *eikos* argument, hardly enough to give him the reputation for innovation and originality he later possessed. Aristotle, however, presents a different version of this same case, which he attributes to Corax, and his version is more interesting. In the *Rhetoric* Aristotle cites an example of a case where the likely is in fact not likely (1402a18–23):

> ἄν τε γὰρ μὴ ἔνοχος ᾖ τῇ αἰτίᾳ, οἷον ἀσθενὴς ὢν αἰκίας φεύγει (οὐ γὰρ εἰκός), κἂν ἔνοχος ᾖ, οἷον ἰσχυρὸς ὤν (οὐ γὰρ εἰκός, ὅτι εἰκὸς ἔμελλε δόξειν). ὁμοίως δὲ καὶ ἐπὶ τῶν ἄλλων· ἢ γὰρ ἔνοχον ἀνάγκη ἢ μὴ ἔνοχον εἶναι τῇ αἰτίᾳ· φαίνεται μὲν οὖν ἀμφότερα εἰκότα, ἔστι δὲ τὸ μὲν εἰκός, τὸ δὲ οὐχ ἁπλῶς ἀλλ' ὥσπερ εἴρηται·

[3] In the orators the most common cognate is the adverb *eikotōs*, "probably." Implicit *eikos* arguments usually take the form, "I would/would not have done something," which may be expressed as a rhetorical question, "Why/How would I have done it?" Gorgias' *Palamedes* is full of such arguments.

> If someone is not open to the charge, for instance a weak man who is accused of assault, he argues that this was not likely. And if he is open to the charge, as would be the case if he were strong, he also argues that this was not likely, because it would seem likely [sc. and therefore he would refrain from doing it]. And the same holds in other cases, for a person must either be open to the charge or not. Thus both positions appear likely, but the one is likely while the other is not likely in a straightforward sense (*haplōs*) but in the way we have explained.

In this version neither argument is false, since they both simply state what is likely, and in each case this may or may not be true. The first argument, that a weak man is unlikely to assault a strong man, is a straightforward *eikos* argument of the sort we have just examined. The second argument is more complex: it acknowledges the traditional *eikos* argument, that a strong man is likely to assault a weak man, but then reverses its effect, arguing that this very likelihood makes the act unlikely. And this reverse *eikos* argument is not only highly original but also a typical product of sophistic thought of the period – a clever, turning-the-tables type of argument, in the form of a pair of opposed *logoi* (words, speeches, arguments) in which the second neatly reverses the first. The reverse *eikos* argument was too obviously clever and sophistic for use in court, and, not surprisingly, the only example of it in forensic rhetoric is in Antiphon's *First Tetralogy* (2.2.6), a fictional case. Traditional *eikos* arguments, on the other hand, are common in forensic speeches, though the word *eikos* itself is used much more often in the earlier part of the fourth century than later.

Before considering actual court speeches, however, it will help to examine Antiphon's *First Tetralogy* and another fictional speech from the fifth century, Gorgias' *Palamedes*, both of which make much use of *eikos* arguments. The *First Tetralogy* is essentially an exercise in the use of *eikos* arguments.[4] The issue in the case is a simple question of fact: did the accused kill the victim or did someone else do it? The reader is given enough facts to cast suspicion on the defendant but these do not provide solid evidence of his guilt. Both sides then marshal an array of *eikos* arguments and counter-arguments, such as that it is likely (or not likely, as the case may be) that the victim was killed by someone else, or was killed in a drunken fight, or that the defendant's known hostility to the victim makes him the likely killer. The defendant also includes a reverse *eikos* argument, that the existence of strong motives for him to kill the victim in fact makes it unlikely that he would do so since he knew he would be the obvious suspect. All these

[4] For more detailed analysis of *eikos* and other issues in the *First Tetralogy*, see Gagarin 2002: 112–18.

arguments provide an interesting display but do not give either side a clear victory.

Over the course of the four speeches that make up the *First Tetralogy*, however, Antiphon also includes discussion of the value or weight of *eikos* arguments relative to other sorts of evidence. This begins in the plaintiff's opening words (2.1.1–2):

> ὁπόσα μὲν τῶν πραγμάτων ὑπὸ τῶν ἐπιτυχόντων ἐπιβουλεύεται, οὐ χαλεπὰ ἐλέγχεσθαί ἐστιν· ἂν δ' οἱ ἱκανῶς μὲν πεφυκότες, ἔμπειροι δὲ πραγμάτων ὄντες, ἐν δὲ τούτῳ τῆς ἡλικίας καθεστῶτες ἐν ᾧ κράτιστοι φρονεῖν αὑτῶν εἰσί, πράττωσι, χαλεποὶ καὶ γνωσθῆναι καὶ δειχθῆναί εἰσι . . . . γιγνώσκοντας οὖν ὑμᾶς χρὴ ταῦτα, κἂν ὁτιοῦν εἰκὸς παραλάβητε, σφόδρα πιστεύειν αὐτῷ.
>
> It is not difficult to obtain a conviction for crimes planned by ordinary people, but when those with natural ability and practical experience commit a crime at that point in their lives when their mental facilities are at their height, it is difficult to know or demonstrate it . . . You should be aware of this, and even if you accept a point as only likely (*eikos*), you should have confidence in it.

With these words Antiphon makes clear from the beginning that *eikos* arguments are necessary when, and perhaps only when, direct proof is lacking; this implies that likelihood has less value in establishing guilt than direct knowledge of the facts, which would commonly come from an eyewitness. Then, at the end of his first speech, the plaintiff concludes that, as a general rule, "someone who plans a crime would never be convicted if he is not convicted either by the testimony of those present or by arguments from likelihood" (οἵ τε γὰρ ἐπιβουλεύοντες ἀνεξέλεγκτοι ἂν εἴησαν, εἰ μήτε ὑπὸ τῶν παραγενομένων μήτε ὑπὸ τῶν εἰκότων ἐξελέγχονται, 2.1.10).

In the next two speeches both litigants refer to these two types of proof, the direct testimony of witnesses and *eikos* arguments. The defendant argues that he should not be convicted just because he is the likely killer, since this does not make him the true killer, whereas the plaintiff, not surprisingly, defends the value of *eikos* arguments when direct evidence is unavailable (2.3.8):

> φάσκων δὲ οὐ τοὺς εἰκότως ἀλλὰ τοὺς ἀποκτείναντας φονέας εἶναι, περὶ μὲν τῶν ἀποκτεινάντων ὀρθῶς λέγει, εἴπερ ἐγένετο φανερὸν ἡμῖν τίνες ἦσαν οἱ ἀποκτείναντες αὐτόν· μὴ δεδηλωμένων δὲ τῶν ἀποκτεινάντων, ὑπὸ τῶν εἰκότων ἐλεγχόμενος οὗτος ἂν καὶ οὐδεὶς ἕτερος ἀποκτείνας αὐτὸν εἴη. οὐ γὰρ ἐπὶ μαρτύρων ἀλλὰ κρυπτόμενα πράσσεται τὰ τοιαῦτα.

> When he says that murderers are not those who are likely to have killed but those who actually did kill, he is correct about those who killed, if it were clear to us who his actual killers were. But if the actual killers have not been revealed, then since his guilt is proven by the *eikos* arguments, this man and no other would be the killer; for such things are not done in the presence of witnesses, but secretly.

In his last speech the defendant responds in similar terms to before, but then, in a complete surprise, he introduces a fact to set against the likelihoods: "I shall show not from likelihood (*eikota*) but in fact (*ergon*) that I was not present [when the man was killed]" (ἐγὼ δ' οὐκ ἐκ τῶν εἰκότων ἀλλ' ἔργῳ δηλώσω οὐ παραγενόμενος, 2.4.8). He then adds that all his slaves will testify that he was home that night.

Because this alibi is introduced at the end of the defendant's last speech, the plaintiff has no chance to respond; thus, the argument for the alibi is never answered, and we are left to conclude, it seems, that the alibi trumps the other arguments. In other words, by placing the alibi at the very end of the *Tetralogy*, Antiphon implies that no valid response can be made.

In the *First Tetralogy*, then, Antiphon clearly indicates that *eikos* arguments in general have limited validity; on the scale of proofs they rank below evidence that is a matter of fact, *ergon*. And because *eikos* arguments are largely absent from his other two *Tetralogies*, which do not concern questions of fact but rather the question of who is responsible for a death when the facts are clear, Antiphon makes clear that *eikos* arguments are only of use where questions of fact are at issue. Over the course of the *Tetralogies*, moreover, Antiphon establishes a set of parallel oppositions: likelihood is juxtaposed with knowledge or truth and also with fact or reality, *ergon*; *ergon*, in turn, is juxtaposed with *logos*, argument, especially in the other two *Tetralogies* but also in Greek thought generally. In other words, the realm of argument, frequently expressed in terms of *eikos*, is contrasted with the realm of fact, truth, and reality.[5]

We find another fifth-century example of this discourse on the value of *eikos* in Gorgias' *Defense of Palamedes*, the only work of his which hinges on a question of fact. The question is, did Palamedes betray the Greeks, as Odysseus charges, or has Palamedes been framed, as he argues? Palamedes establishes this issue early in his defense (5):

> ὅτι μὲν οὖν οὐ σαφῶς εἰδὼς ὁ κατήγορος κατηγορεῖ μου, σαφῶς οἶδα· σύνοιδα γὰρ ἐμαυτῷ σαφῶς οὐδὲν τοιοῦτον πεποιηκώς· οὐδὲ οἶδ' ὅπως

[5] For this metadiscourse about fact and argument, see further Gagarin 2002: 115–18, 122–27, 132–33.

ἂν εἰδείη τις ὂν τὸ μὴ γενόμενον. εἰ δὲ οἰόμενος οὕτω ταῦτα ἔχειν ἐποιεῖτο τὴν κατηγορίαν, οὐκ ἀληθῆ λέγειν διὰ δισσῶν ὑμῖν ἐπιδείξω τρόπων· οὔτε γὰρ βουληθεὶς ἐδυνάμην ἂν οὔτε δυνάμενος ἐβουλήθην ἔργοις ἐπιχειρεῖν τοιούτοις.

I know clearly that my accuser accuses without certain knowledge, since I know that I personally have not done any such thing; and I do not know how anyone could know something that did not happen. If he makes his accusation thinking this is so, I have two arguments to show he is not speaking the truth: I would not have been able to undertake such deeds, if I had wished; and I would not have wished to do them if I had been able.

Since Palamedes has no direct evidence to support the fact that he did not commit the crime, his arguments throughout are almost all indirect *eikos* arguments, often phrased as rhetorical questions: How could I have communicated with the enemy since I could not speak their language? What would I have gained from betraying the Greeks? Or, to put it differently, it is not likely that I communicated with the enemy since I did not speak their language; it is unlikely that I committed the crime since I had nothing to gain.

Odysseus' case, on the other hand, would have been based on facts which, according to tradition, were planted by him in order to frame Palamedes.[6] Tradition also has it that Palamedes was convicted despite his innocence. If we accept this traditional account, we can then say that Odysseus' direct evidence actually supports a false case, whereas Palamedes uses *eikos* arguments in support of the truth. Gorgias does not reveal the outcome of the case, but since he could assume that his audience was familiar with the traditional version of the story, we are probably meant to conclude that Palamedes lost. Thus, as in the *First Tetralogy*, factual evidence is more powerful than *eikos* arguments, even though it is being used to support a false case. In other words, *eikos* arguments are not a better guide to the truth than direct evidence, but sometimes they are the only means available for supporting a true case.

The same sorts of *eikos* arguments, together with the same sorts of assumptions about the relative value of facts and *eikos* can be found in actual court speeches. Take, for example, a passage from Antiphon 6 *On the Chorus Boy* (18):

[6] In the version of Odysseus' speech written by Alcidamas, probably in the early fourth century, Odysseus calls witnesses to testify to these facts.

> αἰτιάσασθαι μὲν οὖν καὶ καταψεύσασθαι ἔξεστι τῷ βουλομένῳ· αὐτὸς γὰρ ἕκαστος τούτου κρατεῖ· γενέσθαι μέντοι τὰ μὴ γενόμενα καὶ ἀδικεῖν τὸν μὴ ἀδικοῦντα οὐκ ἐν τοῖς τούτων λόγοις ἡγοῦμαι εἶναι, ἀλλ' ἐν τῷ δικαίῳ καὶ ἐν τῷ ἀληθεῖ. ὁπόσα μὲν γὰρ λάθρα πράττεται καὶ ἐπὶ θανάτῳ βουλευθέντα, ὧν μὴ εἰσι μάρτυρες, ἀνάγκη περὶ τῶν τοιούτων ἐξ αὐτῶν τῶν λόγων τῶν τε τοῦ κατηγόρου καὶ τοῦ ἀποκρινομένου τὴν διάγνωσιν ποιεῖσθαι καὶ θηρεύειν καὶ ἐπὶ σμικρὸν ὑπονοεῖν τὰ λεγόμενα, καὶ εἰκάζοντας μᾶλλον ἢ σάφα εἰδότας ψηφίζεσθαι περὶ τῶν πραγμάτων.
>
> Anyone who wants can make accusations and level false charges, for these things are in each person's control. But I do not think it is in the power of these men's words to make it so that something happened that did not happen, or that someone committed a crime who did not, but these depend on justice and truth. Now when a murder is planned in secret and there are no witnesses, you are forced to reach a verdict about the case on the basis of the prosecutor's and defendant's words alone; you must be suspicious and examine their accounts in detail, and your vote concerning the facts of the case will necessarily be cast on the basis of likelihood rather than clear knowledge.

The passage has clear echoes of both the *First Tetralogy* and *Palamedes*, but here it is intended as a foil for the defendant, who quickly notes that his case is different: he has witnesses on his side as well as arguments. As he says later in summarizing his case: "I am now presenting you with likely arguments (*logous eikotas*), with witnesses who support these arguments, with facts (*erga*) supporting the witnesses, and with deductions (*tekmēria*) derived from these very facts" (ἐγὼ τοίνυν τούς τε λόγους ὑμῖν εἰκότας ἀποφαίνω, καὶ τοῖς λόγοις τοὺς μάρτυρας ὁμολογοῦντας καὶ τοῖς μάρτυσι τὰ ἔργα, καὶ τεκμήρια ἐξ αὐτῶν τῶν ἔργων, 31). Though the rhetoric here is (as often) a bit of overkill – the facts are essentially the testimonies of witnesses, and the deductions are essentially *eikos* arguments – it boils down to the combination of facts and likely arguments that are drawn from those facts and thus consistent with them. The speaker thus has a complete case that combines arguments, particularly *eikos* arguments, with facts, particularly those derived from witness testimony.

Now, this analysis of forensic discourse as combining facts and arguments did not arise simply out of Antiphon's fertile mind; it is inherent in the structure of Athenian forensic pleading. In court each litigant was given the same set amount of time for his speech, but when he wanted to introduce the evidence of a witness deposition or some other document, he stopped speaking, the water-clock that was timing his speech was stopped (in private cases), and a clerk read out the testimony or the document to the court. When the clerk finished, the litigant resumed his pleading and the clock began again.[7] This procedural structure in essence divided the litigant's case into his own pleading and the clerk's reading out of the documentary evidence – the very combination of argument and fact that the speaker of Antiphon 6 analyzes.

This duality also underlies the important distinction Aristotle draws near the beginning of his *Rhetoric* between artistic and nonartistic proofs, *pisteis entekhnoi* and *pisteis atekhnoi* (1355b35–39):

> τῶν δὲ πίστεων αἱ μὲν ἄτεχνοί εἰσιν αἱ δ' ἔντεχνοι. ἄτεχνα δὲ λέγω ὅσα μὴ δι' ἡμῶν πεπόρισται ἀλλὰ προϋπῆρχεν, οἷον μάρτυρες βάσανοι συγγραφαὶ καὶ ὅσα τοιαῦτα, ἔντεχνα δὲ ὅσα διὰ τῆς μεθόδου καὶ δι' ἡμῶν κατασκευασθῆναι δυνατόν, ὥστε δεῖ τούτων τοῖς μὲν χρήσασθαι, τὰ δὲ εὑρεῖν.
>
> Proofs are either nonartistic or artistic. By nonartistic I mean proofs we do not supply ourselves but that already exist, such as witnesses, interrogations under torture, contracts, and the like; by artistic I mean proofs that can be devised by our own skill and methods. Thus, we need only to make use of the former, but we must discover the latter.[8]

Aristotle's analysis may puzzle a reader unfamiliar with forensic oratory, but it corresponds precisely to the basic duality in forensic pleading, in which the litigant's speech is crafted with (rhetorical) skill or art (*tekhnē*), and thus is "artistic" (*entekhnos*), but the documentary evidence is simply read out by the clerk; it requires no *tekhnē* and is thus "nonartistic" (*atekhnos*). By describing the nonartistic proofs as already existing, so that we need only to make use of them, Aristotle suggests that these written documents have an objective existence independent of the speaker's rhetorical skill; they simply exist and thus can be considered facts as opposed to arguments.[9]

This analysis of a case as a combination of facts (*erga*), which have an objective truth-value, and arguments (*logoi*, including *eikos* arguments), which are needed to support the facts, is widespread in Athenian forensic oratory, though it may take a variety of forms. In any form, however, the truth of established facts is taken for granted: documentary evidence may be rejected as forged, or misleading, or irrelevant, but no litigant ever maintains that arguments in general should overcome established facts or

[7] In the fifth century it appears that witnesses presented their own testimony orally in court, but by early in the fourth century, this testimony, together with all documents (including laws, contracts, wills, etc.) was submitted in writing before the trial and read out to the court by the clerk.

[8] Later Aristotle adds laws and oaths to these nonartistic proofs (*Rh.* 1375a24).

[9] Aristotle is clearly aware that the creation and use of documentary evidence often required just as much artistic skill as the oral pleading; but his analysis nonetheless contains an important insight and has been deservedly influential.

that a particular argument should take priority over a valid document. And yet, a common view among modern scholars is that forensic speakers had just the opposite view of the relative truth-value of facts and arguments. Guthrie, for example, asserts, "The essential theoretical basis of rhetoric was . . . that 'they held the probable (*eikota*) in more honour than the true'."[10] Guthrie is quoting directly from the *Phaedrus* where Plato ridicules the work of earlier and contemporary rhetoricians (267a6–8):

> Τεισίαν δὲ Γοργίαν τε ἐάσομεν εὕδειν, οἳ πρὸ τῶν ἀληθῶν τὰ εἰκότα εἶδον ὡς τιμητέα μᾶλλον, τά τε αὖ σμικρὰ μεγάλα καὶ τὰ μεγάλα σμικρὰ φαίνεσθαι ποιοῦσιν διὰ ῥώμην λόγου.
>
> We will let Tisias and Gorgias rest in peace [says Socrates], who saw that probabilities should be more honored than truths, and who make small things appear great and great things small by the power of speech.

It may not be surprising that Guthrie, like most modern philosophers, is willing to follow Plato's lead, despite Plato's evident bias against rhetoric, but it is difficult to see why a scholar such as Vickers, who sets out to defend rhetoric against attacks such as Plato's, would echo this view: "probability . . . in Greco-Roman law – strange though it may seem to modern ears – . . . was deemed more credible than the 'truth' of witnesses or evidence."[11] Strange indeed, but apparently the grip of Plato and the subsequent tradition that derives from him has a powerful hold, even on otherwise skeptical minds.

If these scholars were to test their conclusions against actual forensic speeches instead of blindly following Plato, they would find that they do not hold up. As we have seen, Antiphon's speakers regularly assert the priority of truth over *eikos*, facts over arguments, though if possible, of course, they seek the support of both types of proof rather than rely on one to the exclusion of the other (e.g. Antiphon 6.31, quoted above). But not every case provides both clear facts and strong arguments, and litigants must sometimes rely primarily on one or the other. More interestingly, not every fact is a fact; that is, not every document that is presented as a fact is accepted as valid. Such cases may hinge on whether a fact is established or a document is authentic, and the arguments on this point will commonly be *eikos* arguments. This is most easily seen in inheritance cases which hinge on the validity of a will. Although a will is a clear example of a nonartistic proof that establishes an objective fact, namely the deceased's intentions, wills are regularly challenged on various grounds, as several of

[10] Guthrie 1971: 180. [11] Vickers 1988: 69–70.

Isaeus' speeches demonstrate. And those challenges are generally cast in the form of *eikos* arguments.

A good example is Isaeus 9 *On the Estate of Astyphilus*. After Astyphilus' death, his cousin Cleon produced a will stating that Astyphilus had adopted Cleon's son, who thus became Astyphilus' direct heir. This claim is challenged by Astyphilus' brother (the speaker) on the ground that the will is a forgery. The main issue in the case is thus the genuineness of the will, and the case for forgery is heavily dependent on explicit *eikos* arguments, which occupy the main body of the speech (7–26). The first of these concerns the making of the alleged will (7–9):

> Now it is likely that he not only desired to adopt a son to leave behind but also considered how to make his will absolutely valid and ensure that whoever he adopted should possess the property, attend the family altars, and perform the customary rites for him after his death and for his ancestors; and he would be absolutely sure that all these things would not happen if he made his will without his family being present, but only if he summoned first his relatives, then the members of his phratry and deme, and finally as many as possible of his other close friends. In that way anybody who might claim the estate on the ground of either kinship or bequest would be easily proved a liar. Now Astyphilus clearly did nothing of the sort and did not summon any of these people when he made the will my opponents claim he did, unless, that is, anybody has been suborned by them to agree that he was present. But I will myself produce all these people to you as witnesses.

In other words, it is common, and indeed expected, that in making a will a man summons his family and other fellow demesmen and phratry members as witnesses; Cleon did not do this; therefore, it is unlikely that he made a will.

Other *eikos* arguments follow: if Astyphilus wanted the will to be secret, he would not have had any witnesses, but he allegedly had several who were unknown to him. It is not likely that Astyphilus made a will before leaving on this campaign, since he had not done so in the past before leaving on other campaigns. It is not likely that he adopted Cleon's son since he was on bad terms with Cleon. And so forth. And all these *eikos* arguments are supported by the testimony of witnesses. The speaker thus supports his claim with both *eikos* arguments and facts, though he denies the main "fact" in the case, the will. Cleon's opposing speech does not survive, but he must have relied heavily on the will, which he presumably had read out by the clerk. He must also have responded to at least some of the speaker's *eikos* arguments with arguments of his own. For example, he may have given reasons why it was reasonable or likely that Astyphilus made a will before leaving on this campaign even though he had not done so before. And he almost certainly supported his arguments with witness testimony and perhaps other documents.

The case as a whole, then, might appear at first to set a speaker relying on *eikos* arguments against a speaker relying on facts (the will), but a fuller analysis shows that this central fact is actually the main issue: is the will a matter of fact or not? On this point the speaker – and almost certainly Cleon too – presents the same sort of combination of facts and *eikos* arguments as we find in most forensic speeches. This produces a complex interaction between facts and *eikos* arguments, in which arguments are used to validate or invalidate a fact (the will), while facts in turn (in the form of witnesses) are used to support arguments. Facts and arguments thus turn out to be mutually dependent; neither, it seems, can exist in isolation from the other.

Similar combinations of facts and arguments are found throughout the fourth century, even as written documents gradually became more common in many areas of Athenian life. Explicit *eikos* arguments, however, become less frequent as the century progresses, though implicit arguments remain common.[12] There is a particularly sharp drop-off in the use of the term *eikos* itself and a corresponding shift to *eikotōs* ("probably"), especially in the speeches of Demosthenes. My sense is that as time went on, arguments with the word *eikos* were felt to be too simplistic, whereas *eikotōs* seemed unobjectionable. As written documents became more common in Athenian courts, they also became more susceptible to challenge. Most questionable are wills, which are routinely challenged. Contracts, which were normally acknowledged by both parties in front of witnesses, are less often challenged as forgeries, but in one case at least, Hyperides 3 *Against Athenogenes*, the plaintiff argues that a contract that he agreed to in front of witnesses was deceptive and therefore invalid.

The one type of document whose authenticity is never challenged is laws. We are told (Dem. 26.24) that the penalty for citing a nonexistent law was death, and no litigant ever accuses another of this crime or argues that a law that has been introduced is not likely to be authentic. This does not mean that laws were never questioned. The meaning of a law, its applicability to the present case, and the legislator's intent in enacting

[12] If we divide the orators into two groups, before 380 (Antiphon, Andocides, Lysias, Isocrates, Isaeus) and after 370 (Demosthenes, Aeschines, Hyperides, Lycurgus, Dinarchus) we find almost the same number of instances of *eikos* and *eikotōs* or *eikota* in the two groups (181 vs. 198), even though the forensic speeches of the later group occupy more than twice as much text as those of the earlier group.

the law are commonly up for discussion. But as evidence in support of a litigant's arguments, laws are often considered especially certain and objective. Particularly interesting in this regard are cases brought by the *graphē paranomōn* procedure, in which a decree and/or its proposer is indicted on the ground that the decree violates one or more laws. In such cases, the law or laws that were allegedly violated had to be written down on a tablet next to the decree, and this tablet was then posted at the entry to the court.

The most famous case brought by this procedure was Aeschines' prosecution of Ctesiphon for his decree proposing a crown for Demosthenes (Aeschin. 3), to which the response was delivered by Demosthenes on behalf of Ctesiphon in his speech *On the Crown* (Dem. 18). Although Aeschines devotes most of his prosecution speech to the broad policy question, whether Demosthenes deserved to be awarded a crown for his public service, he clearly thinks that a particularly strong part of his case is his claim on the narrower point, that the decree violates two specific laws: that an official cannot receive a crown until he has passed the audit at the end of his term (which, according to Aeschines, Demosthenes had not), and that crowns should be awarded in the assembly, not in the theater as proposed by Ctesiphon. Although most modern scholars agree that Aeschines has the stronger case on these two points, Harris has argued strongly that he does not.[13] Be that as it may, Aeschines expresses concern that in his defense speech Demosthenes may try to avoid discussing the question of whether the decree violates these laws, and he appeals to the jury not to let Demosthenes avoid addressing this issue.

For our purpose, the most interesting part of Aeschines' argument on this point is the striking simile he employs in defense not only of the certainty of laws as evidence, but also of the pure objectivity of legal decision-making (3.199–200):

> οὐ γὰρ ἀόριστόν ἐστι τὸ δίκαιον, ἀλλ' ὡρισμένον τοῖς νόμοις τοῖς ὑμετέροις. ὥσπερ γὰρ ἐν τῇ τεκτονικῇ, ὅταν εἰδέναι βουλώμεθα τὸ ὀρθὸν καὶ τὸ μή, τὸν κανόνα προσφέρομεν, ᾧ διαγιγνώσκεται, οὕτω καὶ ἐν ταῖς γραφαῖς ταῖς τῶν παρανόμων παράκειται κανὼν τοῦ δικαίου τουτὶ τὸ σανίδιον, τὸ ψήφισμα καὶ οἱ παραγεγραμμένοι νόμοι. ταῦτα συμφωνοῦντα ἀλλήλοις ἐπιδείξας κατάβαινε.

[13] Harris 1994, 2000. For the more common view (that Aeschines' case is stronger), see Gwatkin 1957. Since Athenian law provided no means for deciding issues like this authoritatively, it is futile to ask which side is correct.

> For the issue of justice is not something imprecise but something defined by your laws. It is just like in carpentry: when we want to know what is straight and what is not, we apply a ruler (*kanōn*) by which we decide the issue. In the same way, in indictments for illegal decrees we have at hand as the ruler of justice this tablet here containing the decree and the laws written side by side. Just demonstrate that these agree with one another and then step down.

The simile presents a view of forensic argument and judicial decision-making that is purely rational and objective. A simple comparison of two texts, the decree and a law, will provide a conclusive decision: either the two agree or they do not; either Ctesiphon's decree is valid or it is not.

The image of a ruler of justice was not Aeschines' invention. Not long before the case "on the crown" another prominent Athenian, Lycurgus, had prosecuted Leocrates for treason because he had left Athens before the battle of Chaeronea and had spent the next six years abroad. Apparently Lycurgus could find no law expressly prohibiting Leocrates' departure, and so he must argue (as litigants often do) that Leocrates' conduct was indirectly addressed by several laws. But first Leocrates notes the contrasting situation, where a crime is explicitly covered by a law: "In cases where a law has indicated what crimes are subject to punishment, it is easy to use this as a guide (*kanōn*) and punish those who have broken the law" (ὅσα μὲν γὰρ τῶν ἀδικημάτων νόμος τις διώρικε, ῥᾳδιον τούτῳ κανόνι χρωμένους κολάζειν τοὺς παρανομοῦντας, 1.9).

The idea that the law is a straightforward ruler or guide that can be applied to human conduct to produce a more-or-less automatic verdict about the lawfulness of that conduct is a comforting thought. Lycurgus would have had a much easier time convicting Leocrates if there had been a law explicitly prohibiting leaving the city before a battle. And Aeschines would like the jurors to think they could simply place Ctesiphon's decree next to the laws he cites and reach an automatic decision that the two do not agree and thus that Ctesiphon is guilty. But the law is not so simple – not in Athens, or even in the United States, where a similar geometric image has played a role in the history of our constitutional law. I refer to the so-called "T-square rule," enunciated most clearly by Supreme Court Justice Owen Roberts in 1936.[14] In an opinion that declared unconstitutional one of Roosevelt's New Deal proposals, Roberts wrote,

[14] *United States v. Butler*, 297 U.S. 1 (1936).

> When an act of Congress is appropriately challenged in the courts as not conforming to the constitutional mandate, the judicial branch has only one duty – to lay the article of the Constitution which is invoked beside the statute which is challenged to decide whether the latter squares with the former.

The T-square rule was directed against scholars associated with the legal realism movement, who at the time were arguing that the tenets of legal positivism and judicial objectivism, which had dominated the court's thinking for more than half a century, served primarily to justify the justices' personal views regarding the sanctity of private property and free enterprise. As it happened, within two years the composition of the Court had changed and the T-square rule was abandoned, though it occasionally resurfaces.

As for Aeschines and his ruler, although he would have been happy if the jurors had used a simple fact-based rule to decide the case "on the crown," he was enough of a realist to know that he had to devote most of his speech to the larger, more problematic matter of assessing Demosthenes' career as a public figure. For his part, Demosthenes dismisses Aeschines' ruler as confusing and overly complex (18.111):

> τῶν μὲν οὖν λόγων, οὓς οὗτος ἄνω καὶ κάτω διακυκῶν ἔλεγε περὶ τῶν παραγεγραμμένων νόμων, οὔτε μὰ τοὺς θεοὺς ὑμᾶς οἶμαι μανθάνειν οὔτ' αὐτὸς ἐδυνάμην συνεῖναι τοὺς πολλούς· ἁπλῶς δὲ τὴν ὀρθὴν περὶ τῶν δικαίων διαλέξομαι.
>
> Since he thoroughly jumbled the arguments he made about the laws that are written side by side, by the gods, I don't think you understood most of them, nor was I able to make sense of them either. But in a simple fashion (*haplōs*), I will discuss the correct understanding of the issue of justice in the case.

Demosthenes' point is that the aim of this trial is justice and the route to a correct understanding of justice is an argument that clarifies the issue, not an allegedly simple and objective comparison of texts, which leads only to confusion. Demosthenes does not reject the texts of the laws themselves, only the use Aeschines makes of them. In fact, he himself cites a law a few sections later (18.120) as well as several decrees in disputing Aeschines' claim that the decree conflicts with existing law. But he insists that a simple juxtaposition of the decree and the laws will not lead to justice; only a clear discussion will do that.

To put all this in terms of our discussion of *eikos* arguments, Demosthenes accepts the fact that specific texts of laws exist but argues that it is not enough simply to cite these texts; one must use *logos* to explain their meaning and their relevance to a specific question of justice. The ruler of justice, like the T-square rule, suggests a direct and automatic path from objective facts to legal decisions, in which arguments, including *eikos* arguments, have no role. Demosthenes counters that facts and arguments are inextricably linked, and that the meaning of a law depends on argument.

In sum, Antiphon's *First Tetralogy* and other early texts establish a clear distinction between *eikos* arguments and direct, objective truth and give priority to the latter, when it is available, though litigants prefer if possible to draw support from both direct evidence and argument. Over the course of the following century, however, it seems that the more common written documents became as a potential source of direct, objective truth, the more problematic these documents became in forensic argument. Subjective arguments, which are almost always based on some degree of likelihood, continued to be necessary in order to establish the meaning, or the relevance, or the very existence of the reality allegedly embodied in the documents. Forensic discourse continues to commit itself to a clear polarity of fact and argument, but facts almost always become elusive and unreliable, either because their authenticity or very existence is challenged, or because their meaning and applicability to any specific case is debatable. *Eikos* arguments, which at first are often acknowledged to be subordinate to direct evidence, come to be seen as controlling that evidence and manipulating it into supporting whatever end the litigant is trying to achieve. In the end, then, fact and argument are not only mutually dependent but inseparable.

CHAPTER 2

Eikos in Plato's Phaedrus

Jenny Bryan

Introduction

References to *eikos* reasoning are found throughout the Platonic corpus. Most famously and perhaps most puzzlingly, the *Timaeus* offers a 'likely account' (*eikōs logos*) of the likeness (*eikōn*) that is the cosmos. In the *Theaetetus* we find what looks like an explicitly critical assessment of likelihood as inferior to proof (162e2–163a1):[1]

> And whatever the crowds want to hear, you give it to them, telling them that it would be strange if there was no one who was wiser than any old cow; and keep your lips sealed about proof and necessity and rely instead on likelihood (*tōi eikoti*). Mind you, if Theodorus or some other geometer were to employ likelihood in his calculations, he'd be pretty shoddy. So you and Theodorus should give some thought to whether, in such important matters, you are prepared to accept arguments which are persuasive and likely (*pithanologiāi te kai eikosi*).[2]

This passage, along with what Plato has Socrates say about *to eikos* in the *Phaedrus*, is often taken as evidence of Plato's hostility towards the apparently sophistic or rhetorical habit of relying on 'mere likelihood' in attempts to persuade. Plato presents *to eikos* as opposed to a respectably philosophical interest in truth and proof. We can see why the *Phaedrus* might provide further grist to this interpretative mill. At 267a6–7, for example, Socrates claims that Tisias and Gorgias 'saw that likelihoods are

Formative versions of this chapter were presented at the St Andrews Research Seminar and the Institute of Classical Studies Ancient Philosophy Seminar. I am very grateful to the audiences at both of these meetings for their challenging and illuminating comments.

1 *To eikos* and its cognates can be translated in a variety of ways, each of which is necessarily loaded. I have chosen 'likelihood' to render *eikōs* in the *Phaedrus* (following Nehamas and Woodruff 1995) in preference to 'probability' (as adopted by Rowe 1986 and Yunis 2011) or 'plausibility' (as preferred by G. R. F. Ferrari 1987) because it seems to avoid the (more recent) mathematical undertones of the former and the implication of subjectivity suggested by the latter.

2 Except where noted, translations are my own.

to be more honoured than truths' (*pro tōn alēthōn ta eikota eidon hōs timētea mallon*). The implication here certainly seems to be that rhetoric as taught and practised by Tisias and (more recently) Gorgias is more concerned with persuasiveness (such as is produced by presenting likelihoods) than with truth.[3] This passing reference to *to eikos* is rightly connected to the more extensive discussion found later in the *Phaedrus*, at 272d2–274a5. This later discussion serves to elaborate the critique of the sophistic analysis of arguments from likelihood as indifferent to truth. Such arguments are apparently tailored to the beliefs of their audience and the need to persuade, with no regard for the truth of the matter. At 272d7–e2, for example, Socrates gives an account of the disregard for truth to be found in the law courts:[4]

> In the law courts no one cares at all about the truth of such things, but rather for what is convincing (*tou pithanou*). And this is what is likely (*to eikos*), which is what the man who intends to speak with skill (*tekhnēi*) must concentrate on.

What Socrates goes on to say about *to eikos* has been subjected to relatively little scrutiny. This is, in part, because the section of the *Phaedrus* dealing with the details of rhetorical theory (apparently taken from contemporary rhetorical handbooks) has itself received relatively little interest from philosophers.[5] It is also, I suspect, partly because scholars already think they know Plato's attitude towards *to eikos* and are thus more than willing to assume that this passage fits that preconceived reading. This chapter takes a closer look at Socrates' discussion of *to eikos* at 272d2–274a5. In particular, it considers the significance of the claim made at 273d2–6 that 'this likelihood comes about for people because of a similarity to the truth'.

[3] One might think that the disparaging talk at *Phdr.* 229d–e of 'experts' reducing myths to likelihoods (*prosbibāi kata to eikos hekaston*) offers a contrast to this attitude. Those who disbelieve myths and seek to explain them in terms of likelihood would presumably do so on the grounds that their analysis is *more likely to be true* than the myth itself. However, Yunis 2011: 93 suggests that the gist of this passage is to establish that, since the 'probable' is merely what the masses believe, and 'the probable is the norm for which mythical interpretation strives, mythical interpretation is doomed to be a vain, unproductive pursuit'. In what follows I will suggest that Plato is not himself committed to the sophistic understanding of *to eikos* as independent of truth.

[4] The fact that Socrates singles out the law courts here is significant. It reminds us of and contrasts with the claim made at *Phdr.* 261a–e that *scientific* rhetoric will be a 'means of leading the soul (*psukhagōgia*) through speeches, not only in the law courts and all other kinds of public meetings, but also in private' (261a7–9).

[5] See, for example, Hackforth 1952: 155, who dismisses 272b–274b as offering little more than a recapitulation of the deficiencies of sophistic rhetoric, although he does admit that 'so far as there is anything new, it is the substitution of the probable (*to eikos*) for the plausible (*to pithanon*) . . . and Socrates' ridicule of this by an illustration taken from Tisias' own manual'.

This issue of 'similarity to the truth' is one which deserves careful consideration not least because it implies that Plato's attitude towards *to eikos* in the *Phaedrus* is rather more nuanced than has generally been recognized. In fact, it seems that the *Phaedrus* suggests the possibility of a positive, philosophical use for *to eikos* within the philosophically respectable, dialectically informed kind of psychagogic rhetoric described in the second half of the *Phaedrus*.[6] Just as the *Phaedrus* argues for a good version of scientific rhetoric as being an appropriate tool for the philosopher trying to guide the soul of his auditor towards philosophy and truth, so it hints at a role for *to eikos*, as having a genuine connection to truth, within this new kind of philosophical rhetoric. This philosophically respectable version of likelihood, with its essential connection to truth, is a redefinition of *to eikos* developed in contrast to what is presented as the standard sophistic attitude to likelihood as entirely divorced from truth. In what follows I will argue that it is important to distinguish the attitude towards *to eikos* attributed to the sophists from that of Socrates. This distinction turns on their different understanding of the relationship between *to eikos* and truth. Plato is not himself dismissing *to eikos* as unrelated to truth. Rather, he is criticizing sophistic rhetoricians for failing to realize that likelihood's persuasive power lies in its similarity to truth.

Tisian *eikos*

Phaedrus 259e2–274b6 presents Socrates' analysis of rhetoric, both as he claims it is practised by 'experts' in rhetoric and described in technical rhetorical manuals and as he claims it should be practised, namely as a medium for dialectically informed persuasion.[7] At 259e7–260a4, in response to Socrates' claim that a speaker needs to know the truth about his subject matter, Phaedrus introduces the suggestion that knowledge of the truth need have no role in persuasion:

> I have heard about this, my dear Socrates, that it is not necessary for someone intending to be a rhetor to learn what is really just, but rather what will seem so to the crowd who will act as judges, and not to learn what is really good or fine, but whatever will appear to be. For persuasion (*to peithein*) comes from these things and not from the truth.

Socrates counters this claim with an attempt to demonstrate that knowledge of truth and reality is actually what best equips someone who wishes

[6] See Asmis 1986 and Moss 2013 for a discussion of the nature and significance of *psukhagōgia* in the *Phaedrus*.

[7] See Yunis 2009.

to persuade scientifically, that is, someone practising the genuine science (*tekhnē*) of persuasion (260a5–262a4). He then goes on to consider a variety of rhetorical tricks such as are set out in contemporary manuals of rhetoric (266d5–269d1), with the clear implication that such artifices, despite being presented as part of a 'science' of rhetoric, actually belong to a practice that is no more than a pseudo-science (precisely because it seems to be based on no kind of genuine understanding).[8] At 271c10–272b4 Socrates concludes that a true expert in scientific rhetoric must have an understanding of both the nature of speeches and the nature of souls and, further, of how the former should be applied to the latter in order to produce the maximum (i.e. maximally beneficial) persuasive effect. When Phaedrus notes that the acquisition of such understanding is no mean feat, Socrates admits as much and suggests that it is necessary to double-check whether any less demanding path to rhetorical expertise is available (272b7–d1). It is at this point that Socrates returns to the question raised by Phaedrus' original suggestion at 259e7–260a4 that a competent rhetor has no need for the truth. Socrates now offers a close scrutiny of the teachings of Tisias, who has, as noted above, already been named (at 267a6–7) as a champion of *to eikos*:

> In the law courts no one cares at all about the truth of such things, but rather for what is convincing (*tou pithanou*). And this is what is likely (*to eikos*), which is what the man who intends to speak scientifically (*tekhnēi*) must concentrate on. For again, they say that sometimes one should not say what actually happened, if it was not likely to have happened, but rather say what is likely, both in prosecution and defence, and that indeed one ought to pursue the likely whenever one is speaking, often saying goodbye to the truth. For when this happens throughout one's speech, it brings about the entire science. (272d7–273a1)[9]

Those who discuss this account of *to eikos* in detail tend to focus on the question whether Plato is giving a truthful account of the rhetorical use of likelihood.[10] For my purposes, however, what matters is that Plato

[8] This is, of course, very close to the critique of rhetoric offered by Plato's *Gorgias*. See, for example, 463a6–c7. The *Phaedrus* seems to expand on a position hinted at in the *Gorgias*, that if rhetoric can be separated from its explicit political motivations, it may serve as a useful handmaiden for philosophy (see *Grg.* 503a5–b3).

[9] As Yunis 2011: 219 observes, the claim that *to eikos* brings about the entire science of persuasion is a direct challenge to Socrates' claim that it requires extensive training in dialectic and knowledge of truth and psychology. If persuasion simply involves telling people what they already think, it is easily achieved.

[10] See Gagarin 1994, Kraus 2006, and Nehamas and Woodruff 1995: 75 n. 170. For an extended discussion of rhetorical/sophistic *eikos* see Tindale 2010: 69–82. See also Gagarin in this volume for a detailed discussion of *eikos* arguments in Athenian forensic rhetoric.

is clearly representing *to eikos* as somehow opposed to truth. Faced with a choice between the two, the sophistic rhetor is encouraged to prefer likelihood. In fact, truth and likelihood stand so far opposed that what is true is often unlikely and therefore, it seems, unpersuasive. Before moving on, it is worth observing that this characterization of *to eikos* and its relationship to truth is not offered as Socrates' own. The entire description is attributed to an anonymous 'they'. 'They say' that in the law courts, no one cares about truth and that sometimes one should prefer the likely to what is true.[11] Indeed, Phaedrus agrees that Socrates has given an account of what is said by 'those who profess to be experts about speeches' (273a2–3). This attribution should not be overlooked. Clearly, Socrates is not recommending that rhetors should say goodbye to the truth. Equally, I will suggest, there is no reason to assume that he considers *to eikos* and truth to be separable in this way (indeed, there is good reason to think that he does not).[12]

Socrates labours this issue of sophistic rhetoric's separation of truth and likelihood by giving an example apparently taken from Tisias' own manual.[13] First, however, he asks Phaedrus to clarify what Tisias means by *to eikos* (273a6–c5):

SOCRATES: But you've studied at least Tisias himself carefully. So let Tisias tell us this too: doesn't he say that likelihood is nothing other than what the masses think?

PHAEDRUS: What else?

SOCRATES: It's likely (*eoike*) that it was when he discovered this wise and skilful technique that he wrote that if a feisty weakling is sued for mugging a cowardly muscle-man and stealing his cloak or something, neither should tell the truth. The coward should say that the feisty man had help to beat him up and the latter should retort that they were alone and employ that common tactic: how could someone like me attack someone like that? The muscle-man will not admit to his cowardice but will concoct some other lie and may thus help his opponent to refute him. And in other cases the way of speaking scientifically will be much the same.

[11] Note the *phasi* ('they say') at 272d2. Compare the translation of Rowe 1986 with that of Nehamas and Woodruff 1995. Rowe's repetition of 'they say' reminds us that Socrates is 'giving the wolf's side' (272c10–11) in representing the view that knowledge is not required for scientific persuasion.

[12] A similar point can be made about the passing reference to *to eikos* at 267a6–7. Socrates is attributing the view that 'likelihoods are to be more honoured than truths' to Tisias and Gorgias. Whilst he says nothing about his own view of *to eikos*, it is surely not this. Note that even the discussion of *to eikos* at *Tht.* 162e2–163a1 is put into the mouth of Protagoras.

[13] Aristotle attributes a version of this example to Corax at *Rh.* 1402a. See Gagarin 1994: 51.

Again, it is worth emphasizing that the definition of *to eikos* as 'nothing other than what the masses think' is explicitly attributed to Tisias; it is not presented as Socrates' own view.[14]

Socrates concludes his discussion of Tisias as follows (273d2–e4):

> Ὅτι, ὦ Τεισία, πάλαι ἡμεῖς, πρὶν καὶ σὲ παρελθεῖν, τυγχάνομεν λέγοντες ὡς ἄρα τοῦτο τὸ εἰκὸς τοῖς πολλοῖς δι᾽ ὁμοιότητα τοῦ ἀληθοῦς τυγχάνει ἐγγιγνόμενον· τὰς δὲ ὁμοιότητας ἄρτι διήλθομεν ὅτι πανταχοῦ ὁ τὴν ἀλήθειαν εἰδὼς κάλλιστα ἐπίσταται εὑρίσκειν. ὥστ᾽ εἰ μὲν ἄλλο τι περὶ τέχνης λόγων λέγεις, ἀκούοιμεν ἄν· εἰ δὲ μή, οἷς νυνδὴ διήλθομεν πεισόμεθα, ὡς ἐὰν μή τις τῶν τε ἀκουσομένων τὰς φύσεις διαριθμήσηται, καὶ κατ᾽ εἴδη τε διαιρεῖσθαι τὰ ὄντα καὶ μιᾷ ἰδέᾳ δυνατὸς ᾖ καθ᾽ ἓν ἕκαστον περιλαμβάνειν, οὔ ποτ᾽ ἔσται τεχνικὸς λόγων πέρι καθ᾽ ὅσον δυνατὸν ἀνθρώπῳ.
>
> Tisias, for a long time before you even arrived we were saying that this likelihood (*touto to eikos*) comes about for people because of a similarity to the truth (*di' homoiotēta tou alēthous*). And just now (*arti*) we explained that in every case the man who knows the truth (*ho tēn alētheian eidōs*) best understands how to discover similarities (*tas homoiotētas*). So if you have anything else to say about the science of speaking, we shall listen. But if not, we will believe what we have just explained, that unless someone enumerates the natures of those who will listen and is able to divide the things that exist according to their kinds and to embrace each thing individually under one form, he will never master the science of speaking, to the extent that is possible for man.

Socrates' claim that 'we were saying that this likelihood comes about for people because of a similarity to the truth' is particularly intriguing for several reasons. The first is that it seems to present a direct challenge to Tisias' definition of *to eikos* as unrelated to the truth. The second is that it is not altogether clear where or even that Socrates *has* been making this claim. The third is that it is equally unclear just what it means to say that 'likelihood comes about for people because of a similarity to the truth'. I turn now to a consideration of these questions.

Persuasion and similarities

In discussing Socrates' claim that 'we were saying that this likelihood comes about for people because of a similarity to the truth', scholars generally look

[14] Yunis 2011: 220 notes that the definition of *to eikos* 'in relation to the beliefs of the masses is precisely what makes the probable useful in sophistic rhetoric, which aims at mastery in the mass institutions of the polis'. Since this is obviously not Socrates' motivation, there seems no reason to assume that he shares this definition.

for a connection to an earlier passage in the *Phaedrus*.[15] So, for example, Ferrari tells us that 'Socrates' riposte to Teisias' parry is dismissive . . . Socrates simply refers back to his earlier principle that the plausible attains its plausibility by resemblance to the truth, and the argument he built upon it: that he best deceives who knows the truth.'[16] Ferrari does not specify precisely where Socrates has stated this principle but he is presumably referring to the analysis of persuasion at 261e6–262c3.[17] Certainly, there is good reason to assume a connection between the two passages. The discussion of Tisias' teachings on *to eikos* was introduced in response to a worry that Socrates was setting the bar too high in suggesting that only the man fully trained in psychology and dialectic could achieve the science of persuasion (271c10–272c4). This reminds us of Phaedrus' suggestion at 259e7–260a4 that 'it is not necessary for someone intending to be a rhetor to learn what is really just' which is, as noted above, followed by Socrates' attempt to show that knowledge is actually necessary for persuasion (260a5–262a4).[18] Both 273d2–6 and 261e6–262c3 share the context of a refutation of the claim that persuasion can be achieved without knowledge.

Perhaps the most obvious point of connection between these two passages is their shared concern with the role of 'similarity' (*homoiotēs*) in persuasion. The mention of similarity to the truth at 273d2–6 will be considered in detail below. At 261e6–262c3 the notion of similarity is placed at the heart of Socrates' analysis of persuasion. In response to the suggestion that knowledge is not a necessary condition of the science of persuasion, Socrates argues that all rhetoric (both public and, more controversially, private) is a kind of 'leading of the soul through speeches' (261a7–b2). At 261e2–4 he goes on to claim that a speaker in possession of the science of speaking will be able to persuade his audience equally well of either side of an argument – he will possess the skill of *antilogic* (*antilogikē*).[19] Thus, he concludes, if there is a science of speaking, 'it will be that by which someone will be able to make everything which can be made similar to something else similar to everything else to which it can be made similar,

[15] In fact, the *ara* at 273d3, as an '*ara* of realization' (see Denniston 1954: 38–39) would seem to indicate that Socrates is both referring back to a previous discussion and realizing its implications for *to eikos*. I am grateful to Mike Trapp for this observation.

[16] G. R. F. Ferrari 1987: 82. As we shall see, Socrates has nowhere stated the principle quite as Ferrari describes it, i.e. in terms of persuasion having a resemblance *to the truth*.

[17] See also W. H. Thompson 1868: 132 and Yunis 2011: 221, who suggests that 273d2–6 'recapitulates' the argument at 261e6–262c3 with a new and justified focus on *to eikos*.

[18] However, 273a2–5, in which Phaedrus notes that 'we did touch briefly on this sort of thing previously', seems to be a reference to the passing mention of *to eikos* at 267a6–7, rather than to the discussion of persuasion at 261e6–262c3.

[19] See Kerferd 1981: 59–67 for a discussion of antilogic as a feature of sophistic rhetoric.

and to bring to light when someone else is making things similar and hiding it' (261e1–4).[20] Socrates goes on to clarify the role of similarities within persuasion and thereby to establish that knowledge is necessary. It is worth quoting his argument in its entirety (261e6–262c3):

SOCRATES: Does deception occur more often regarding things that are very different or things that are only a little different?
PHAEDRUS: Regarding those that are a little different.
SOCRATES: Then you are more likely to escape detection if you proceed from one thing to its opposite if you take small steps than if you take large ones.
PHAEDRUS: Absolutely.
SOCRATES: So someone intending to deceive someone else and to avoid being deceived himself will need to have precise knowledge of the similarity and dissimilarity of the things that are.
PHAEDRUS: Necessarily.
SOCRATES: So if he is ignorant of the truth of a thing, will he be able to distinguish in other things greater and lesser similarities to the thing of which he is ignorant?
PHAEDRUS: That is impossible.
SOCRATES: Clearly, therefore, those who have beliefs contrary to what is the case and are deceived get into this state through certain similarities.
PHAEDRUS: That's how it happens.
SOCRATES: So is there any way in which someone who does not know the nature of each of the things that are will be expert in shifting the position of others little by little through similarities away from what is the case to its opposite on each occasion and himself escape this process?
PHAEDRUS: No, never.
SOCRATES: So, my friend, the man who doesn't know the truth but rather chases after opinions, will, so it seems, present a ridiculous and unscientific 'science' of speaking.

Socrates claims that knowledge of similarities is a necessary condition of deception (which is, it seems, essential to persuasion).[21] This is because persuasion involves imperceptibly manoeuvring one's audience away from their original belief. The imperceptibility of this process relies upon shifting them away from their original position step by step, moving between things, presumably propositions, which 'differ little' until one has persuaded one's audience to adopt a position opposite to that with which they started. It is the knowledge of similarities that allows one to structure one's argument

[20] See Yunis 2011: 186: 'rhetoric is an art that demonstrates propositions by means of likenesses'.
[21] Yunis 2011: 187: 'persuasion of this kind is called deception because it takes place without the auditor's conscious assent, and thus is a means for the speaker to impose his will on the auditor. It is immaterial from the perspective of the art whether the auditor is persuaded to hold true or false beliefs.' See Murray 1988.

in a persuasive – that is, deceptive – fashion. But Socrates' claim is not simply that knowledge of similarities is a necessary condition of successful scientific persuasion. He also claims that one cannot recognize similarities (and thus which similarities are closer than others) unless one knows 'the nature of each of the things that are'. One cannot, it seems, know that *x* is similar to *y*, or that *x* is more similar to *y* than to *z* unless one knows the true natures of *x*, *y* and *z*. Therefore, it is not just knowledge of similarities but also, and more fundamentally, knowledge of the nature and truth of things (to which knowledge of similarities is secondary) which is the necessary condition of successful scientific persuasion.

There is a clear connection between this account of the necessity of knowledge for scientific persuasion and what Socrates says about *to eikos* at 273d2–6. In fact, it is clear that Socrates has the earlier discussion in mind when he says 'and just now (*arti*) we explained that in every case the man who knows the truth (*ho tēn alētheian eidōs*) best understands how to discover similarities (*tas homoiotētas*)'. Both passages assert that, thanks to his pursuit of knowledge of the nature of things, it is the dialectician who is best placed to produce successful scientific persuasive accounts.

There is, however, a subtle difference between these two passages. This difference does not serve to undermine the clear connection between 273d2–6 and 261e6–262c3, but it should prompt us to reconsider the nature of this connection. As has been noted, the primary reason for drawing a link between 273d2–6 and the earlier analysis of persuasion is their shared concern with similarities. Indeed, scholars tend to proceed as if Socrates' claim that 'this likelihood comes about for people because of a similarity to the truth' is a mere restatement of the preceding discussion of the role of similarities in persuasion. However, when we look more closely, it becomes clear that the match between the two passages is not quite so neat. Let us look again at what Socrates has to say about the role of similarities in persuasion at 262b5–8:

> Ἔστιν οὖν ὅπως τεχνικὸς ἔσται μεταβιβάζειν κατὰ σμικρὸν διὰ τῶν ὁμοιοτήτων ἀπὸ τοῦ ὄντος ἑκάστοτε ἐπὶ τοὐναντίον ἀπάγων, ἢ αὐτὸς τοῦτο διαφεύγειν, ὁ μὴ ἐγνωρικὼς ὃ ἔστιν ἕκαστον τῶν ὄντων;
>
> So is there any way in which someone who does not know the nature of each of the things that are (*hekaston tōn ontōn*) will be expert in shifting the position of others little by little through similarities (*dia tōn homoiotētōn*) away from what is the case (*apo tou ontos*) to its opposite on each occasion and himself escape this process?

This passage certainly talks of persuasion through similarities and, presumably, 'what is the case' is equivalent to 'truth', but the relationship between these two elements is rather different from that expressed in the definition of *to eikos* at 273d2–6. There *to eikos* is said to come about 'because of a similarity to the truth (*di' homoiotēta tou alēthous*)'.[22] Even if we substitute persuasion for *to eikos* at 273d2–6, this will not be sufficient to establish a clear equivalence between the two passages. In the first case Socrates seems to be talking about similarities between propositions. The truth or falsehood of these propositions is mentioned only to suggest that scientific persuasion enables one to persuade people *away* from the truth (*apo tou ontos*). Of course, it is emphasized throughout the argument at 261e6–262c3 that knowledge of truth will enable one to recognize these similarities, but this is not equivalent to claiming that one is thus recognizing (and thus harnessing the persuasive power of) a similarity *to the truth*. In fact, the passage as a whole implies that the scientific speaker is seeking to determine similarities and differences to the beliefs held by one's audience, so as to construct an imperceptibly persuasive argument. This looks rather similar to the definition of *to eikos* as 'nothing other than what the masses think' attributed to Tisias at 273a7–b1. In both cases, it seems, one structures one's argument according to the beliefs of one's audience, but in the case of genuinely scientific persuasion, this structure is informed by knowledge of how things really are (and thus of what the steps between similar propositions should be).

At 261e6–262c3 Socrates talks of achieving persuasion *through* (*dia* + accusative) similarities. These are not offered as similarities to the truth, but rather to the beliefs held by the audience. At 273d2–6, on the other hand, Socrates describes *to eikos* as coming about *because of* (*dia* + genitive) a similarity to the truth. I want to emphasize that I am not denying a connection between these two passages. It is clear that one exists and, indeed, is signalled in the text. However, I do want to claim that the connection between these passages is not as straightforward as is often assumed and, further, that a closer scrutiny of the claim made at 273d2–6 is worthwhile for what it can tell us about Plato's attitude towards *to eikos*.

At this point, it is worth considering one potentially deflationary objection to my emphasis on the apparent difference between the two passages. Those who want to champion the reading of 273d2–6 as a recapitulation of

[22] The use of the genitive *tou alēthous* to express the target of similarity is striking. One would more usually expect a dative (see LSJ 1225). But Plato uses the genitive to express the target of resemblance elsewhere, for example at 262b2–3, 'So if he is ignorant of the truth of a thing, will he be able to distinguish in other things greater and lesser similiarities to the thing of which he is ignorant (τὴν τοῦ ἀγνοουμένου ὁμοιότητα σμικράν τε καὶ μεγάλην ἐν τοῖς ἄλλοις διαγιγνώσκειν).' I am grateful to Nicholas Denyer for pointing out this parallel. He suggests that the cases for similar object and target of similarity may be considered interchangeable on account of the symmetrical nature of similarity.

the earlier analysis of persuasion might suggest that, whilst the claim that 'likelihood comes about for people because of a similarity to the truth' is apparently different in form from Socrates' previous suggestions about the role of similarities in persuasion, this difference only amounts to the fact that it is some kind of abbreviation of that earlier argument. So perhaps it is possible to explain 'because of a similarity to the truth' as simply offering a compact allusion to 261e6–262c3. Such a reading would presumably run along the following lines. In the earlier argument Socrates analyses persuasion as a form of deception, wherein the deceiver alters his auditor's position by gradually moving them through imperceptible stages away from their earlier opinion to an agreement with its opposite. Now, at 262b5–8, Socrates describes how such a speaker leads his audience away from 'what is the case' (*apo tou ontos*), i.e. from the truth, implying that their original belief was indeed true. Presumably, the first step in such an argument would be to determine (and make the case for) a proposition imperceptibly different from, i.e. similar to, that truth. The next step will, of course, be to determine and make a case for a second proposition imperceptibly different from, i.e. similar to, the first imperceptibly different, i.e. similar, proposition. The strength (and science) of the argument lies in correct judgment over the similarity which holds between these propositions. Further, if proposition *y* is similar to proposition *x* and proposition *x* is similar to the truth, one could justifiably claim that proposition *y* is itself somehow similar to the truth. In fact, if each proposition, including the opposed conclusion, is similar to the one before up to and including the original truth, one may want to claim that this false conclusion can also reasonably be claimed to be similar to the truth. Indeed, one may think that this is Socrates' point at 261e2–4 when he claims that the science of speaking 'will be that by which someone will be able to make everything which can be made similar to something else similar to everything else to which it can be made similar'. The truly artful speaker can make even falsehood similar to the truth. If this is an appropriate reading of 261e6–262c3, it can be analysed as more or less equivalent to 273d2–6. When Socrates says that 'likelihood comes about for people because of a similarity to the truth', he means that persuasion (such as that induced by *to eikos*) is induced *by means of* similarities (to the truth) *because* they are similar to the truth. Thus, 273d2–6 is just an abbreviation of the previous argument about the role of similarities in persuasion.

If this (or something like it) is the explanation of how these two passages fit together, then I do not feel too discouraged. In fact, it seems to me that any reading of 261e6–262c3 which wishes to claim that it is equivalent to

273d2–6 will, like the one just mooted, have to assume that the argument is to be read in terms of manipulating not just any similarities, but, specifically, similarities *to the truth*. Only this will explain the apparently causal status of similarity to the truth at 273d2–6. However, even if this relationship between the two passages holds, and I am happy that it should, I want to suggest that there is something particularly significant about Socrates' decision to single out *to eikos* in the *Phaedrus* and that this significance lies in its relationship to truth.

To eikos and similarity

At this point it is worth reminding ourselves that Socrates characterizes Tisias' attitude towards *to eikos* as something entirely divorced from truth; indeed, he appears to have suggested that it is preferable to truth. Tisias' likelihoods are analysed as deceptive tricks unconnected to truth. As I have been emphasizing, Socrates offers a contrasting analysis of *to eikos* as coming about *because of* a similarity to truth, so that the man who knows the truth is best placed to employ *to eikos* because he alone is able to recognize similarities. Above, I looked at the connection between the earlier analysis of persuasion and this claim. Now I want to consider a further possible (and perfectly compatible) explanation for Socrates' redefinition of *to eikos* in terms of a similarity to truth.

When we look at the development of the terminology of *eikos*, we can see that Socrates' analysis of likelihood in terms of similarity (to the truth) is not altogether novel. Indeed, it is rather more common than might be supposed by those who concentrate on Plato's analysis of sophistic *eikos*. For example, we can turn to the *Homeric Hymn to Hermes*. This text offers what Gagarin has suggested may be the earliest extant argument from probability. Hermes has stolen Apollo's cattle and Apollo has demanded retribution. In defending himself before Zeus, Hermes appeals to the implausibility that he should have done such a thing (265–66):[23]

> But I am not like (*eoika*) a cattle-rustler or a strong man,
> And this is not my work; rather other things are my concern.

Here *eoika* refers explicitly to similarity: Hermes is pointing out his dissimilarity, as a mere babe, from the kind of person who tends to steal cattle. But Hermes' claim that he is not similar to a cattle-rustler also appears to be adducing the notions of both appropriateness and, significantly,

[23] Gagarin 1994: 51.

likelihood. The sense of the argument seems to be that it is unlikely that Hermes would have behaved in a manner so unlike that appropriate to an infant.[24]

Perhaps the most famous and problematic exploitation of this connection between likeness and likelihood belongs to Plato himself. At *Timaeus* 29b5–c3, Timaeus draws a connection between likeness and likelihood:

> So, accounts of what is permanent and stable and evident with the help of reason are stable and permanent... but accounts of that which is created to be like that [which is stable and unchanging], being of a likeness, are likely (*ontos de eikonos eikotas... ontas*) and stand in proportion (*ana logon*) to those others. As Being is to Coming-to-be so truth is to convincingness (*hotiper pros genesin ousia, touto pros pistin alētheia*).

Scholars have worried about precisely why Plato should think that accounts of likenesses should be likely.[25] Most explain it as an expression of some kind of deficiency, either of truth or of certainty. In fact, it is possible to read this claim as a positive one, setting up a standard at which descriptions of likenesses should aim.[26] For current purposes, however, what matters is the clear connection being drawn between likeness (*eikōn*) and likelihood (*to eikos*). It is also significant that this connection is glossed in terms of a relation (*ana logon*) to truth. Likely accounts stand in some kind of relationship to the true accounts which describe stable originals.

I do not want to claim that the *Phaedrus* is offering or assuming the same explanation of *to eikos* as what I take to be the relatively technical account found in the *Timaeus*. I simply want to draw attention to the fact that Plato is clearly aware of and interested in the possibility of considering and explaining likelihood in terms of its relationship to truth, at least by the time that he writes the *Timaeus*.[27] I want to emphasize this for two reasons. The first is because it implies that Plato's attitude towards *to eikos* is rather more complex than is often recognized.[28] The second is that I think

[24] It is possible to identify a connection between likeness and likelihood (or a manipulation of the etymological connection between these ideas) in other early (and explicitly philosophical) contexts, for example Xenophanes B35 and Parmenides B860. See Bryan 2012 for a general account of the development of *eikos* terminology.

[25] See Rowe 2003b: 28 for a succinct summary of the puzzle.

[26] See Bryan 2012: 114–60 and Burnyeat 2005.

[27] Although my account assumes that the *Phaedrus* predates the *Timaeus*, the relative dating is more or less immaterial; I want to claim only that both dialogues can be read as evidence that Plato was alive to the issue of the relationship of likelihood to truth.

[28] In fact, it seems to be that what is said about *to eikos* as separate from the truth in the *Phaedrus* encourages scholars to interpret the *eikōs logos* of the *Timaeus* as being somehow deficient.

it hints at a possible alternative reading of Socrates' statement at 273d2–6. My suggestion is that Socrates does not offer the redefinition of *to eikos* as coming about because of a similarity to truth purely as a restatement of his earlier argument at 261e6–262c3. Rather, Socrates wants to redefine *to eikos* through its relationship to truth precisely in order to *challenge* the Tisian definition of *to eikos* as nothing other than 'what people think', with no apparent concern for the truth of the matter. If this is the case, Socrates is not rejecting *to eikos* outright as a means of persuasion. Rather, he is offering a rehabilitation of *to eikos* as a persuasive tool available *only* to the dialectician, for only he can recognize what is similar to the truth and thus present an account that is appropriately similar.[29]

Against the background of a traditional conceptual connection between likeness and likelihood such as that identifiable in the *Hymn to Hermes*, I suggest that it is possible to read 273d2–6 as adducing that traditional connection (to which Plato also turns in the *Timaeus*). It is worth looking once more at that passage in detail:

> Tisias, for a long time before you even arrived (*palai . . . prin kai se parelthein*) we were saying that this likelihood (*touto to eikos*) comes about for people because of a similarity to the truth (*di' homoiotēta tou alēthous*). And just now (*arti*) we explained that in every case the man who knows the truth (*ho tēn alētheian eidōs*) best understands how to discover similarities (*tas homoiotētas*).

Socrates claims that they have been drawing the connection between *to eikos* and the truth 'for a long time before you arrived (*palai . . . prin kai se parelthein*)'. He then says that they have 'just now (*arti*)' established the superiority of the dialectician in discovering similarities. As noted above, this reference to a previous conversation is generally and rightly taken to refer back to 261e6–262c3. However, the claim that they have been discussing this topic for a long time is particularly puzzling. Is the earlier discussion really extensive enough to justify this claim? My tentative suggestion is that 273d2–6 also refers to a general, perhaps even traditional, conceptualization of *to eikos* as persuasive through some kind of relationship of similarity to the truth. It is significant, I think, that the first claim is something that has been said for a long time (*palai*) whilst the second

[29] On this reading, 273d2–6 asserts the ability of the dialectician to perform a version of the art of persuasion described at 261d10–e4. The persuasive speaker is someone who can make anything resemble anything; the speaker who produces an *eikos* argument is making his position resemble the truth. See Cooper 2004: 67–71.

is something explained recently (*arti*).[30] My suggestion is that the *palai* clause refers, in part, to the time before 'Tisias came along' not just in the dialogue, but in the history of rhetorical theory. Socrates is referring back to a time before the sophistic separation of persuasion and truth, a time such as that of the *Hymn to Hermes*, where the conceptual and etymological connection between *to eikos* and similarity to truth was explicit. This is, of course, the connection that Plato will go on to exploit in his *Timaeus*.

So, my suggestion is that in this section of the *Phaedrus* Plato is doing more than recapitulating his earlier arguments against sophistic rhetoric. He is redefining *to eikos* in such a way as to reject both the persuasive value of *Tisian* likelihood (as Tisias himself would diagnose it) and the concomitant claims that persuasion is or can be unrelated to the truth.[31]

To eikos and truth

This returns us to the question of what Socrates is claiming when he says that *to eikos* comes about in the minds of people because of a resemblance to the truth. One possible explanation is simply that, in creating a likely argument, a persuasive speaker can quite reasonably be thought to be making his argument look like it is true. After all, when we say (in an approving fashion) 'that seems likely', what we are really saying is 'that seems likely to be true' or 'I am inclined to believe that'. This is a relatively obvious point, but it is one worth making against the analysis of *to eikos* attributed to Tisias in the *Phaedrus*. If Tisias really did claim that likelihood 'is nothing other than what the masses think' and, because of this, it has no relationship to truth, he seems to have overlooked precisely this relatively obvious point. For he has failed to see that to hold an opinion is to commit to its truth. So, for example, if a jury is convinced of some falsehood, this is because they think it likely to be true. Indeed, in the case of the feisty weakling and the cowardly muscle-man described at 273a6–c5, it may well be that they are convinced of an account that is untrue, but the point is that it is considered *likely to be true*. In fact, it is apparently considered more likely to be true than the truth itself. However, the fact that the truth can be considered unlikely or that what is false can be considered likely

30 *Palai* can sometimes stand as equivalent to *arti*, but it is also true that the two can stand in opposition, as at Pl. *Tht.* 142a1–2: Have you been here a while (*palai*), Terpsion, or have you just (*arti*) arrived from the country?

31 See Allen in this volume for a consideration of Aristotle's view of the connection between truth, similarity to the truth, and persuasion.

is no justification for Tisias' claim that likelihood can be understood and explained without reference to truth.

It is worth emphasizing that I do not want to suggest that Plato is denying that Tisian *eikos* arguments lack likelihood.[32] Rather, I think, Socrates is intimating that Tisias has misunderstood the nature of likelihood, or perhaps misdiagnosed the way in which it exercises its persuasive effect on the audience.[33] So rather than suggesting that Plato is making the rather counterintuitive claim that Tisias' suggested lines of argument are not genuinely likely, it is possible to reassess them as examples of *to eikos* properly understood as deriving their persuasive power from their perceived resemblance to the truth.[34] If the *eikos* nature of an account lies in its similarity to the truth, we can see that only the man who knows and looks to the truth will be able to produce *eikos* arguments. Plato is not rejecting *to eikos* as something entirely divorced from truth. He is attacking what he claims is the sophistic definition of *to eikos*. In so doing, he offers an analysis of *to eikos* as something perfectly compatible (indeed, only really compatible) with the project of philosophical psychagogic rhetoric informed by truth set out in the *Phaedrus*.

Conclusion

At the very least, I hope to have made a case for holding back from attributing to Plato what he presents as the sophistic definition of *to eikos* as unrelated to truth. I have tried to show that there is a way to read Socrates' claim that *to eikos* comes about through its similarity to truth as expressing a rather more positive attitude to this feature of rhetoric and, indeed, that this is a better reading than simply assuming it to be a recapitulation of the argument at 261e6–262c3. This positive attitude towards *to eikos* would, I think, be perfectly in keeping with the positive role that the *Phaedrus* ascribes to rhetoric as the potential handmaiden of

32 Werner 2012: 156–59 suggests that Socrates recognizes that the sophists will be persuasive because they do offer arguments which appeal to the beliefs of their audience. On my account, this appeal is not in itself sufficient to attain genuine likelihood, for this is a quality which an argument will possess only if it both appeals to its audience *and* is constructed in accordance with knowledge of the true nature of things.

33 Reinhardt 2010 suggests that Plato's emphasis on the connection between truth and likelihood in the *Phaedrus* may be a direct response to contemporary theory on *to eikos* as represented by the *Rhetorica ad Alexandrum*.

34 Some may object to the suggestion that Plato is using 'similarity to the truth' as equivalent to 'perceived similarity to the truth', but this emphasis on the perception of similarity induced in the audience seems perfectly in keeping with the description of persuasion as making something similar to something else presented at 262d10–e5.

philosophy and the philosopher's psychagogic endeavours. I have shown that there is a tradition of understanding likelihood as somehow connected to the idea of similarity to truth and that Plato is not averse to playing with this connection between likelihood and likeness elsewhere in the corpus. If, as Socrates seems to be suggesting, *to eikos* gains its persuasive force from its similarity to the truth, rather than from the fact that it is a mere statement of the beliefs of the majority, there certainly seems no reason that it should not have a role in philosophical persuasion. As Socrates reminds us more than once, the dialectician is equipped with knowledge of reality and truth and best placed to recognize what is similar to or different from that truth. He is therefore the man best placed to employ *to eikos* and rhetoric in general for philosophically respectable purposes. In the mouth of the sophists, *to eikos* is unpredictable and dangerous; employed by the philosopher, however, it can be a respectable tool of respectable persuasion.

CHAPTER 3

Aristotle on the value of "probability," persuasiveness, and verisimilitude in rhetorical argument

James Allen

A family of terms used by Aristotle and other ancient Greek authors may serve as a useful point of entry into this inquiry: *eikos* (likely), *pithanon* (plausible, persuasive), and *endoxon* (reputable). They are not synonyms nor are they etymologically related. That they have something in common nonetheless is suggested by the fact that, beginning with Cicero, they were all translated as *probabile*, meaning roughly "what invites or lends itself to approval," and words derived from it. In Cicero's hands *probabile*, though also used to render other terms, corresponds most often to *pithanon*, while *verisimile*, like or resembling the truth, corresponds more closely to the Greek *eikos*. He introduces the two terms together and treats them as near synonyms (*Academica Priora* 31).[1] Plato had already brought their Greek counterparts, *pithanon* and *eikos*, into close connection in the *Phaedrus* (272d–e), where, drawing on an association between likelihood and likeness or resemblance to the truth that was already old at the time, he argues that the persuasive power of likely argument is due to a similarity or likeness to truth (273d1–e4).[2] Aristotle sometimes treats *pithanon* as though it were interchangeable with *eikos* (e.g. *Rh.* 1400a5–14), and likewise with *endoxon* (e.g. *Rh.* 1356b26–33). Modern European languages are, through the Latin *probabile* and *verisimile*, ultimately dependent on *pithanon* and *eikos* for the terms they use for probability: *probabilité*, *vraisemblence*, *Wahrscheinlichkeit*.

The term "probable" and the ideas it has been used to express have had a long and eventful history. The discovery and rapid development of

I am indebted to Victoria Wohl and an anonymous reader for valuable comments and criticisms, and to Jamie Dow for illuminating conversation and correspondence about some of the topics tackled here.

[1] Cf. Görler 1992; Glucker 1995. The terms rendered as *probabilis* by Cicero also include *eulogos* (reasonable), which does not, however, figure in the discussion of rhetoric. On Aristotle's use of the term, which also occupies a prominent place in Stoicism, see Le Blond 1938.

[2] On *eikos* in the *Phaedrus*, see Jenny Bryan's contribution to this volume. On the ancient background of the idea of the *eikos* as likeness to truth, see Bryan 2012: esp. ch. 1.

the modern mathematical theory of probability during and after the seventeenth century, in particular, transformed our understanding of probability to such an extent that it has been argued we should regard "the emergence of probability" in this period as something entirely new.[3] Some older uses of the term are likely to puzzle the modern reader. A sentence from Gibbon, "such a fact is probable but undoubtedly false," for instance, sounds strange to contemporary ears.[4] Although there may be continuities between still-current uses of "probable" and related words, on the one hand, and the ancient family of terms from which they are descended, on the other, it is older and ancient uses of "probable," closer to their Greek originals, that are relevant to this essay and which I mean to signal by "probable" and related terms.

Without pretending to have exhausted the topic, which is full of interesting complications, we can distinguish roughly between an approbative use of *pithanon* (and *probabile*) and a contrasting pejorative or disparaging use. The first often serves to endorse a view with reservations or render a judgment with the proviso that one may after all be mistaken. This is not all that different from the way in which we say "it's probable that," "probably," or "it's likely that" or "very likely." In contexts where an argument has been presented, it means that it is compelling or is to be reckoned with or the like, even if it is not conclusive. In the contrasting case, the point of calling a view *pithanon* or probable is less to say that there are positive reasons for rejecting or doubting it than to represent it, or the grounds that have been offered in support of it, as only superficially compelling, as specious or merely plausible, as we would have to say instead of "probable but false." This is the kind of persuasiveness against which we ought to be on guard and by which we should strive not to be taken in. To put matters somewhat grandly, to speak of "probability," the *pithanon* or the *eikos*, in the first way is, roughly speaking, to say that it would be a proper and correct exercise of reason to accept a view on the strength of the available considerations (in a certain context, for this can change should more considerations become available). To speak in the second way, on the other hand, is to offer an explanation, or part of one, for a defective exercise of reason, actual or potential.

That the same terms can be used to these opposite ends in this way is not surprising. Whether being *pithanon* – persuasive or plausible – counts

[3] The title of a celebrated book, Hacking 1975. For an alternative account laying greater stress on continuities between new developments and what went before, see Franklin 2001.

[4] Cited in Hacking 1975: 19.

in favor of a view or not will depend on to *whom* it is persuasive and *how*. Likewise whether being *probabile* should recommend a view or not will depend on by *whom* it is such as to be approved and *why*. Typically one takes oneself to have exercised one's faculties in a judicious way, by sifting the evidence, giving due attention to the relevant considerations, consulting trustworthy witnesses and the like. And the fact that a view tends to persuade one or invites one's acceptance in these conditions or tends to persuade people who are like one in these ways, or human beings in general to the extent that they are, speaks in its favor. It is different when one has others in view (or sometimes oneself at other times and in other conditions). The fact that a view has features that make it persuasive to other people who are undiscriminating, credulous, or foolish, or to people in general to the extent that they are such, and invites acceptance by them no longer counts in its favor. By calling an argument plausible or persuasive – explicitly or implicitly – to *such* people, one intends to disparage it. If one recommends a view or argument to someone else that one presents as persuasive or plausible in this sense, it is not for acceptance but as a means of persuading a third party or parties.

Likeliness conceived of as resemblance to the truth participates in the same double character. The semblance of truthfulness can be found in the absence of truth and will explain why gullible and undiscriminating people take falsehoods for truths. What is more, it can be made to belong to false or dubious propositions by skilled practitioners of persuasion. On the other hand, when securely grasped truths are impossible to come by, even for the most thoughtful and discriminating, likelihood or resemblance to truth may be the best that can be hoped for and can serve as a (suitably cautious) recommendation.

Rhetoric is the art of persuasion – "the manufacturer of persuasion" or "a device for producing persuasion"[5] – and above all the persuasion of crowds or large public bodies. Detractors of rhetoric lean heavily on the pejorative uses of *pithanon* and *eikos*. They can derive support not only from Plato's depiction of the master rhetorician Gorgias in the dialogue of the same name, but also from Gorgias' own words in the advertisement for the power of speech that he includes in his display piece, the *Encomium of Helen* (8–14), and the advice Tisias and Gorgias are said in the *Phaedrus* to offer the orator, namely to give preference to the *eikos* or the *pithanon*

[5] Pl. *Grg.* 453a, e; 459c–d; 452c–d; 455a. On this formula, which was widespread, and its possible origin as a coinage of Plato's, see Mutschmann 1918.

over the true (*Phdr.* 267a; cf. 260a, 272d).[6] The picture that emerges from these testimonies is of a power that puts all convictions on a level in point of persuasiveness, regardless of whether true or false, good or evil. The persuasiveness it imparts to its arguments is only ever accidentally related to the truth of their conclusions or the soundness of the case it presents for them, if any. Conceived of in this way, rhetoric appears to be nothing more than a system for the organized exploitation of human beings' cognitive weaknesses and defects of intellectual character.[7]

After Gorgias has identified persuasion as the business of rhetoric in Plato's dialogue, Socrates induces him to agree that there are two kinds of persuasion, one that produces conviction without knowledge, the other that imparts real knowledge or genuine understanding (*Grg.* 454e; cf. *Rh.* 1355b29). Since even rhetoric's stoutest defenders would never claim that the orator, constrained as he is by speaking in law courts and at other public gatherings and before intellectually limited audiences, can achieve persuasion of the second, knowledge-imparting kind, the only alternative appears to be to admit that rhetorical persuasion is of the first kind, which implants conviction apart from knowledge. Since Gorgias has also agreed that the law courts and other deliberative bodies before whom the orator speaks tackle issues of justice and injustice and the beneficial and harmful, he comes perilously close to conceding that rhetoric is persuasion of the ignorant by the ignorant about matters of the utmost importance (*Grg.* 459b, d; cf. *Phdr.* 260c7).

The choice between alternatives offered by Socrates is too stark, however. It is exhaustive only if quite disparate conditions are lumped together as ignorance. The account of the value of persuasive argument in rhetoric defended by Aristotle is in part a reaction to Plato that is meant to overcome this excessively narrow choice. But though it is fair to say that Aristotle has a more favorable attitude toward the use of persuasive argument in rhetoric, which stands in sharp contrast to Plato's in the *Gorgias*, I shall also suggest that the differences between Aristotle's favorable view and the more favorable attitude toward rhetorical persuasion that Plato adopts when he returns to the subject in the *Phaedrus* are at least as important

[6] Tisias' name is often coupled with that of Corax in ancient testimonies (see Radermacher 1951: 28–35). Aristotle attributes to Corax an example of *eikos* argument like that ascribed to Tisias by Plato (*Phdr.* 267a, 273b–c; *Rh.* 1402a17–23). On Corax and Tisias and the place of *eikos*-based argument in rhetoric, see Michael Gagarin's contribution to this volume.

[7] Cf. Pl. *Grg.* 464d2, where Socrates says that rhetoric hunts by means of pleasure for folly (*anoia*) in audiences the better to deceive them.

and illuminating. Though he does not lean as heavily as Plato does on the term *eikos*, which he sometimes uses for one variety of persuasive argument among others (*Rh.* 1357a30–b2, 1402b13–1403a2; cf. *An. Pr.* 70a2–11), Aristotle too makes use, though in a different way, of the idea of similarity to the truth in his account of persuasive argument in rhetoric.

If Aristotle's view of rhetoric was more favorable than Plato's in the *Gorgias*, it was not because he was kindly disposed to rhetoricians active before and during his lifetime, who are the object of a scathing indictment in the first chapter of the *Rhetoric*. Aristotle's perspective is, however, in important ways different from that of the *Gorgias*. His principal charge is that rhetoric, such as it then was, occupies itself with playing on the emotions of the audience at the expense of presenting arguments to them about the matters in contention. He complains that the rhetoricians have little or nothing to say about the enthymeme, the rhetorical argument or syllogism, which in his view ought to be the central concern of rhetorical theory.[8] So harsh is his condemnation of appeals to emotion that it may come as a surprise when, in the second chapter, he presents emotion as one of the three artistic *pisteis*, types of proof or ways of securing conviction, alongside the presentation of the speaker's character and the argument itself (*Rh.* 1356a1–20).

For our present purpose, we can put aside the interpretive issues raised by the introduction of the non-argumentative *pisteis* – they do not alter Aristotle's emphasis on the centrality of argument – to focus on Aristotle's objections to the use of emotion in actually existing rhetorical theory and practice. He maintains that appeals to the emotions do not bear on the matter at issue or affair in contention – they are *exō tou pragmatos* – instead they relate to the hearer and work by affecting him or her (1354a18, 28; 1356a3–4, 14–15). The orator intent on rousing the feelings of his hearers is, he maintains, like someone who bends a straightedge (*kanōn*) before using it (1354a25). By contrast, in order to come to grips with the affair at issue, argument is essential (1355a15–18, 22), ideally argument confined to establishing that it, the issue or *pragma*, is or is not the case, has come to be or has not (1354a26–27; cf. 1356a3–4, 15–20) or will come to be or not (1354b13–15). It would be surprising if there were not a substantial overlap between the gratification of an audience by an orator, to which Socrates objects in the *Gorgias*, and the corruption or perversion of their capacity for judgment through emotion, against which Aristotle inveighs in the

[8] Cf. Burnyeat 1994.

Rhetoric.[9] But the opposition between the defects and failings of rhetoric and the alternative that would remedy them that Socrates' argument sets up is different from the one that emerges from Aristotle's account.

In a celebrated passage in the *Gorgias* arts that care for the soul and those that care for the body are presented as counterparts to each other (463a–466a). Two pursuits are counterparts in the relevant way by standing in the same relation to different objects or domains. Thus the arts of medicine and justice are counterparts by caring in the same way for the body and the soul respectively, of whose natures they possess genuine understanding and whose good condition – health and virtue – are their respective objects. In Socrates' system of analogies, as false likenesses or counterfeits of justice and medicine respectively, rhetoric and cookery are likewise counterparts, aiming at pleasure and not the good of the soul and the body, of which they lack knowledge, and to which as a result of these facts they are each related in a different but less creditable way than the corresponding true arts. Rhetoric and cookery are, in a word, impostors, and the alternatives to them – the remedies for their defects – are the real things, the arts of justice and medicine.

The alternative to playing on the emotions of the audience for Aristotle is not, as it is in the *Gorgias*, the care of the soul grounded in the substantive understanding of its nature, but argument about the matter at issue. The orator who concentrates on argument has already taken an important step in the right direction. Aristotle acknowledges that argument has an important didactic function that it can be made to discharge only by the master of a substantive discipline (*Rh.* 1355b28–32), and therefore presumably in the care of the soul through education. If he thought this were the only function of argument, his critique of rhetoric would resemble that of Socrates, who late in the *Gorgias* envisages a true but unexemplified art of rhetoric that is hardly to be distinguished from the art of the statesman, who teaches people to be just (504d, 508c, 517a–b). But Aristotle's object, revealed in the opening words of his treatise – "rhetoric is the counterpart of dialectic" – is to vindicate rhetoric by comparing it to another kind of argumentative discipline occupied with another kind of argument.

Dialectic enjoys a higher reputation than rhetoric, yet according to Aristotle does not have a substantive subject matter of its own about which

[9] According to Socrates, rhetoric caters to the auditors' gratification (*kharis*) and pleasure (*Grg.* 462c; cf. 464d). Aristotle notes that emotional appeals are less effective in deliberative than in forensic oratory because, while the audience's own interests are at stake in the former, in the latter they look only to their own gratification (*Rh.* 1354b11, 33).

its masters have knowledge and are in a position to instruct others. The language of "counterparts" is the same as that used in the *Gorgias*, and it is likely that Aristotle intended his words as an implicit rejoinder to Plato.[10] If comparing rhetoric to dialectic in this way elevates the former, however, dialectic as Aristotle understands it, between which and rhetoric there is in his view an analogy, is a less elevated business than it was for Plato, to whom dialectic was always the method that leads to philosophical truth and who sometimes comes close to identifying it with philosophy itself, whose coping-stone he calls it in the *Republic* (534e).

For Aristotle the two disciplines are counterparts by furnishing wide-ranging *practices* of dialectical and rhetorical argument with the appropriate methods. These are roughly characterized at the beginning of the *Rhetoric* as upholding and examining an argument, on the one hand, and defending and accusing (to which Aristotle later adds praising and blaming and counseling for and against when he distinguishes the three genres of oratory at 1358a36–b8), on the other. Though shared to an extent by everyone, they can be developed and practiced as arts of dialectic and rhetoric respectively (1354a3–11; cf. *Soph. El.* 172a30–36). Like dialectic, rhetoric equips its practitioners to construct arguments about any and every subject without the substantive knowledge that is the mark of the expert in each of the special sciences.[11] To set up either as a discipline with a special subject matter of its own is to miss its true nature, which is precisely to be a faculty of *logoi* (arguments, speeches) and not a science with its own defined subject matter (*Rh.* 1358a23–26, 1359b12–16),[12] which was precisely the possibility that it was the object of the first sustained argument in the *Gorgias* to exclude (449c–451d).

Part of the burden of Aristotle's argument in defense of rhetoric, as it ought to be, is that the lack of a substantive subject matter, together with the other features shared by the two disciplines, can be a reproach to rhetoric only if it is also a reproach to dialectic. In what follows I aim to demonstrate how his analogy between rhetoric and dialectic extends to the help each is able to lend us in our efforts to discover and grasp the truth, where, in the case of rhetoric, talk of the *pithanon* and the *eikos* in the positive sense is at home.

Aristotle has the most to say about the value of persuasive argument in rhetoric in and around a passage dedicated to showing that rhetoric is *useful* (*khrēsimos*) – henceforward "the passage" or "the main passage"

[10] On this point, see Brunschwig 1994, Rapp 2002: 20–21.

[11] *Rh.* 1354a1–3, 1357b7–9, 1356a31–34, 1358a21–25, 1359b11–16.

[12] Cf. *Rh.* 1355b3, 1356a33.

(1355a20–b9). It is full of comparisons with dialectic, and it resembles and refers to a passage in the *Topics* that sets out the purposes in relation to which that treatise and the method of dialectic expounded in it are *useful* (101a25–b4).[13] The usefulness of both arts requires special discussion because of the distinctive peculiarities that set them apart from the ordinary run of arts and sciences. When they make use of argument for didactic or other purposes, each of those arts and sciences is confined to the fixed set of principles proper to its domain and the consequences deduced from them. Each is of value to the extent that knowledge of that domain is of value, which in the case of practical and productive disciplines will itself in turn depend on the value of the products or actions made possible by that knowledge.

As we have already observed, rhetoric and dialectic are not confined to arguing about matters belonging to the domain of one special art or science. They are not even confined to making the case for true conclusions. Alone among disciplines they argue on opposite sides of the question (1355a29–36). The invention of a persuasive or reputable argument for a false conclusion can be just as much the exercise of the art of rhetoric or dialectic as the invention of an argument for the truth that is its contradictory. Rhetoric is defined by Aristotle as "the faculty of discerning the potentially persuasive about each matter" (*Rh.* 1355b23; cf. 1355b10–14). The object of the *Topics*, the treatise on dialectic, is "to discover a method by means of which we shall be able to construct syllogisms about every proposed problem from reputable premises (and by which, when it is our task to respond, we will avoid saying anything contrary to the thesis we defend)" (100a18–21). Evidently such pursuits stand in need of justification in a way that the special arts and sciences do not.

Here is a translation, starting with 1355a14–18 (§ 0), which comes before the official beginning of the passage at 1355a20ff. (= § 1) and is separated from it by a short ellipsis:

> 0. For it belongs to the same faculty (*dunamis*) to see the true and what is like the truth (*to homoion tōi alēthei*), and at the same time human beings are sufficiently well orientated toward the true by nature and mostly hit on truth. For which reason being good at aiming for reputable opinions belongs to the person who is also disposed in the same way toward the truth. . . .
>
> 1. Rhetoric is useful (*khrēsimos*) because the things that are true and the things that are just are naturally stronger than their opposites, so that if the

[13] Discussed in Allen 2007: 98–100.

judgments are other than they should be, defeat is one's own fault, which is a cause for reproach.

2. What is more, with regard to some auditors, not even if we had the most exact knowledge would it be easy for us to persuade them by speaking on the basis of it. For the account in accordance with truth is instruction (*didaskalia*), and this is impossible, but it is necessary rather to make our proofs and arguments on the basis of common items, just as we also said in the *Topics* about encounters with the many (cf. *Top.* 101a30).

3. Furthermore it is necessary to be able to persuade auditors of opposites, just as in syllogisms, not so that we should do both (for it is not necessary to persuade people of base things), but so that how things are should not escape us, and so that when someone else makes unjust use of arguments we should be able to dissolve them. None of the other arts, then, argue for opposites, but dialectic and rhetoric alone do this, for they are in the same way of opposites.

4. Yet the underlying matters are not like this, but the things that are true and the things that are better are by nature always more readily argued and more persuasive, to speak without qualification.

5. And furthermore it would be strange if it were a disgrace not to be able to defend oneself bodily, but being unable to do so in speech or argument (*logos*), which is more proper to human beings than the use of the body, were not. And if someone making unjust use of such a faculty of arguments (*dunamis tōn logōn*) would be responsible for especially great harms, this belongs in common to all the goods apart from virtue, above all the most useful.

Like the broader stretch of text to which it belongs, the passage is characterized by abrupt and puzzling transitions. What follows is a highly selective reading that ignores some problems in order to focus on the light the passage and its context throw on the positive value of rhetoric and the persuasive arguments that are its stock in trade.

The passage from the *Topics* to which Aristotle refers at 1355a28 reads: "having catalogued the opinions (*doxai*) of the many, we shall converse with them, drawing not on views alien to them but on their own, inducing them to shift their opinion (*metabibazontes*) when we think they speak wrongly" (101a30–34). As Aristotle notes, since the persuasive is always persuasive to someone, but no art studies the particular, rhetoric will not study what is reputable to, for example, Socrates or Hippias, but what is reputable to people of such and such a kind (*Rh.* 1356b26–57a1). Rhetoric, with its mastery of what is persuasive and to whom, will be in an excellent position to invent arguments persuasive to different types of audience. Many of these groups will be far less discriminating than those to which Socrates (or even Hippias) belonged, and elsewhere in the *Rhetoric* Aristotle emphasizes the limitations of the audiences to whom, and the constraints under which, orators must speak (1357a3, 1419a18). There is, then, room for a wide range of possible relations in which the considerations put forward by the orator can stand to those by which he is moved himself, from being substantially the same to being entirely different.

Nonetheless the picture of human epistemic powers that emerges from the account of the usefulness of rhetoric is surprisingly rosy. Human beings, Aristotle asserts, are orientated by nature to the truth, which they mostly hit; things true and things good are naturally stronger than their opposites and more readily argued for and persuasive, to speak without qualification. This picture is lent support elsewhere in Aristotle's work. Talk of inducing people to shift opinions is echoed in an especially revealing passage on method in the *Eudemian Ethics* (1216b26–36). There Aristotle says "best of all would be if all human beings were seen to agree with what shall be said but, failing that, then in a certain way all, which they will do being induced to shift their opinions (*metabibazomenoi*), for every one has a certain kinship with or affinity for the truth (*oikeion ti pros tēn alētheian*).[14] To judge by the passage on the usefulness of rhetoric, like the figure described at *Topics* 101a30–34, the orator has a part to play in assisting human beings to do what comes naturally to them, namely to grasp and appreciate the truth for sound reasons.

In the passage under examination Aristotle adopts the point of view of a speaker who is better informed or more knowledgeable than his auditors and is set on persuading them of the truth. To win their assent, he must adapt or tailor his argument for them by exploiting considerations that they find persuasive, where these may differ from those that would figure in a properly didactic, scientific demonstration of the same conclusion, if any, and from the reasons by which he is most moved himself. There is, however, another non-didactic use of argument available to those lacking expert knowledge, which likewise depends on and assists the human affinity for truth. It is discussed explicitly, albeit briefly, in the *Topics* but is, I suggest, also implicit in the *Rhetoric*, where it constitutes yet another way in which rhetoric is the counterpart of dialectic.[15]

[14] On this and similar passages, see Barnes 1980.

[15] Unless this is meant to be covered by "so that how things are should not escape us" (1355a32 in § 3), which may refer not to things in general, however, but to the immediate circumstances of the rhetorical dispute.

The *Topics* first refers to the dialectical version of this use of argument in I 2, as the last of three uses for the treatise specified there, directly after its use in "encounters with the many," on which we have already touched. It is "in relation to the philosophical sciences" (101a27–28, 34–36).[16] "Being able to raise difficulties on both sides of the question, we shall more easily discern the truth in each." The "we" who will be aided in our search for the truth are, in the first instance, the parties to a dialectical debate conducted in the right spirit of fair and open-minded inquiry, perhaps the audience if there is one and, it also seems, the solitary philosophical inquirer who, as Aristotle envisages him, will in effect take both parts in a dialectical discussion in his quest for the truth (cf. *Top.* 155a7–16). It is a beneficiary of the last kind whom Aristotle has chiefly though not exclusively in view when he returns to the issue in the last chapter of the treatise (163b9–16):

> With regard to knowledge and philosophical wisdom, to be able to see and to have seen the consequences that follow from each hypothesis is no small instrument (*organon*), for it only remains to choose one of them correctly, and for this a good nature is needed, that is, the good nature that relates to the truth, to be able to choose truth and flee falsehood well, precisely what those who are well endowed by nature do, for loving and hating well what is presented to them, they judge what is best.[17]

The appeal to a good nature here is yet another reference to the natural human affinity for the truth. That the art of rhetoric, apart from enabling the orator to tailor his message to his audience, can also be of use in the sphere in which it operates in something like the way in which dialectic is of use in its sphere is suggested by another passage in the *Rhetoric*, a few pages after the one cited above. After observing, in another passage also cited above, that rhetoric does not grasp the particular any more than any other art does and is, therefore, not occupied with what is persuasive or reputable to individuals but, like dialectic, with what is so to types (1356b28–34), Aristotle continues: "for [dialectic] also does not syllogize from any chance assumption, but from the beliefs of those who require argument, while rhetoric syllogizes from the beliefs of those accustomed to deliberate, for its function concerns the kinds of things about which we deliberate and regarding which we do not possess arts" (1356b34–57a2; cf. 1359b18–20).[18]

[16] At *Top.* 101a36–b5 Aristotle adds a further point (*eti de*): "in relation to the first principles of each science." On the question whether this is a fourth use, different from those already announced, or a further specification of the third use, see ad loc. Brunschwig 1967: xii, 116–17; R. Smith 1997: 52–54.

[17] On this passage in relation to *Top.* 101a34–36, see Barnes 1991.

[18] Translating the text printed by Kassel.

Aristotle's observation in the earlier passage, "not even if we had the most exact knowledge, would it be easy for us to persuade some by speaking on the basis of it" (1355a24–25), then, was a concession for the sake of argument. Vital as its tailoring function may be, the art of rhetoric typically finds its application in relation to matters of which there is not exact knowledge and which require deliberation. The audience's limitations and constraints of time and occasion are not the only reasons why "instruction is impossible" in the sphere where rhetoric is of most use (1355a25–27), or why the arguments that it employs do not and should not meet the didactic standards erected by Socrates in the *Gorgias*.

If dialectic can be applied to especially good effect in the preliminary investigation of issues falling under the philosophical sciences – domains in which there is a fixed system of settled and immutable truths, the understanding of which is the mark of the master of the corresponding science – rhetoric is especially useful in spheres where immutable truths are few and prone to exceptions, where constantly renewed deliberation on the basis of shifting and inconclusive evidence is required to make the decisions on which action is based and where there is no such thing as a fixed and settled condition of the intellect in which all the truths proper to the domain are grasped. Aristotle's remarks here invite comparison with his account of deliberation in the *Ethics*, where we learn that it is concerned with things that are to be done and up to us, but which permit of being otherwise and do not fall under any of the exact and self-sufficient sciences, in this way resembling medicine and navigation more than gymnastics or the art of writing (*Eth. Nic.* 1112a33–b10; cf. 1141b8–12). If philosophical inquiry conducted by a solitary investigator can sometimes resemble an internal dialectical discussion in which the philosopher takes both parts, rhetorical debate in a deliberative assembly will sometimes resemble the deliberations of an individual agent, in which, however, the tasks of making cases for and against different courses of action are assigned to different orators.

If the case that I have been assembling is on the right lines, then, rhetoric has a positive contribution to make not only to the orator tailoring his argument for the truth as he knows it to an audience, but also, by enabling orators arguing on opposite sides of a question to make the most persuasive case possible for each, to the process of collective deliberation undertaken in pursuit of truth. Of course these uses do not exclude each other. Both depend on human beings' natural affinity for the truth, which rhetoric, well used, assists. With these thoughts in mind, I want to return to the passage on the usefulness of rhetoric to discover what light it throws on the persuasive as it is employed by rhetoric when it is being put to use in these ways.

Among several statements in the main passage remarkable for their epistemic optimism, the assertion at 1355a36–38 (§ 4) stands out: "things true and things better are by nature more readily syllogized and more persuasive to speak without qualification." By "more readily syllogized" Aristotle presumably does not mean that the case for something is more readily put into the form of a syllogism (whether by "syllogism" we mean something strictly answering to the theory of the *Prior Analytics* or, as we probably should, something looser), but rather that a good case can be made for it more easily than the alternative because there are more and more compelling grounds supporting it to hand. Aristotle's statement, "the true and the better are more persuasive," makes the best sense when viewed in relation to the remarks about human beings' natural affinity for the truth, which are abundant in the context, especially the assertion at 1355a14–18 (§ 0): "human beings are sufficiently well orientated toward the true by nature and mostly hit on truth." When human beings' natural capacity to discover and appreciate the truth is given the chance to work as it should and is provided with the help it needs (sometimes by dialecticians or orators), the true and the better will be more persuasive to them.

A characteristic, F, belongs to something without qualification (*haplōs*) if it is true to say that it is F in abstraction from particular circumstances or with qualifications mentally removed, for example qualifications such as *to* or *in relation to someone* or *something, at a time* or *in a place*.[19] The qualifications from which abstraction needs to be made here will refer to the factors that hinder or impede this capacity, for example strong emotion, partiality, prejudice, inattentiveness, and the like on the side of the listener, deceit or incompetence and the like on the side of the speaker. When the orator serves the interests of truth and justice, then, he does so precisely by exercising his faculty for the discovery and presentation of persuasive or plausible arguments.

This way of conceiving of the matter may be contrasted with the way of thinking that opposes the persuasive to the true that is prominent in Plato's *Gorgias* and *Phaedrus* and, it seems, in Tisias' and Gorgias' own advertisements for their discipline. That the capacity to discover and grasp the truth is not to be separated from the ability to assess and appreciate the persuasiveness of arguments is apparently the burden of 1355a14–18 (§ 0), where Aristotle asserts that it belongs to the same faculty to see the

[19] Cf. *Top.* 115b29–35.

truth and what is like the truth.[20] Aristotle's view can be usefully compared with two alternatives in the *Phaedrus*. There as we have seen, rhetoricians like Tisias are said to have advised the orator to pay no attention to the truth, but to cultivate the persuasive, which is what seems true to the many, which is the likely (272d–273a; cf. 259e–260a). The point is illustrated by the notorious story of the attack by a weak (but brave) man on a strong (but cowardly) one (267a, 273b–c; cf. *Rh.* 1402a17–23). Each is advised not to tell the (implausible, unlikely) truth, but a (likely) falsehood instead. To be sure, the truth may occasionally be likely, but if so it is only by accident.

To this Socrates opposes a different picture, in which the persuasiveness or likeliness of a case, instead of being only accidentally related to the truth and capable of being manufactured by people who are wholly ignorant of it, can be produced – in an artful, systematic way – only by those with knowledge of the relevant truths. He fastens on the idea of the likely as the source of persuasiveness, and he argues that the orator who wishes to deceive without being deceived works by exploiting likenesses to the truth or reality. This orator passes by degrees through them to the likelihood that he will present as the truth to his auditors in order to induce them to shift their opinions (*metabibazein*) as he would. There can, then, be no art of rhetoric apart from knowledge of the truth (261e–262c).

Aristotle agrees with Plato in rejecting the idea that the persuasive or likely is related to the truth only by accident, but with a crucial difference. To Plato's way of thinking, the capacity systematically to invent persuasively likely arguments is inseparable from a grasp of the truths which they are like, but, though there is a moment in the *Phaedrus* when rhetoric and dialectic seem about to converge (266c7–9), the moment passes and the task of discovering and grasping truth is reserved for dialectic (266c, 269b). Instead rhetoric is conceived of as an ability to lead souls (*psukhagōgia*) that makes use of knowledge won by other – dialectical – means (261a, 271c).

On the interpretation I am defending, however, Aristotle would object, and perhaps in the passage under consideration does object, to this division of labor. On his view, the two counterpart disciplines, though operative in different spheres, stand in the same relation to truth. Although it is not the function of either to discover or secure the truth by itself, each, by enabling its practitioners to make the most persuasive or reputable case possible for a thesis, can be highly useful to this end. By bringing out what can be said on both sides of a question, they provide assistance to human beings'

[20] The relation of 1355a15–18 (§ 0) to their context is puzzling. See the appendix to this chapter for a discussion of the problem and a suggested solution.

capacity to discern the true and the false. And this will have been the point of Aristotle's statement that it belongs to the same faculty to discern the truth and what is like the truth, which so resembles Plato's view about the relation between the true orator's ability to argue persuasively by likelihood and knowledge of the truth (*Rh.* 1355a14–15, *Phdr.* 273d).[21]

Aristotle's statement that it belongs to the same faculty to discern the truth and what is like the truth (1355a14–15) makes the best sense if we understand talk of the *verisimile*, or that which is like the truth, not as meaning that for each *verisimile* or likelihood there is a truth to which it is similar – the way in which artistic verisimilitude is sometimes understood. Rather something will be likely or *verisimile* by exhibiting the features, characteristics, or marks that belong to the truth, for example, implying or being implied by known truths, being compatible with them or conforming to well-established patterns among similar previous occurrences, or in some contexts, the power to illuminate or explain other truths, and so on. The reason why it belongs to the same person to be both good at aiming at reputable opinions and disposed in the same way towards the truth is that the reputable opinions are similar to the truth in virtue of displaying these marks or features, and the task of assessing the reputable credentials of an opinion, by attending to them, cannot be divorced from that of pursuing the truth.

As a result, though Aristotle grants rhetoric more autonomy than Plato does in the *Phaedrus* – it can be practiced without knowledge or specialized expertise – this is not because he stands with rhetoricians like Tisias and Corax, who appear to have thought that the orator can, and will be well advised to, practice his trade without paying any regard to the truth at all. By contrast, Aristotle held that rhetoric, like dialectic, is useful because of the aid it is able to lend human beings in search of the truth.

This is the conclusion that I have been most concerned to establish. I have done so, however, by dwelling on some programmatic remarks near the beginning of the *Rhetoric*. There is an obvious question that is raised by those remarks, namely how well the rosy picture of the discipline they imply, especially on the interpretation defended here, squares with the more detailed advice offered to the orator in the rest of the treatise. This is too large an issue to be tackled here, but the conclusions reached so far may throw some light on a puzzling observation of Aristotle's about the persuasive or plausible; by attending to this we will also be able to look

21 Cf. Cooper 1994: 203–208, who calls these lines the "Phaedran" passage and speaks of the relation here between the abilities to see the truth and to see what resembles it going in the opposite direction from Socrates' in the *Phaedrus*.

at how Aristotle uses the three terms whose affinities were our point of departure: *pithanon*, *endoxon*, and *eikos*.

Shortly after the passage on the usefulness of rhetoric Aristotle states: "it belongs to rhetoric to discern both the persuasive and the apparently persuasive just as it does to dialectic to discern both the syllogism and the apparent syllogism" (1355b15–17). This is puzzling because, though we are used to opposition between reality and appearance, the persuasive or the plausible often occupies the same place as the apparent in a like contrast with the truth. To speak of the "apparently persuasive," then, might seem to be like speaking of the "apparently apparent" in opposition to the really or truly apparent. We could perhaps interpret the passage as involving an implicit reference to two perspectives. It might appear to a poor practitioner of rhetoric that an argument would be persuasive to his audience when in reality it would not be. This is no doubt true, but it seems less plain that it is the point Aristotle has in view. For one thing, it does not fit the parallel with dialectic especially well. Behind that parallel, I suspect, is the accomplished dialectician's power to recognize an opponent's fallacious argument for what it is. Others may be taken in by it; he understands why they are, but also why they should not be (which will ideally enable him to persuade them to correct their mistake).

What is more, it does not seem to be a point of this kind that Aristotle wants to make when he contrasts the reputable and the apparently reputable in the *Topics* or when he contrasts the likely and the apparently likely later in the *Rhetoric* (*Top.* 100b23–101a1; *Rh.* 1402a2–27). The *Topics* passage is a difficult one, but its aim seems to be to distinguish between considerations that have some force or weight and those which superficially appear to do so but whose grip on us dissolves when they are properly scrutinized. Alexander of Aphrodisias offers among other examples the statement "he who sees has eyes," which would permit us to deduce that the one-eyed person has more than one eye (*In Top.* 20, 10, 10–19 Wallies).[22]

The reference to the apparent likelihood occurs in the chapter of the *Rhetoric* devoted to *topoi* of the apparent enthymeme (II 24). According to Aristotle, the whole method of Corax consists in this fallacy, the abuse of likelihood (1402a17). He cites a version of the notorious encounter between the strong man and the weak man as an example (1402a17–23; cf. *Phdr.* 267a, 273b–c). He treats this kind of argument by likelihood as a species of the fallacy of *secundum quid* as it is analyzed in the *Sophistical Refutations* (cf. 166b37–167a20, 168b10–16, 169b9–12, 180a23–31). This approach is

[22] See the note ad loc. Brunschwig 1967: 114–15.

not without problems,[23] but for our present purpose it is notable that Aristotle, unlike Plato, calls this a merely apparent likelihood, so implying a distinction between merely apparent and real likelihoods with a different character. It seems fair to conclude that these are grounds or considerations which afford good if not conclusive reasons for a conclusion, considerations by which it is reasonable to be moved and which can be regarded as such even if in the end we find the considerations assembled on the other side of the issue more weighty and compelling.

If Aristotle's remarks about the persuasive and the apparently persuasive are to be understood along the same lines, then here at least we find a view of the persuasiveness that it is the object of rhetoric to produce that squares well with the rosy view of the discipline's usefulness in the pursuit of truth that it has been the chief object of this essay to illuminate. Despite the abuses of which rhetoric is capable,[24] and for all the defects of actually existing rhetorical theory and practice, well-practiced rhetoric can, through its mastery of probable argument, serve human beings' natural affinity for the truth.

Appendix: *Rhetoric* 1355a14–18

These lines are separated from the passage proper, §§ 1–5, by a back-reference to an earlier assertion that the professional rhetoricians of Aristotle's day favor judicial oratory because it is easier to persuade an audience by rousing their emotions when their own good is less obviously at stake (1354b26–1355a1). The "for" (*gar*) with which it begins should mean that it lends support to the preceding sequence of assertions (1355a3–14): that the artistic method of rhetoric is properly concerned with *pisteis*; that the *pistis* is a demonstration; a rhetorical demonstration is an enthymeme, and the enthymeme a syllogism of a kind; that the syllogism is the concern of dialectic, with the result that he would be best able to invent enthymemes who adds to the knowledge of how and from what a syllogism arises, which comes from dialectic, a grasp of the sort of things about which enthymemes are (on this passage, see Burnyeat 1994: 12–15). But it is hard to see how. It has been argued that the truth and what is like it are the concerns

[23] Cf. Allen 2001: 58–62.

[24] The fact that rhetoric is capable of doing great harm when used unjustly is a feature it shares with all the goods apart from virtue, especially the greatest, e.g. strength, health, wealth, and strategy, Aristotle observes, here agreeing on another point with the Gorgias of the dialogue, who maintains that though rhetoric, like boxing and other competitive accomplishments, can be used to do harm, it is not the fault of the teachers of these skills, but of those who abuse them (Pl. *Grg*. 456c7–457c3).

of dialectic and rhetoric respectively (e.g. by Rapp 2002: 78). The point would then be to support the view, asserted in the preceding lines, that the two disciplines are related. The problem, acknowledged by the view's proponents, is that it is not obvious that Aristotle thought of dialectic and rhetoric in this way. Elsewhere he often takes dialectic to be concerned with opinion by contrast with the special sciences, which are concerned with truth (cf. Bonitz 1870 *ad* 203b47–49).

What is more, it is not clear how the relation between dialectic and rhetoric that is asserted in the preceding, namely that they require an understanding of the syllogism, which is the special study of one of them, is supported by the existence of a faculty that discerns both truth and verisimilitude. Admittedly, rejecting this view leaves the relation with the foregoing as much of a mystery as ever. In view of these difficulties, I wonder if these lines might not have been displaced. They are of a piece with the epistemological optimism of §§ 1–5 and would, I suggest, fit especially well after § 4, 1355a36–38, where they would lend support to the statement that "the things that are better and the things that are true are by nature more readily argued and more persuasive, to speak without qualification" (1355a36–38) *because* (*gar*) "it belongs to the same faculty to discern what is true and what is like the truth and human beings are well orientated to the truth and mostly hit on it . . . " (1355a14–18). If lines 1355a14–18 do not belong where they are found, the sequence of statements about enthymemes and syllogisms and the need for the orator to learn from dialectic about the latter can be viewed as a well-motivated digression that helps to contrast the proper focus of rhetoric's artistic method on argument with the rhetoricians' misguided obsession with emotion and consequent neglect of deliberative oratory, to which 1355a18–20 returns.

CHAPTER 4

"Likely stories" and the political art in Plato's Laws

Ryan K. Balot

Likelihood, what is likely or possible, likely stories: this closely related nexus of ideas plays a central role in a remarkable number of Platonic dialogues. The reason is that "the likely" is related to "the true" in a highly ambiguous way. Plato understandably found these ambiguities worthy of philosophical investigation. The present essay examines the role of likely stories, or plausible myths, in Plato's *Laws*. In this dialogue we find Plato's most searching exploration of "the likely" and "the true" in their political dimensions. Clarifying their relationship will shed light on Plato's vision of the limits, necessities, and possibilities of political life. In order to set the stage, it will be useful to begin with the rhetoric, politics, and metaphysics of likelihood in two other Platonic treatments.

In the *Phaedrus*, first, Socrates lays bare the interrelations of rhetorician, subject matter, and audience, in the context of debates that were familiar to and contentious among his contemporaries. Socrates criticizes Tisias and other rhetorical teachers for pandering to conventional beliefs because, in the process, they obscure the appropriate position of "the likely" in the art of rhetoric (272d–273b). These professors misguidedly equate "the likely" with "what is accepted by the crowd" (273b1, trans. Nehamas and Woodruff 1995). This is why they have come to believe that "the likely" must be esteemed more highly than "the true" (267a), an opinion that Socrates works hard to reverse in his conversation with Phaedrus. In the latter part of the dialogue Socrates argues that, whether he wants to teach or persuade, a speaker should acquire two kinds of systematic knowledge: first, of the truth about his subject matter and, second, of the soul-types of his audience (260d–e, 270b, 277b–c). In doing so, he elaborates the positive account of didactic rhetoric found in the *Gorgias* and points toward other positive treatments in the *Statesman* and in Aristotle's *Rhetoric*.

In the course of giving this topic the careful attention it deserves, though, Socrates does not ignore or dismiss the possibilities of intentional deception, secrecy, and purposeful obscurity. Knowledgeable speakers, he says, are better than the ignorant at producing similarities – at moving, little by little, from what is true to what is not (263b). Hence, deceiving others is easier for those who genuinely understand similarities and dissimilarities (262a–b). Moreover, and more importantly, those with knowledge speak differently to different people, knowing when to speak and when to remain silent (276a), and knowing how to instill words (*logous*) and customary practices (*epitēdeuseis nomimous*) into each particular soul so as to produce the kind of "persuasion" and "virtue" that "you want" (270b). The possibility of knowing "the whole" (270c2), of course, is by no means taken for granted; but, even assuming that such knowledge is possible, Socrates makes it clear that using plausible fictions will be within the knowledgeable speaker's repertoire: simple speeches for simple people (277c).

Plato entertained a different vocabulary of likelihood, and a different set of purposes, in the *Timaeus*. Here too, however, he continued to reveal the tight connections between likely stories, rhetoric, and political life. In the *Timaeus* Socrates characterizes his interlocutors – Timaeus, Critias, and Hermocrates – as people who by nature and education participate in both philosophy and politics (19e–20a). Was Timaeus' cosmology, then, implicitly intended to have political significance? The dialogue itself draws attention to this question, because it begins with Socrates' and Critias' lengthy discussion of politics (17c–26d) and promises to finish with Critias' exposition of Athenian political life, which is to be based on the accounts of Socrates and Timaeus (27a–b). As an account of the origins of human beings (27a), among other things, Timaeus' exposition is acknowledged to be metaphysically and epistemologically prior to Critias' political *logos*. The metaphysical account appears unmistakably to be prior to the political account. Yet there is no reason to suppose that Timaeus' "likely" metaphysics is not crafted, even in a decisive way, with a view to politics (Critias' subject) and human excellence (Socrates' subject); indeed, the framing device suggests as much.[1]

Timaeus offers an account of likelihood that is not at odds with Socrates' account in the *Phaedrus*. At the beginning of his cosmology Timaeus draws a set of apparently familiar distinctions between being and becoming, truth and convincingness, reality and likeness (29b–c). Instead of disparaging "the likely story" by comparison with the truth, however, Timaeus proposes that as human beings we should be satisfied with an account that is no less likely than others (29c–d). On this provisional basis, Timaeus proceeds

[1] This is the central theme of Lampert and Planeaux 1998, who argue convincingly that "By pairing Timaeus and Critias, and by fitting Timaeus's cosmology within the account of the city in motion, Plato puts cosmology into the service of political philosophy" (119).

to describe the creation of a providential and rationally ordered universe, which is truly living and possesses soul and intelligence (30b–c, 92c). As he plumbs these metaphysical depths, Timaeus repeatedly calls this account a "likely story," whether he describes it as *muthos* or *logos* (e.g. 30b, 44d, 48d, 55d, 56a, 68b, 72d, 90e). As many have seen, in fact, Timaeus connects the vocabulary of "likelihood" to the idea that things that "come to be" (such as the cosmos) are likenesses of stable realities or "Being." Hence, our accounts of them are necessarily also merely "likely," as opposed to the "stable" and "unshifting" accounts of "fixed" and unchanging Being. But this epistemic distinction does not imply that as human beings we will ever be able to discover any such "stable" or "unshifting" account of Being as such. No: as Timaeus stresses, "We must do our very best to make these accounts as irrefutable and invincible as any account may be" (29b7–c1, trans. Zeyl 2000). Even if the ultimate realities are *per se* knowable with clarity, they are knowable in that way only to the god "and to any man he may hold dear" (53d7).[2] As the dialogue unfolds, it emerges that this man is the rhetorically and philosophically gifted statesman.

Throughout his long and complex speech, readers can see Timaeus fitting his "likely" account of nature to the purposes of ethical teaching and political governance.[3] These points emerge, in part, through Timaeus' representation of human psychology. Timaeus works hard to explain the origins of sense perception, pleasures and pains, fear and spiritedness, and the other emotions (42a–b). Sense perception arises from "forceful disturbances" – the transformations brought about in bodies by "things coming to them and leaving them" (42a4, trans. Zeyl 2000). Having been created in a mixing bowl (41d–e), human souls are internally discordant with themselves, filled with conflicting emotions (42a–b) and subject to "toilsome transformations" (42c4). Timaeus emphasizes self-mastery as the key to justice and the good life (42b). He particularly concerns himself with explaining the punishments due to those who fail to master fear, pleasure, and pain: in a dialogue framed by stories of war and "manly" political activity, these akratic souls would be born again as women and then as wild animals (42b–c, 76d–e; cf. 86b–c). Faulty bodily constitutions conspire with defective political constitutions to produce vicious

2 Cf. Johansen 2004, Burnyeat 2005.

3 Too few Platonic scholars have grappled with the dialogue's political framing device. or discussed the metaphysics and psychology of works such as *Timaeus* and *Laws* in connection with the politics of these dialogues. For the conventional view of Plato's use of *muthos* in this and other dialogues, see J. E. Smith 1985. Noteworthy exceptions are Lampert and Planeaux 1998, Kalkavage 2001, and Howland 2007. Howland 2007: 12 argues that "Timaeus' cosmic poem seems designed to answer to the needs of political animals."

anger, cowardice, and stupidity in the soul (86e–87b). The "naturalness" of humanity's capacity for rational self-governance remains highly uncertain. Yet, as a philosophical statesman (cf. 20a), Timaeus accepts the task of justifying the city's commitment to order by referring to the natural order of cosmic intelligence – and, as Howland has observed, to the need to use force in order to produce order from disorder (30a, 35a–b, 47e–48a).[4]

But this commitment remains difficult, if not impossible, to achieve in practice. Despite the god's own dedication to rational order, Timaeus' account emphasizes the difficulty of preserving measure or balance within the human world (47d–e, 87d–88d). This prevailing instability, in fact, is why the statesman will find it crucial to rely on beliefs about the gods' "unstraying revolutions" (47c3), on divine justice and punishments, and on the hierarchical ordering of humanity to make his audience moderate and law-abiding. Despite initial appearances, *imitatio dei* truly involves establishing a civil religion and ruling over men with prudence, by means of either persuasion or force. Invented by past men "who claimed to be offspring of gods" (40d8, trans. Zeyl 2000), and accepted without any adequate proofs (40d), the traditional gods themselves are caring and providential, yet they are free of any responsibility for humankind's evils and mistakes (42d–e). Only the gods and a small group of human beings have a share in *nous* (rational intellect), while all other human beings have a share only in true belief (51e; cf. 53d). Timaeus explains that the gods hierarchically separated immortal and mortal souls within humanity, in order to ensure that the "best part" is in charge and rules by force (71b–c), while the "ambitious part" is located in the chest, ready to obey the citadel's commands to vanquish appetite (69d–70b; cf. 72e–73a). It is a short step to declaring the justice of philosophical rule and the goodness of a life dedicated to our most sovereign and divine element – our intelligence (89d, 90a–d).

In light of these divergent accounts of "likelihood," similitude, and plausible fictions, I propose to explore two questions about "likely stories" in Plato's *Laws*. First, what is the purpose of myths and fictions, namely "likely stories," in this most explicitly political of Plato's dialogues? Is it to obscure the truth for political purposes, for example to cultivate moderation or to tame the aggressiveness of certain spirited citizens, or is it to maximize true belief among all citizens – or are there still other possible motivations? Second, is it problematic to base the political world of Magnesia on "likely stories," however these are to be construed? What consequences do the Athenian's "likely stories" have for the political theory of Plato's *Laws*?

[4] Howland 2007: 13–15.

The character of the Athenian's likely stories

Initially, at least, the idea that the Athenian's "likely stories" would be actively deceptive, though politically salutary, might seem deplorable.[5] Yet perhaps this moralizing outlook is anachronistic, the product of Enlightenment commitments to transparency and publicity. Consider Kant's transcendental formula of public right: "All actions affecting the rights of other human beings are wrong if their maxim is not compatible with their being made public."[6] Did Plato's characters express similar commitments to publicity, similar qualms about spreading falsehoods publicly? Seemingly not: and the topic surfaces repeatedly in his dialogues, not only in the *Phaedrus*.

The most obvious example is the so-called "noble lie" of the *Republic*, Socrates' own "myth of the metals" (414b–415d). Socrates liberally sprinkles this section with the vocabulary of "falsehood" (e.g. 414b, 414c; cf. 414e).[7] He also refers back to the earlier discussion of falsehoods that are useful, he says, against enemies or to prevent the ignorant from wrongdoing, or to explain ancient things that are unknown (382c–d). The rulers of the city will be like doctors who use falsehoods for the good of the city (389a–d). Despite these discussions of falsehood, to be sure, it is arguable that his "myth of the metals" shows Socrates communicating what he takes to be the truth about Callipolis' citizens – namely, that they are, in fact, born into a hierarchy of natural talent or merit that the city should do its best to respect, through testing children and making occasional intergenerational adjustments (415b–c). This is the sort of myth (*muthos*, e.g. at 415a2) that makes it easier for everyone to accept what is, according to the dialogue's political provisions, just, namely political inequality.

By contrast, Socrates does not equivocate about the use of carefully designed sexual lotteries in Callipolis. These lotteries are meant to insure that inferior people will not complain about their dismal sex lives, when superior young men are given sexual opportunities more frequently, in order to father more children (460a–b, trans. Grube 1992; cf. *Ti.* 18d–19a). This is simply a lie – and probably not a very effective one at that, since even the "inferior" will come to see that the lotteries invariably give prizes to top warriors and mathematicians. This may be just or not. We are given

5 Stalley 1983: 173–78; Bobonich 2002: 111–12.

6 "Perpetual Peace, Second Supplement: Secret Article of a Perpetual Peace" (Reiss 1991: 114–15) and "On the Agreement between Politics and Morality According to the Transcendental Concept of Public Right" (Reiss 1991: 125–26).

7 Notably, Socrates also refers to the secret handling of children and their upbringing in the framing prologue to Timaeus' long speech (*Ti.* 18c–19a).

no argument to address that question. Socrates argues only that the lotteries are useful to the city's breeding practices: that is, they are supposedly good for the city as a whole without necessarily being good for each, or even most, of its members.

What about the Athenian Stranger's own use of myths, plausible fictions, and even deception in Plato's *Laws*? Why does the Athenian not only use myths and plausible fictions, as he says, but also raise questions and doubts about the relationship of the true and the just to the useful? Is it possible to escape the worry that, despite his emphasis on the truth, truthfulness, and his own search for the truth (e.g. 634c, 730c), the Athenian also recommends to Kleinias the use of salutary fictions in order to moderate or control the future citizenry of Magnesia?

These questions matter because they are directly connected to the purposes of the Athenian's political regime, as well as his vision of what is politically possible. They have given rise to highly divergent interpretations. Bobonich, for example, holds that Magnesia is an educative regime designed to enable all its citizens to flourish to the greatest extent possible, through rationally grasping the goodness of virtue and embodying that goodness in their lives. As a result, the Athenian's preludes, his chief innovation, "are not intended to inculcate false but useful beliefs or to effect persuasion through non-rational means."[8] Others propose that the Athenian's teachings, and particularly the preludes, are meant to encourage virtuous behavior, albeit through emotional appeals to praise and blame, or through a kind of "preaching," rather than through strictly rational persuasion.[9]

Zuckert, by contrast, maintains that in Magnesia philosophy supports a "specific political order" by offering ideological teachings that reassure citizens of the essential connections between human happiness and rationally discoverable cosmic order – even though the "'truths' the philosophers would teach are not coherent."[10] Along the same lines, albeit more insistently, Strauss holds that the Athenian is aware that his likely stories are false; his theological and ethical teachings are merely politically salutary, designed to cultivate piety and to moderate the otherwise disruptive human passions, particularly *thumos* (spirit, anger) and the desire for pleasure.[11]

[8] Bobonich 2002: 104, cf. 91–123; Bobonich 1994: 272.

[9] Morrow 1960: 556–60; Stalley 1983: 42–44; Laks 1991; Nightingale 1993: 291–93; Yunis 1996: 227–36.

[10] Zuckert 2009: 51–146; cf. Benardete 2000: 288–314.

[11] Strauss 1975: esp. 2–3 and 26–37, with 140–56 and 179–86; cf. Pangle 1980: 388–91, 400–402, 413–17, 495–96, 503–504, 507–10, with Pangle 1976. Pangle 1980: 504 writes, "The city's piety is based on the truth inasmuch as its piety provides an avenue for philosophic pursuit of truth." On Platonic metaphysics as rhetoric, in general, see Parens 1995.

In order to make progress on this question, it is useful to examine several prominent likely stories in the *Laws*, both in their particular settings and with a view to their general dramatic and argumentative functions. From the outset, the Athenian's likely stories come into view in a studiously indeterminate way: their status is unclear and they resist straightforward categorization. The question of telling the truth or crafting fictive narratives (or even lies) is present on the text's surface, but it is by no means definitively resolved on that surface. I will eventually suggest that even if these ambiguities have made readers uncomfortable, they were not unintentional; in a sense yet to be determined, the ambiguity itself is the point.

Early on, to take one example, the Athenian likens the human being to a "divine puppet, put together either for their play or for some serious purpose – which, we don't know" (644d8–e1).[12] The *logos* declares that our souls consist of gold and iron cords, gentle reason and powerful passions (644e–645b); and this *logos* somehow supports or "saves" a *muthos* – "the myth of virtue, the myth about us being puppets" (645b1–2). The epistemic status of this analogy is never clarified, even if the Athenian immediately uses these psychological principles to address the "pleasures, pains, the spirited emotions, and the erotic emotions" that are crucial to education, that are intensified by drinking wine (645d6–8).[13] The Athenian briskly arrives at the conclusion that the art of politics is the care of the soul, which depends upon an educative regime anchored in correct understanding of "the natures and habits of souls" (650b7). The image of the puppet continues to be useful later in the discussion, when the Athenian tries to define humanity's humble, and yet paradoxically dignified, position in relation to the gods (804b; cf. 803b).

Given the Athenian's stress on developing an education to virtue (643e–644a, 705d–706a, 765d–766b, 770c–e, with Stalley 1983: 34–40), readers might forget that the Athenian does not argue carefully for this particular anatomy of the soul (by contrast, of course, Socrates does argue carefully for a very different anatomy of the soul in the *Republic*; cf. Frede 2010: 118). He does, however, emphasize that each person should follow one of the cords: "this cord is the golden and sacred pull of calculation, and is called the common law of the city" (645a1–2), a characterization that tightly links our rationality to the city's laws. This cord is "gentle rather than violent" and requires "helpers" if "the race of gold" – the law – is to

[12] All translations from the *Laws* are from Pangle 1980 unless otherwise indicated.

[13] According to Stalley 1983: 61, "Overall it is difficult to avoid the conclusion that the Athenian has not thought through his account of choice in these passages [discussing the 'puppet of the gods']."

control the other cords (645a6). In its most perfect condition, this golden cord embodies "prudence" (*phronēsis*) or "intelligence" (*nous*), which is the leader of the divine goods (631c–d) and helps to ensure that all the city's virtues, including moderation, justice, and courage, are "knit together" in the best way possible (632c). The Athenian's particular representation of human psychology grew out of the initial conversation about the law and is fitted closely to the city's need to develop the civic virtues by means of intelligent "guards" (632c; cf. 688a–b, 906a–907a, 961a, 962c, 964d) and "helpers," namely punishments and the use of force, as well as religious ideology, that will support the golden cord against the encroachments of the passions.

The Athenian's psychology is impressive and may even be correct; at this stage, however, it is not only provisional, but also specifically designed with the city's legal, ethical, and educational needs in view. At the end of the *Laws*, as if to confirm the provisionality and simplification characteristic of the puppet image, the Athenian states that the city's future philosophers will need to reconsider all the key questions of human goodness *ab initio*: they will need to examine whether virtue is one or four or somehow both (963c–965e) and how the beautiful and the good are both many and one (966a). Kleinias had grasped that virtue was the aim of the regime, and that *nous* was the leader of the four virtues, but he did not recognize the need to explore any additional questions about virtue (963a). The Athenian even initiates a short Socratic examination of the unity and multiplicity of virtue with Kleinias (963d–964a), although this dialogue within the dialogue, meeting with apparently insuperable difficulties (963d–964a), quickly transforms itself into a proposal to institutionalize an educational system for the city's philosophers (964d–965b). Even then, the Athenian is emphatic that "it isn't easy to discover what things should be learned" (968d3). These educational topics are "incapable of being described beforehand" (968e4). This is why he considers it so risky and impossible, albeit paradoxically necessary, to offer an account of "the education and upbringing that have now again become the topic of the discussion" (969a2–3). At the end of Plato's longest and most prescriptive dialogue, the very basis of the regime is up in the air, all over again, although the images and arguments seemed settled, even reified, in the mind of Kleinias.[14] The Athenian has offered the puppet analogy merely as a likely, although perhaps the

[14] As Bobonich 2002 has argued, Plato's representations of human psychology change significantly from dialogue to dialogue – if one wishes, from "early" to "late" texts.

best available, and even the most politically necessary, story about human psychology.[15]

At the beginning of Book 2, to take another example, the Athenian explicitly raises the possibility of using falsehoods in the training of the city's young. For the characters themselves, the immediate question is whether the just life is also necessarily the most pleasurable life – or, if these lives are mutually exclusive, then which of the two is happier (662c–d). The Athenian indicates that he and Kleinias might come to agree about justice and pleasure, despite their present differences, "if some god were to give us consonance" (662b1–2). Immediately thereafter, he invokes Zeus and Apollo, the divine legislators of Knossos and Sparta respectively, as though their arrival will settle the question; but after verging on the impiety of interrogating the gods' views, the Athenian shifts his cross-examination from them to human "fathers and lawgivers" (662c–e). He sees that harmonizing the pleasant and the just, the good and the noble, is crucial for persuading many citizens to live "the pious and just life" (663b2). Then, through a rapid series of exchanges, the Athenian secures the tentative, not to say grudging, agreement of Kleinias, at least, that the unjust life is not only more shameful but also less pleasant than the just life (663d). Yet Kleinias has enough political awareness to recognize that few will accept that justice is so pleasant (663e), even if many poets and others say so, under pressure of harsh penalties (662b). For his part, the Athenian may genuinely believe that the just, noble, pleasant, and good all move as one (662b), but he recognizes that his short argument is inadequate to establish even the harmony of pleasure with justice, not to mention piety (663b, 663d) or nobility (663a). He needs the help of a god, but this help will not be forthcoming until Book 10, where the just are shown to receive pleasure in the afterlife and the unjust are appropriately punished (904d–905c; cf. 662c–d, 664b–c).

This impasse explains why the Athenian now turns to the possibility of deception. He immediately adds that, whatever its truth-value, his argument equating the just and pleasant lives would help to convince citizens to live justly: "Even if what the argument has now established were not the case, could a lawgiver of any worth ever tell a lie more profitable than this (if, that is, he ever has the daring to lie to the young for the sake of a

15 The image has certainly given rise to divergent interpretations. Zuckert 2009: 114–15 argues that the psychology based on the puppet is incomplete because the Athenian has not taken into account the citizens' anger; Bobonich 2002: 258–92 regards the image as expressing Plato's later, unitary psychology.

good cause), or more effective in making everybody do all the just things willingly, and not out of compulsion?" (663d6–e2; cf. 663b). The Athenian concludes this stretch of the conversation by emphasizing the need to find advantageous beliefs, not necessarily true ones. By referring to the myth of the "sown men" (663e–664a), he assures his interlocutors of their capacity to persuade the citizenry of useful beliefs, true or not. Yet his arguments for the truth or justice of these useful ideas are either nonexistent, or too narrow and circumscribed to bear the political weight placed upon them.[16]

Likely stories, or useful fictions, also surface later in the dialogue, particularly in the Athenian's intentionally frightening "likely stories" about punishment.[17] In Book 9 the Athenian points out the shamefulness – and yet the necessity, given that they are human beings legislating for other human beings – of laying down punishments for criminals (853b–c). After offering incisive and moderately subversive reflections on criminal responsibility, the Athenian descends to the worst types of crime, such as the "voluntary and wholly unjust murders of kinsmen," for which the punishment is carried out not only on earth but also by "watchful Justice, the avenger of the blood of kinsmen," who ordains a strict *quid pro quo*: if someone intentionally kills his mother, for example, he will be reborn as a woman and killed by his own child (872d–873a). This type of "myth" (872d–e), along with other stories of punishment, retribution, and demonic luck, is likely to be more useful than true (cf. 865d–e, 870d–e, 877a–b, 880e–881b). As Stalley points out, "The Athenian repeats stories about the vengeance of the gods and of the souls of the dead even though he officially regards vengeance as irrational (729e, 865d–e, 870d–e, 872e–873a)."[18] So too with the prelude addressing temple robbers, which explains their evils as motivated by "a certain gadfly that grows naturally in human beings as a result of ancient and unexpiated injustices" (854b3–4). The laws exist in order to help citizens live in friendship and moderation and to curb those with an intrinsically tough nature (880d–e).

In Book 10 the Athenian also offers a "likely story" designed to explain the ordered rationality of the cosmos to disbelievers. He delivers an impressive philosophical theology based on arguments about self-moving motion and the priority of soul to body, all intended to rebut materialist philosophers whose metaphysical theories have helped to corrupt the souls of the young (cf. 885d, 890a). Along with these arguments about the existence of the gods, the Athenian develops a metaphysics obviously designed to promote

16 Stalley 1983: 62–64 agrees, calling the reference to lies the Athenian's "disreputable argument."

17 On punishment, see especially Stalley 1983: 137–50; Saunders 1991.

18 Stalley 1983: 148.

justice within the city: the god, as craftsman and supervisor, has taken care that acting for the good of the whole does not conflict with but rather simultaneously constitutes the good of each part, including human beings (903c–d). The "good" is virtue: the leader of the gods constructed the entire cosmos and each of its parts "so that in the whole, virtue would very much triumph, and vice meet defeat, in the easiest and best way" (904b4–6). This implies that, although we are responsible for ourselves, we nonetheless inhabit a providential universe where criminals will pay just retribution either on earth, in Hades, or somewhere even more extreme (904c–d, 905a–b).

As we discovered in the framing narrative of the *Timaeus*, the Athenian's appreciation of cosmic order conforms surprisingly closely to his political purposes and designs. These purposes emerge most forcefully in the section that the Athenian calls "mythic incantations" (*epōidōn . . . muthōn*, 903b1), although his evocation of a cosmic battle over justice and injustice (906a) and his system of punishments (907d–909a) run a close second.[19] These passages develop the Athenian's exhortation to new colonists, where he argues that the "avenging god" and "Zeus, god of strangers" will supervise relations with outsiders (729e–730a). It is striking that those sent to the "Moderation-Tank" must learn to be "moderate," having had discussions with members of the Nocturnal Council (909a); that is, they must practice the city's religion and keep their opinions to themselves, without being forced to agree that Magnesia's civil theology is correct or adequate to the cosmos itself. Perhaps it is not accidental, but rather ideologically significant, that the *Laws* begins with the word "god" (624a1) and that Timaeus begins his long speech with an invocation of the god (*Ti.* 27c).

Diverse audiences

How to adjudicate between the competing scholarly positions on the Athenian's "likely stories"? My own view is that we need not, and indeed we

[19] See Bobonich 2002: 101–104, 108–109 for the view that these arguments are "elaborate and philosophically sophisticated." Kraut 2010: 64–65 argues that Book 10 demonstrates "cosmological truths." Schofield 2003: 12–13 believes that Book 10 constitutes a "reasoned justification of the religious assumptions which underpin his whole approach to the question of the best *politeia*," even though he then leaves open the question of whether Book 10 is nothing more than a "civil theology." Many commentators agree that the Athenian aims to embody his own understanding of the cosmos in his religious teachings, albeit at a reduced level appropriate to the pedagogical purposes of the city and the diverse intellectual capacities of its citizens: see Schofield 2003: 13; Schofield 2006; Van Harten 2003. For discussion of some of the problems with the Athenian's theological arguments, see Strauss 1975: 148–50; Pangle 1976; Stalley 1983: 169–78; Halper 2003: 261–63; Zuckert 2009: 121–29, 143–46.

cannot, make a final determination between these positions. The reason is that the city is a community of members with diverse inclinations, talents, and needs.[20] In addition to Kleinias and Megillos, in fact, the dialogue shows us a variety of other figures who would hear the Athenian's legislation in diverse ways. This variety of characters cuts across the familiar Platonic distinction between philosophers and non-philosophers. There are those who think freely and "follow reason alone" (835c). Others refuse to believe the city's religious mythology and refuse to participate in its prayers and sacrifices (887d–888c). The dialogue's "cast" ranges from the philosophers of the Nocturnal Council (964d–968d), to civic-minded individuals who serve on the Nocturnal Council and in the city's magistracies (951d–952b, 961a–b), to atheistic philosophical inquirers (821a–b, 899c–d, 908b–d), to other types of criminal who need punishment in order to become moderate (862d–863c).

Other characterizations are more general. While the Athenian expresses optimism about the pedagogical efficacy of the legal preludes (722e–723b, 857c–e; Stalley 1983: 42–44), and about the city's future generations (769a–e, with Bobonich 2002: 395–408), he also expects certain citizens to be incapable of an education to virtue. Necessity compels each city to compromise the standards of genuine – that is proportional – justice, and to introduce "equity," which weakens the city's conformity to justice (757d–e). Most citizens will merely abide by the laws without much reflection (966c, though cf. 846d), perhaps out of fear or piety (838d–839c). They will be ignorant of anything more than the most rudimentary education, which is exemplified by the Athenian's reference to simple mathematics (818a–b; cf. 735a, 809e–811b; Pangle 1980: 485–91). The Athenian likens the "populace and majority in the city" to the "part that feels pain and pleasure in the soul," the part that refuses to obey "knowledge, or opinions, or reason – the natural rulers" (689b2–3).[21]

20 See Strauss 1975: 180–81: "One may also think of the fact that the common good has a ceiling (the genuine virtue of all citizens who are capable of it) and a flooring (survival), and that these two ends may make opposite demands on the statesman." Cf. Halper 2003: 263–64, who, in discussing the peculiarities of the Athenian's arguments about self-moving soul and cosmic justice, imagines a diverse audience consisting of ordinary citizens, philosophers, and "well-educated citizens whom we might call the professional class" (264); and Rowe 2010, who distinguishes between philosophers and non-philosophers on his way to making the obvious point that experienced readers of the dialogue are imagined to be experienced in reading Plato's other works: I believe that Rowe is mistaken, however, in thinking that Plato's other dialogues provide conclusive arguments for the Athenian's positions. Schofield 2010 holds, rather, that the *Laws* contains two different political projects: one of maximizing virtue, the other of constraining those "whose appetite for virtue is limited" (22).

21 Kraut 2010: 67 agrees that "in any large population a variety of psychological types can usually be found" but argues that most Magnesians will possess at least a "fragile" form of (demotic) virtue.

Political founders must, nevertheless, communicate effectively with citizens of all types. Recall, therefore, Socrates' advice to Phaedrus that philosophers – such as the Athenian Stranger – will be best at understanding how to create likely stories through the manipulation of similarities and subtle differences. Socrates had maintained that the most effective speakers would be aware of the diverse "soul-types" within their audiences. They would choose to modulate their speeches accordingly, even occasionally simplifying or obscuring important points in order to instill whichever virtues and beliefs they might find desirable. In Plato's *Timaeus*, Timaeus himself staged an elaborate performance along these lines – one that provided a template of sorts for the Athenian Stranger.

If the dialogue as a whole is designed to speak differently to different people, then we can understand why the Athenian's presentation of the puppet, of the coincidence of just and pleasant lives, of punishment, and of theology is studiously indeterminate. Some citizens, such as those "tested and hardened by only a small education" (735a4), will take comfort in a providential, albeit occasionally punitive, universe designed by beneficent gods with human "puppets" in mind. The puppet image itself will help them grasp that their most important virtue consists in self-mastery, or controlling their powerful urges, including, ultimately, their spirit as well as their desire. They will identify above all with the city's traditional religious rituals and sacrifices, where they will mix and "become more friendly to one another," that is seek peace, at least, if not virtue (738d–e; cf. 628c–d, 688a–b, 697b–c, with Stalley 1983: 37–40).

Others will question and explore the Athenian's "likely stories" and find in them an invitation to think philosophically about the problems he raises. They will pick up on the Athenian's oblique hints that we should resist having excessively powerful convictions about the world above and the world below, especially when our beliefs conflict with our development of excellence in the soul (727d). Some will be rationally educated by the city's preambles and learn to think through questions of justice, convention, and religious ritual for themselves; others will learn to obey the laws for more or less good reasons and, at the limit, they will obey out of fear; others still will be frightened into submission or, though rulers, be "enslaved" to the laws (715d).

The diversity of the citizenry becomes especially important when we recall that the Athenian recommends using the *Laws* itself as a source and model for the city's educational "curriculum" (811c–812a, 957c–e). Quite possibly, Magnesia's citizens are meant to peruse the lawgiver's arguments and other writings on their own (cf. 822e–823a, 858c–859a, 890e–891a, with

Bobonich 1996 and Nightingale 1993).[22] Their readings of the dialogue might well mirror the interpretations advanced by scholars, because, as readers, we ourselves are ranged alongside the officials, or even citizens and students, of Magnesia. We must make a judgment for ourselves about which characters to identify with and which lines of argument make most sense. Should we identify with Kleinias' new convictions about the rational order and providence of the cosmos, and the harmony between cosmos and individual? Or should we identify rather with the Athenian's uncertainty and even skepticism about the nature of virtue and the gods?

Despite these intentional ambiguities, the dialogue offers astute readers at least one important signal that we are supposed to identify with the Athenian's own philosophic uncertainty. To see what I mean, consider one particularly well-known example of a "likely story." In describing the city's educational regime, the Athenian declares that "nowadays" all Greeks tell falsehoods about the "great gods, Sun and also Moon" (821b6). The falsehood is that these and other stars are "wanderers": they never follow the same path but wander through the heavens randomly. The Athenian insists that this falsehood is offensive to the gods, because "each of [the stars] always moves in the same circular path, which is one and not many – though each appears to move in many" (822a5–8). The Athenian says that he regards this astronomical mistake as blasphemous – but why?

The answer comes later, when the Athenian argues that "the entire path and motion of heaven and of all the beings in it has the same nature as the motion, revolution, and calculations of Intelligence, and proceeds in a kindred way" (897c4–7); this motion is circular (898a). Readers of *Timaeus* will recognize instantly that the circular path was ordained for these bodies by the god. In Timaeus' speech, circularity signifies rational order and intelligence itself (38c–d, 40a–b, 47b–c). Hence, in this context, the Athenian means to correct the traditional Greek religious teaching about "wanderers" in order to help future Magnesians appreciate that the cosmos is rationally ordered, as both he and Timaeus vigorously maintain. In this way the city's religious teachings form part of an educational "ascent" designed to encourage all citizens, to the maximum extent possible, to respect and appreciate, if not worship, the goodness of both rational order and the activities of the rational intellect (*nous*).[23]

[22] As Bobonich 1996: 259–73 has shown, there are many interesting questions to ask about the relationship between written law and the critique of writing found in Plato's *Phaedrus*. This relationship, however, is beyond the scope of the present investigation.

[23] See Bobonich 2002: 107–109, citing earlier scholarship; Halper 2003: 264–65.

Put differently, this education is meant to encourage the citizens to regard the rational intellect (*nous*) as leader of the soul and rational intellect as such (capital N-*Nous*, if you like) as the god of gods. At the very center of Book 10, in fact, the Athenian calls *Nous* "god, in the correct sense, for the gods" (897b2).[24] It is here, however, that we appreciate the suspiciousness of the Athenian's theology, on two grounds. First, it is suspiciously convenient that *nous* rules in the cosmos, since this provides a theological justification not only for the philosophical life of the individual but also for rule by the philosophical elite of the Nocturnal Council. We discerned a similar pattern in Timaeus' own representation of the cosmos and its relationship to human psychology.

Second, the Athenian's conception of *Nous* as the god of gods looks very much like a projection of what the Athenian considers to be humanity's own highest concerns and activities.[25] Was the Athenian unaware of what Xenophanes had once said? "But if horses or oxen or lions had hands, or could draw with their hands and accomplish such works as men, horses would draw the figures of the gods as similar to horses, and the oxen as similar to oxen, and they would make the bodies of the sort which each of them had" (Xenophanes, fr. 15, trans. Lesher 1992). It is not difficult to believe that the Athenian's cosmology was developed out of the activity of philosophy itself: *Nous* is the god of gods, just as human *nous* is the leader of the human virtues. If the traditional gods were anthropomorphic, so too is the untraditional Platonic god *Nous*.

Public teachings, political challenges

Politics is the realm of the "likely story," the indeterminate grey region of opinion. It is appealing to envision the Athenian as a benefactor of humanity, who uses "likely stories" in order to educate non-philosophers and to make possible a decent, moderate political life. For example, the Athenian gradually moves his interlocutors from their unreflective Dorian outlook to a more thoughtful position on the gods and the virtues. From the very start, he skillfully transforms the Dorians' conception of courage from an exclusively militaristic one to one that combats both pleasures and fears (633c–634c). He grasps their starting-points and incrementally

24 This is of course a contentious passage; the reading in the text is accepted not only by Pangle 1980, but also by Mayhew 2010: 206, although Mayhew does not offer a political reading of Platonic cosmology; cf. Mayhew 2008 for more attention to the political context and meaning of Book 10.

25 Cf. Halper 2003 for a different, but careful and interesting, interpretation of Plato's use of the motif of *imitatio dei*, in the theological sections of the dialogue.

enlarges their understanding. Although the Dorians have misgivings about his criticisms of their lawgivers (630d, 666e–667a, 806c–d), he convinces his companions that the *logos* requires them to limit the Dorians' respect for narrow and traditional expressions of courage and to view *nous* as the highest of the soul's virtues (687e, 963a). By the end of the dialogue we can see that Kleinias is convinced by the Athenian's arguments for the ordered rationality of the cosmos and the gods' supervisory care over human beings (907b; cf. 821c). Both beliefs help to ground his newly discovered commitment to founding a city dedicated to virtue, and to his (now) unwavering belief in the leadership of *nous* (962e–963a).

Even so, the dialogue is intended less to appeal than to lay bare the chilling and irresolvable difficulties of political life. These stem from humanity's inability to grasp the complete and final truth about our condition. As the Athenian often emphasizes (e.g. 925e–926b), he does not adequately grasp "the whole." This is why he continually reminds his interlocutors that he is neither a true legislator nor a possessor of god-like wisdom. While the Athenian does present himself as an accomplished philosopher (892d–893a, 968b) – for example, he is highly adept at deploying the Socratic paradoxes within the legal code (e.g. 860c–e) and at using other advanced philosophical methods such as "collection and division" (815b–816d) – he also states explicitly that he is, like his companions, a man in search of, rather than in possession of, knowledge (859c; cf. 641d). He knows that his treatment of what is most important – human psychology, the theory of punishment, theology, the virtues – has been inadequate.

It is not always, or even often, easy to know why the Athenian treats these topics inadequately: perhaps because he lacks full knowledge, but also perhaps because political life demands that his accounts be radically foreshortened, that his public teachings (true or not) be designed primarily to persuade, moderate, and even control a diverse citizenry. The Athenian finds himself in the highly awkward position of having to legislate for citizens who need diverse benefits from the city's institutions and laws, when he himself, in truth, does not adequately grasp what is good for the citizens. The dialogue is intended to dramatize just this difficulty, which is intrinsic to political life.

Despite the Dorians' changed perspective on certain central questions, then, we are left with two serious difficulties. First, although the Athenian was explicit in his decision to simplify the argument for the "non-swimming" Dorians (892d–893a; cf. 900c), his interlocutors go beyond accepting his presentation as a merely provisional statement of his worldview – a worldview that, even if expressed more fully and completely, would still be radically inadequate. By the end of the dialogue Kleinias and Megillos have become forceful exponents of the Athenian's stated ideas (857e, 858b–c, 885e–886a, 887b–c, 890d, 899d, 907b, 963a, 965e, 966b), even though they often fail to understand his meaning (820d, 963c–d) or forget obvious lessons (714d–e, 792b–c). They are convinced that they need the Athenian's help in founding the city (969c–d; cf. 892d–893a) and that they must install philosophers in the central ruling body of the city, the Nocturnal Council (968b–c). Kleinias wants to hold on firmly to the Athenian's "truths" – understandably, because he wants those truths to be functionally effective in the life of Magnesia. As much as Kleinias and Megillos evolve over the course of the dialogue (e.g. 687e, 820a), though, they remain deaf to the Athenian's self-criticisms and qualifications of his own theses (896b–c, 907b–d).

Kleinias, in particular, believes that the Athenian offers the "pious" and "correct" argument and therefore trusts him (898c, 890d, 903a). He also believes that the laws proposed should be "set down in writing" and "entirely stable" (891a1–2). Kleinias as a practical lawgiver will reify the Athenian's arguments, rendering them more stable than they were intended to be – more stable than they in fact are. Given the Athenian's explicit and emphatic commitment to truth and his authoritative manner of presenting arguments (e.g. 870a–d, 896b–c), Kleinias can be forgiven for believing it both correct and beneficial to render the fluid, oral *logos* concrete in written law, even though he thereby reverses the apparent "lessons" of the *Phaedrus* (890e–891a, with Bobonich 1996). Be that as it may, for Kleinias, the Athenian's arguments are true, stable, even immutable: in Kleinias' view, the only question is whether even the "slow learner" can examine the laws and the lawgiver's writings often, so as to arrive at a proper understanding of them. In short, by the end of the dialogue the Dorians have internalized the Athenian's *logos*, reified its implications for the law code, and expelled any religious doubts from their souls (907b). Many modern scholars have taken the perspective of Kleinias.

Second, and on a related note, it is disturbing that the Athenian's ideas will function as a controlling ideology for Magnesia, even though he does not have fully convincing answers to the fundamental questions that he himself raises. On the basis of likely stories, the Athenian is willing to exercise coercive power over life and death – which makes him uncomfortable and helps to explain his endless deferral of legislation.[26] When he arrives,

[26] The Athenian's emphasis on the deferral of legislation: 737d, 768c, 768e with Laks 2000: 263–66. At 781d–e he has to "go way back" to make appropriate legislation for women, which he can do

after endless deferral, at one of the central areas of any legislative code – the penal code – he remarks upon its messiness and even shamefulness (853a–c). At its heart politics is an ugly, messy, uncomfortable, and typically dangerous activity, at least in the Athenian's view. Yet in emphasizing the discomfort of his protagonist, Plato also intends his readers to "feel," quite acutely, the discomforts involved in actualizing political theories in real legislation. The nature of political activity, as opposed to philosophical dialectic, is that we must act decisively in this thoroughly disordered and largely unknowable world.[27]

Conclusion

It is fair to conclude that Plato teaches us, his readers, that disorder and "remainders," and our own discomfort with them, are characteristic features of "the political" as such. The *Laws* clarifies the tensions involved in political life – not in order to provide any resolution of those tensions, which is impossible, but simply for the sake of our understanding of the nature of political life.[28] This does not imply, however, that political life, or the Athenian's engagement with it, lacks dignity. Quite the opposite. Though an accomplished philosopher, the Athenian has chosen to forsake private philosophical conversations with his peers, where the participants can "afford" to be playful, experimental, and even erotic.[29] He has chosen

because the three speakers are "enjoying leisure." At 783b they still have not completed the topic of education; cf. 788a. Finally at 796d the Athenian says that his account of the gymnastic art is complete. But even at 796e he still must continue to discuss other parts of education.

27 This is why Laks 2001: 113 is correct to say that "the transformation of legislation into philosophy remains out of the reach of common humanity, if not of humanity as such," but this should mean not only that humanity is incapable of attaining to a full and complete grasp of the truth, but also that legislation and philosophy are two intrinsically different activities. Philosophical ideas cannot be seamlessly transformed into legislation without loss or remainder – far from it; and the dialogue means to render this issue vivid.

28 It is confusion over this issue, I think, that leads Schofield to say: "it would be a mistake for those of us who are some species or other of democratic liberal to think we can find very much to identify with in its pages . . . There is a message here for our times – but warning, not beckoning, us" (2003: 13). My own view, as expressed above, extends Strauss 1975 and Pangle 1980.

29 Plato uses the dialogue, in part, in order to bring out the conflict between philosophy and politics in this way. This focus on the conflict between philosophy and politics does not imply, as Rowe believes, that "Socrates is dead and buried" or that "The city of Magnesia, we may say, leaves no room for someone like him" (2003a: 92). Rowe 2003a also finds, oddly to my mind, that Plato himself "still thinks that he is genuinely upholding Socratic ideas as well as continuing to subscribe to Socratic method," when in fact he is not, according to Rowe; Rowe knows Plato better than Plato knew himself. In Rowe's interpretation we see the dangers of anachronistically working with a conception of "a thoughtful, liberal-minded, egalitarian" Socrates (Rowe 2003a: 92). For a more careful attempt to connect the thinking of the *Laws* with that of other ("earlier") Platonic dialogues, see Larivée 2003.

to address the realm of constraint, necessity, and force – and perhaps to make irreversible mistakes in that realm. It is not Kleinias or Megillos, but rather the Athenian, who exemplifies true courage, because he has chosen to take the meaningful risk of legislating even when he knows that his legislation will be misunderstood and will, in any case, always be inadequate to the task of cultivating human excellence. He chose this life, I think, as the clearest possible imitation of the god's own mixed life.

CHAPTER 5

Open and speak your mind

Citizen agency, the likelihood of truth, and democratic knowledge in archaic and classical Greece

Vincent Farenga

In *Democracy and Knowledge: Innovation and Learning in Classical Athens* (2008), Josiah Ober makes an important theoretical contribution to our understanding of Greek citizenship. He hypothesizes that a participatory, deliberative democracy such as Athens relied on a citizenry organized with the goal of "transforming raw data and unprocessed information into politically valuable knowledge" (18). Through the use of this "knowledge in action" (18), this society of "free and equal individuals" somehow "would be able to form . . . compatible intentions, would come to share beliefs about others' intentions, [and] could gain common knowledge" (8). By using a "social epistemology" (34) to understand citizen behavior, Ober points to an essential function of Greek citizenship which has remained obscure: the production of knowledge in forms that were socially, politically, militarily, and economically advantageous to the state.

I do not think it is a coincidence that the democratic citizens of Athens were the most enthusiastic and industrious users of deliberative and forensic arguments based on *to eikos*, usually translated as "what is likely or probably true." As the contributions to this volume explain, this was a potent method of inquiry that did not rely on a sure grasp of what was true or real but on what could reasonably be considered true or real according to various logics of likelihood, plausibility, or appropriateness. In this chapter I ask whether such arguments contributed to the production of politically valuable knowledge by citizens acting as both individual and collective agents. I suggest that one particular use of *to eikos* expanded the horizon of what citizens could claim to know individually and collectively, and that this invigorated two of the "epistemic processes" Ober identifies as instrumental to democratic knowledge. These are "aggregation," or "the process of collecting the right kinds of dispersed knowledge in a timely manner for decision making," and "alignment," which permitted "people who prefer similar outcomes to coordinate their actions by reference to shared values and a shared body of common knowledge" (Ober 2008: 26–27).

I focus on the role that a citizen leader played in providing others with incentives for speaking out during deliberation to share their individual knowledge. I first examine the contributions of the Athenian Solon and of Xenophanes of Colophon and Elea to the early formation of a group agency capable of leading citizens to collective knowledge, which I equate with the archaic term *noos*/*nous* (and its cognates), signifying "mind" or "intelligence." Part of my discussion deals with the difficulty that archaic and classical thinkers faced when conceptualizing the move from individual to plural agency, especially when understanding how the knowledge of individuals could be so organized or aligned. I then discuss how *to eikos* bridges this gap by redefining the validity of what individuals contribute to group knowledge, especially in the Athenian democracy, in effect "opening" citizens' minds to alternative senses of what was true or real. Lastly my argument considers one practice of speech in Athens that was instrumental to aggregation and alignment because, like the use of *to eikos*, it permitted individual speakers to share personal knowledge. This was *parrhēsia*, or the freedom in deliberation to "speak one's mind" not about what was absolutely true but about what one thought *might* be true.

Joint actions, group agency, and Greek citizenship

If an essential function of Greek citizenship was the production of useful knowledge, then states had to coordinate citizen interaction in order to deliberate and make decisions. This requires an awareness of how a citizen, acting as an individual, could contribute and respond to what other individuals say and know, but in such a manner as to produce a plural or group understanding in the form of an assembly's proposals (*gnōmai*), decrees (*psēphismata*), and laws (*thesmoi, nomoi*), or a jury's verdicts (*gnōmai*). To account for this, Ober draws from contemporary theories of individual and group agency and from the contemporary philosophies of joint action and rational choice (2008: 6–7). List and Pettit (2011) provide the basic sense of agency as the threefold capacity: (1) to perceive and represent how things are in an environment; (2) to prefer, or form a motivation about, how things should be; and (3) to process these representational and motivational states to produce an intention and so change a perception into a preference (19–21). A human agent, however, does more than react to an environment and form simple propositions based on a limited set of possibilities: sophisticated propositions are formed by measuring the degrees to which something is or is not the case or by expressing a possibility, necessity, or obligation that something be the case (26). The agent's intentions are therefore based on judgments about what should be the case, incorporating theories of causation, counterfactual conditions, complex scenarios based on contingencies, or projections far into the future (26–27). He or she thus articulates a variety of preferences and then monitors or regulates these, checking for accuracy, consistency, and contradictions (24–31).

The key to coordinating an individual's agency with that of others in a group – for our purposes, when a citizen participates with others in collective deliberation and decision-making – is the alignment of intentions. This occurs in a type of collective action resulting from a "joint intention" and so qualifying as a "joint action." (In other words, this is not an action a group of citizens performs together accidentally or unconsciously.) Using Bratman 1999 and List and Pettit (in progress at the time), Ober sees a joint action result when an individual citizen puts aside personal differences in order to: (1) share the same goal with others; (2) make an individual contribution to it; and (3) know consciously his or her own intentions, others' intentions, and the reasoning others use to complete the task in question.[1] A joint human action also includes the ability to monitor or regulate itself for accuracy, consistency, and contradictions. And so a group agent "is organized to seek the realization of certain motivations in the world and to do so on the basis of certain representations about what the world is like."[2]

To what degree does this linking of individual agents to group agency conform to ways in which the Greeks thought about citizenship? At the start of *Politics* 3 Aristotle's goal is to define the citizen in an absolute, genuine, or full sense: "For we seek to define the citizen absolutely" (*zētoumen gar ton haplōs politēn*, 1275a19). He achieves this by folding the question of citizen identity into the question of citizen agency, and he understands this agency in performative terms, defining what it is that a genuine citizen *does* or how he *functions*: "A citizen in the unqualified sense (*haplōs*) is defined by nothing other than participating in making decisions and in holding office" (*metekhein kriseōs kai arkhēs*, 1275a22–23). He refines the definition so that a citizen is "one who has the ability to share in an office (*koinōnein arkhēs*) that requires deliberation (*bouleutikēs*) or making a decision (or judgment, *kritikēs*)" (1275b18–19), or even more simply one who holds the indefinite "office" of assemblyman (*ekklesiastēs*) and juror (*dikastēs*, 1275a25–30) because his state grants him this power (*exousia*, 1275b15).

[1] List and Pettit 2011: 32–33; cf. Ober 2008: 5–12. [2] List and Pettit 2011: 32.

For Aristotle a citizen is therefore a singular agent whose fundamental actions of deliberation and decision-making engage him in "sharing" or "participating" with others (*metekhein, koinōnein*). But does his analysis attend to the questions of coordination we see in List and Pettit's model of individual and group agency, joint intention and action, and so on? Aristotle represents what it is that citizens do both individually and collectively by the term "partnership" (*koinōnia*), which describes the cooperative activity of political life organized in various ways by various kinds of constitution (*politeia*): "For if in fact the state is a partnership, it is a partnership of citizens in a constitution" (1276b1–2). He also adds the term *sunthesis* (composite structure) as a synonym for the other two (1276b7). So if this understanding of a partnership is also agency-centered, the term should refer to an action that amalgamates what an individual says and decides with what a collectivity says and decides.

Aristotle offers three analogies illustrating how an individual citizen interacts with the collectivity, each underscoring the primacy of collective action and identity over individual (1276b1–29). But his third analogy bifocally acknowledges the distinctness each individual can bring to a collective action. In evaluating that distinctness, he raises two important questions: (1) How does an individual's skill contribute to the common skill required for the partnership or composite structure to succeed? (2) What is the ultimate purpose of each partnership or composite structure? Here Aristotle compares an individual citizen to a crew member on a ship. Each crew member has a function or capacity to fulfill (*dunamis*): one is a rower, another a helmsman, another a lookout, and so forth (1276b20–29). We have, he claims, two options for judging the excellence (*aretē*) of the sailors' skills. The most accurate reckoning would be an individual one (*logos idios*) suited to each man's performance at his specialized task, but there is also a common reckoning (*koinos tis*) that will suit them all. This depends on how well each contributes to the primary task (*ergon*) of sailing: to insure safe navigation (*hē sōtēria tēs nautilias*, 1276b26). Likewise, since citizens share a partnership just as sailors do, a citizen's excellence depends on the kind of constitution defining his participation and his contribution to community welfare (*hē sōtēria tēs koinōnias*, 1276b28–29).

I think this analogy of citizenship to seamanship makes clear how Aristotle's understanding of the function of a citizen conforms to List and Pettit's (and Ober's) concepts of individual and group agency and joint action and joint intention. Each crew member shares the same goal (safe navigation), makes a contribution in accordance with it, and shares his own and others' intentions (and reasoning) in order to reach port safely. The philosopher's notions of partnership, constitution, and composite structure all foreground the need for individuals to "align" (List and Pettit's term) with others their own perceptions and representations of what is the case in their environment, their own desires or motives for changing this, and their intentions for a preferred state of affairs.

But Aristotle's strongest description of citizenship as a form of group agency occurs later in *Politics* 3, when he famously surveys the arguments for why it is better that the mass of citizens be sovereign (*kurios*) than the virtuous few (1281a40–b20). The key "summation" argument is that the numerical accumulation of many contributions, even though most are imperfect in their moral nature and their share of virtue and practical wisdom, nevertheless turns out to be superior to the contribution of the virtuous few. This is because the plurality of individual opinions suddenly, magically, becomes a collectivity represented by a figure whose contours are almost comical. Abandoning his customary analytic vocabulary of terms such as *koinōnia*, *politeia*, and *sunthesis*, Aristotle imagines that "when they come together they become like (*hōsper*) one human being with many hands, feet, and faculties of perception (*aisthēseis*), but also with one personality (*ta ēthē*) and intelligence (*dianoian*)" (1281b5–7). This simile makes the conceptual leap between individual and group agency, producing an imaginary, democratic "Super-citizen" aggregating and aligning all individual perspectives into a collective person and mind. In effect this is an ancient approximation of the modern "emergentist" notion of group agency, one animated by a mysterious force that supervenes to transform individuals into a group agent.[3]

Solon and early Greek citizen agency: opening the citizen mind

Ober's hypothesis that one essential function of Greek citizenship is the production of useful knowledge identifies the era of the Athenian democracy (508 to 322 BCE) as the crucible when democratic knowledge secured Athens' primacy among Greek states (Ober 2008: 70–73). But political history in Athens and elsewhere suggests that the formation of citizens into a group agent whose identity and function produce socially and politically useful knowledge was a gradual process. Starting with early discourses

[3] On the "elimitivist" vs. "emergentist" traditions of group agency, see List and Pettit 2011: 2–4, 9–11, and 73–76.

(poetic and prose) in political and moral thought, we can identify multiple attempts *c.* 600–450 BCE to develop collective forms of deliberation whose shared production of knowledge meets the challenges faced by city-states undergoing strife between elites and commoners and the introduction of new institutions like councils and law courts. Solon (*c.* 630–560 BCE) is a prime example of such efforts. Before Solon, epic poets, prophets, oracles, and judicial chiefs all relied on the privileged sight of a sage (or knowledge expert) whose virtuoso performance of wisdom revealed the nature of the past, the hidden wishes of gods, the fate of humans, and the "straightest" or most just solution (*dikē*) to a dispute.

But in Solon's verses early in the sixth century we detect a break in this knowledge tradition with a call for new ways to represent both the agent and process of a different kind of knowledge. The theory of group agency suggests that Solon's goal was (1) to align the individual capacities of citizens (and the limited capacities of factions among them) to perceive and represent what was the case in their political conflicts; (2) to form joint preferences and intentions about what should be the case; and (3) to articulate these as propositions for change. His verses demonstrate problems in organizing Athenians into a citizen body capable of functioning as a group agent – but also key resources to resolve these difficulties. First, following List and Pettit, contributors to group agency need to self-identify spontaneously as members if joint intentional attitudes are to be carried out. This often requires them to overcome an "identification gap" that "fails to activate" them as members because they cannot recognize themselves as subjects of a joint intention (List and Pettit 2011: 186–87). Second, a group agent must be held responsible for its actions and their consequences in several ways: (1) by facing a "normatively significant" choice about an act that is good or evil, just or unjust; (2) by having sufficient understanding and information to make a normative judgment; (3) by having sufficient control over its circumstances to make the choice (155); and (4) by relying on an individual member who is an "activist" and can remind others about their shared identity, their "we-ness," including past performance and fundamental group attitudes, even incorporating a group that is not fully functioning by "rally[ing] their fellows around the suggestion that the group already has corporate status" and can exercise responsibility (193).

The personae Solon demonstrates in both his elegiac and his iambic verses play the role of such an "activist" struggling to persuade diverse Athenians that as citizens they share in a group agency that can form joint intentions about the dangers to their state's welfare and develop remedies to cure them.[4] It is, I think, of primary importance that Solon isolates for citizens the agency of knowing, especially the cognitive act of thinking, *to noein*, and the faculty they possess that enables them to think, *noos* (*nous*), their "mind" or "understanding."[5] Most of his surviving verses embed the practice of *to noein* and *noos* within deliberative times and places where citizens fight verbally and physically over political control, leadership, lawmaking, the change of constitutions, wealth and poverty, freedom and slavery, juridical offense and punishment, and so forth.

As harangues, the poems issue a new cognitive imperative to Athenians, one aimed at their inability to take responsibility as a group agent and to understand that they themselves (and not the gods) are responsible for their miseries and the impending destruction of their city (11.1–4 and 4.1–8).[6] He, in other words, exhorts them to see that they are both political and moral agents when they act in the public sphere because it is their use of *to noein* (9.6) and *noos* (4.7, 4c.3, 6.4, 11.6, 34.3) that determines (in List and Pettit's terms) their formation of representational and motivational states. Starting with Homer and Hesiod, *noos* and its cognates indicate the ability to use one's vision to discern accurately an object or action in a given situation, often when seeing how to evade danger and formulate the intent to escape. The accuracy of one's *to noein* or *noos* can also include the ability to pierce through appearances to grasp the true nature of an object or true intention of another person.[7] These two terms can also be instrumental in exercising one's responsibility, because they describe a particular individual's ability, or a group's habitual ability, that is distinguishable from others' and makes one liable to blame for intellectual and moral shortcomings.[8]

So when Solon harangues the leaders of the poor (the *dēmos*) to come to their senses (4.7), the elites to moderate their ambitious intention (4c.3; cf. 6.4), or the citizens at large to smarten up (11.6), he appeals to their habitual use of *noos*. But when at 9.6 he assumes the Olympian perspective that includes the foolish or ignorant behavior of both the mighty and the lowly within the polis, he uses *to noein* to issue a universal imperative to act before it is too late: *ēdē khrē panta noein*! ("Now is the time to exercise fully

[4] On Solon's multiple personae (political and poetic), see Farenga 2006: 295–303 and McGlew 1993: 101–11.

[5] On *noos* (*nous*) in Solon, see J. Lewis 2006: 23–41, esp. 31 and 35, and Lewis 2001: 126ff.; see also Almeida 2003: 191–94 and 203 and Mülke 2002: 112–13 and 200.

[6] Solon fragments are numbered following West 1971.

[7] See Von Fritz 1974: 23–25 and Barnouw 2004: 65, citing Schmitt 1990: 182–83.

[8] We find this sense in Hesiod (Von Fritz 1974: 27–28). Sullivan 1989 demonstrates how in the lyric poets *noos* is a "psychic activity" that reveals a person's "true nature" (131) or "inner nature" that is often hidden from others (140–41 and 160).

our power of discernment!"). Or, to translate this into a more colloquial English idiom, he cries out to them to abandon their various hubristic and uninformed habits of thought: "Open your mind!"[9] Here *to noein* must refer to a collective thinking that is coordinated as a joint intention, a shared opening of the collective citizen intelligence. It may be modeled on the superior discernment of a sage like Solon, a speaking subject "who knows," and who feels his own spirit command him to teach all Athenians (4.30). (He even describes himself in the mocking words of a mindless citizen as "Solon the complex thinker and advisor," *Solōn bathuphrōn . . . boulēeis anēr*, 33.1). But, as an "activist" member of a group agency that he struggles to incorporate and hold responsible for its joint intentions, he possesses an intelligence within the grasp of all citizens, a "mind" all can imitate.[10]

Like Aristotle, Solon is aware of how difficult it can be to see how individual thinking participates in a joint cognitive action. In the surviving fragments he refers multiple times to the need citizens have to "fit" or "form joints" between their individual or factional opinions and those of other citizens. The citizen mind should function, he claims, by developing skills like that of a carpenter, more specifically a "joiner," a craftsman who assembles already existing pieces of wood into a new construction by fashioning joints to connect them. He uses this metaphor to describe his own use of state power to transform enslaved Athenians into free citizens ("I fashioned a joint (*xunarmosas*) with my power linking both force and just settlement alike," 36.16–17), but also when he wrote laws and applied them fairly to individual cases ("I fashioned a joint (*harmosas*) when I wrote laws for both the wicked and the righteous," 36.19–20). But the term also describes the broader project of how to distribute privileges justly among diverse segments of the citizenry. Distribution in the right way (*eunomiē*) depends on a structure (*eukosma*) where elements fit together (*artia*) in a manner that is well ordered and intelligently interconnected (4.32–39).[11] Through *harmozō* and *artios* (and their cognates) Solon describes the sort of composite structure, like Aristotle's *sunthesis*, *politeia*, and *koinōnia*, that aligns the perceptions and intentions of individuals and factions so they might accept responsibility for joint actions needed to rule Athens.

9 *Pace* J. Lewis 2006: 140 n. 22, for whom the translation "mind" is anachronistic. As Sullivan observes, Greeks represented *noos* in this sense as a possession (using *ekhein*), because it was within their capacity to control (1989: 146).

10 The need to imitate Solon's "mind" (*nous*) or "intention" (*dianoia*) in creating a specific law or the laws in general became a topos in Athenian forensic rhetoric. See Farenga 2006: 332–40 (esp. 340 n. 104) and Johnstone 1999: 26–27.

11 For a fuller discussion of Solon's cognitive joinery, see Farenga 2006: 287–92.

Nevertheless, Solon does see limits to what *to noein* and *noos* can know, but these are limits to the knowledge a citizen produces as an individual rather than as a group agent. His nonpolitical verses look like the beginnings of moral philosophy because they struggle with questions about what causes individual human happiness or misery, an individual's capacity to have an accurate understanding of his own merits and faults, and how one's present circumstances in life relate to one's ultimate fate. The dividing line between a citizen's public life and private life, and that between what he can discover through use of political discourse and through use of moral discourse mark a boundary separating the capacity of *to noein* and *noos* to produce trustworthy knowledge. Expanding on a Homeric theme, poem 13 notes how citizens' "noetic" powers diminish when as moral agents they struggle to make a living on sea and land, and while trading, farming, manufacturing, or practicing crafts (13.43–58).[12] Just as they do not understand that Zeus will sooner or later inflict punishment on them for their sins, "so do we humans, both the righteous and the wicked, think alike (*noeomen homōs*): each has the belief (*doxan*) that he's doing fine before suffering some disaster" (13.33–34). Likewise a man of low reputation (*deilos*) believes (*dokei*) he is esteemed (*agathos*), and an unpleasant-looking man believes he is handsome (*kalos*, 13.39–40). As John Lewis claims, for Solon, when we contemplate our individual life and lifestyle (*bios*), "Psychically . . . a basic limitation to human foresight makes it impossible to know either what we *are* or what we *should do*, even in situations that are under our control. Failure here is of *to noein*."[13]

Our very "situatedness," Solon seems to be saying, keeps us locked up in a moral ignorance that cannot transcend an individual citizen's life, where collective thought and action and shared ways of knowing may not hold sway. These fragmented worlds, individually experienced, are shot through with what he calls *doxa*, the appearances of present circumstances, and the kind of understanding rooted in appearance (*to dokein*) that an individual applies to the representational and motivational states of his or her life. Unknowability therefore permeates our subjective worlds of seeming; outside the political sphere we lack the cognitive joinery responsible for constructing and maintaining a joint knowledge of our moral situation.

[12] On this theme of "epistemological pessimism" concerning human knowledge as we find it in Homer and other lyric poets before Solon, see Lesher 1990: 225–26.

[13] J. Lewis 2001: 132 (original emphasis). Solon identifies hubris among wealthy citizens as one impediment that produces a "mind that cannot fit in with others" (*mē noos artios*, 6.4); among citizens more generally, the inability to see that they might lose their freedom to a tyrant is symptomatic of a mind "scattered," full of gaps or porous (*khaunos . . . noos*, 11.6).

At best the moral agent is one who should know that he cannot know trustworthy knowledge about himself. The only agent who might hold such knowledge is the epic agency of Zeus's mind (intention) or judgment (13.17, 25, 75) and Moira, Fate itself (13.63). And so Solon declares that "the understanding (*noos*) of the gods is entirely invisible to humans" (fr. 17) and "The invisible measure of genuine wisdom (*gnōmosunēs*) is the most difficult thing to discern (*noēsai*), for it alone holds the limits (*peirata*) of all" (fr. 16).[14]

Xenophanes, *ta eikota*, and the speciousness of truth

Social and political knowledge was, as a kind of joint action, not inherently problematic for Solon. But a citizen's subjective and private self-knowledge was too faulty to contribute to a knowledge one could share with others in a political partnership. How could a citizen describe himself and his everyday struggle in ways that invited others not only to join him in his reasoning but also to construct this information into joint actions suitable for producing collective knowledge? In the era after Solon no "activist" member of a Greek community contributed more to answering this question than Xenophanes of Colophon. Living *c.* 570 to 475 BCE, he addressed the divide Solon saw in the public versus private use of *noos* (and *to noein*) by overhauling its agency.[15] Like Solon, he employed the epic representation of cosmic determination, the mind of Zeus, but he stripped this agency of its anthropomorphic qualities and spliced it with the newer agency of *phusis*, which Ionian thinkers were using to represent macrocosmic forces interacting in harmony and conflict. What results is a universal, divine mind:

> One god is greatest among gods and men,
> not at all like mortals in body or in thought (*noēma*).
> (fr. 23)
>
> Whole he sees, whole he thinks (*noei*), whole he hears.
> (fr. 24)
>
> But without any effort he shakes all things by the thought (*noou*) of his mind (*phreni*).
> (fr. 25)[16]

[14] Lewis cautions that the context of these two fragments is entirely "unrecoverable" (2001: 131 n. 49).
[15] For the dates of Xenophanes' life, see Graham 2010: 130. Some scholars prefer a birth year closer to 560. The fragments are numbered according to Diels and Kranz 1951.
[16] This fragment clearly reworks Homer, *Il.* 528–30, where Zeus's nod makes all Olympus quake. See Lesher 1992: 107 and Bryan's recent discussion (Bryan 2012: 11–12).

> He always remains in the same place, not moving at all,
> nor is it fitting (*epitrepei*) that he change location now here, now there.
> (fr. 26)

In this defamiliarized form, Xenophanes begins to lay bare the agency of efficacious or practical thinking. Reduced to an iconoclastic mind, all that remain of Homer's and Hesiod's Zeus are the basic cognitive functions of agency in List and Pettit's sense: (1) perception and representational states of what is truly the case; (2) a discerning evaluation, or motivational states about what should be the case; and (3) intentional states to bring about desired change through a capacity to energize all things into a kind of developmental motion (presumably linked to *phusis*). This is also an intelligence that is holistic, synoptic, unchanging, and autonomous. In light of group agency theory, this sole divinity functions like a reified mind capable of aggregating all perspectives, aligning them perfectly, producing a knowledge that is firmly "codified" or enduring, and not strictly dependent on what individual agents think. And true to the modern tradition of "emergentist" notions of group agency, the force animating this divinity mysteriously supervenes to cause the energetic "shaking" motion it imparts to everything. In this regard Xenophanes' god looks like an early manifestation of Aristotle's democratic "Super-citizen," whose multiple hands, feet, and perceptions but single mind and personality suddenly emerge from many imperfect contributions by individual citizens.

No small part of the mystery behind this divinity is how Xenophanes could claim to know anything about it in the first place. But he makes a surprising move of great strategic importance when in another fragment he claims:

> Now no man has come to know what is certain (*to . . . saphes*), nor will any one know (*estai eidōs*) about the gods and what I say about all things. For even if he should manage to describe things as they come to pass, he himself would still not know it (*oide*); but opinion (*dokos*) is brought forth over everything. (fr. 34)

Like Solon, Xenophanes recognizes the limitation that appearances (*doxa*) place on trustworthy knowledge. But he expands this flaw in *noos* and *to noein* to include not just the subjective or private knowledge we generate about our own lives but all human understanding, public as well as individual. So is Xenophanes claiming that the problematic "situatedness" Solon attributed to individual moral knowledge pertains to all human knowledge? Not quite, because his epistemological skepticism permits humans access to varying degrees of proximity to the true nature of things. As Jenny Bryan has most recently (and persuasively) argued, the philosopher uses two different notions of similarity to classify one kind of knowledge as definitely false or untrue – for example, what Homer and Hesiod describe, or what humans believe when they claim their gods are "like" (*homoios*) themselves (frs. 10–12, 14–16).[17] But a different kind of similarity characterizes, in a more indefinite relation to truth, something as: (1) resembling the truth; and so (2) possibly true; but also (3) possibly bearing only a "specious similarity" to the truth. As Bryan argues, this is the complex access to truth Xenophanes claims for his own teachings when he exhorts listeners, "Let these things be believed as like (*eikota*) the truth (*tois etumoisi*)" (fr. 35).[18]

By bracketing off divine knowledge as genuine but inaccessible, Xenophanes hits hard at the truth-telling capacity of traditional sources for this knowledge (poets, oracles, prophecy, even the sage). And in their place he opens a vast terrain of what we know we cannot truly know.[19] So while his "fallibilist" pursuit of knowledge may reflect epistemological skepticism and pessimism, it also generates an imperative that humans now proceed to develop a hit-or-miss methodology with at least a chance of approximating the true knowledge that is beyond their ken.[20] And this transformation of acts of knowing (or claiming to know) into acts of risk-taking may be the wiser path:

> Not from the beginning have the gods revealed all things to mortals, but in time by inquiring (*zētountes*) they [mortals] hit upon what is better (*epheuriskousin ameinon*). (fr. 18)

What may at first appear in this epistemology as cognitive loss turns into cognitive gain since risk-taking encourages competition among speakers for a less ambitious but still satisfactory kind of quasi- or hypothetical knowledge – it creates simulacra (*eikones*) of knowledge that may have their place in a joint action to produce knowledge.[21] Just as Aristotle imagined a democratic Super-citizen emerging from many bits of imperfect virtue

[17] Bryan 2012: 36–43. [18] Bryan 2012: 42–48.

[19] In an Aristotelian idiom this is the possibility of knowing what is unknown, or "that it's possible to know that the unknown is unknown" (*esti gar epistēton to agnōston hoti agnōston*, Arist. *Rh.* 1402a7).

[20] On Xenophanes' fallibilism, see Lesher 1992: 164–65, who rejects this characterization, and Bryan 2012: 46ff., who endorses it.

[21] See Graham's comment on this fragment as a "hymn to progress" on the human capacity for social improvement through the pursuit of knowledge (2010: 133). He adduces Lesher's claim that Xenophanes is here praising inquiry as a superior path to truth compared with divination by divine omens (Lesher 1991).

and knowledge (*hekaston morion . . . aretēs kai phronēseōs*, *Pol.* 1281b4–5), so does Xenophanes expand what *noos* and *to noein* can achieve through group agency. Bryan sees his understanding of human cognition as, compared to the divine mind, limited in perspective by our experience: an "essentially perspectival opinion" versus the god's "non-perspectival knowledge" that is "unlimited" and "holistic."[22]

And yet this limitation opens a new cognitive initiative that will, under the right circumstances, contribute to producing democratic knowledge. Ober's chapter on the "aggregation" of knowledge in democratic Athens discusses several institutional innovations that facilitated the more efficient collection of information; it is a question of understanding "how the epistemic horizon was broadened over time."[23] Bryan helps us see Xenophanes' rehabilitation of *doxa* as a giant step in an earlier phase of knowledge aggregation; she calls it a "*broadening*" rather than narrowing of perspective, one capable of producing more "comprehensive" results.[24] Xenophanes' inquiry (*zētēsis*) improves on Solon's because it refuses to eliminate the "situatedness" of *doxa*, seeing this universal quality of human knowledge as generating two different kinds of similarity to the truth. One (what is *eoikōs*), in contrast to the other (what is *homoios*), rescues the cognitive potential of a citizen's contribution to deliberation when he says, *moi dokei* ("it seems a good idea to me . . . ") or *hōs eoike* ("as it seems reasonable [plausible, possible, fitting] . . . ").

If Solon could exhort citizens, "Open your mind!" – but only when they deliberated on political matters as a group agent – Xenophanes could shout the same phrase when deliberating on both public matters and the more intimate moral questions of personal virtue and wisdom.[25] In fact his profile in the *testimonia* is that of an angry, outspoken critic of moral and political matters, a man who felt his *sophia* deserved the citizens' recognition for its contribution to "the right kind of government" (*eunomiē*, fr. 2.19).[26] Lesher even says he "most closely resembles the gadfly of Athens" in this regard.[27] In fact the source preserving fr. 35 ("Let these things be believed as like (*eikota*) the truth") suggests that the ancients understood it as provoking others to contribute their opinions to a philosopher's expression of his own ideas.[28]

22 Bryan 2012: 48–51. 23 Ober 2008: 123. 24 Bryan 2012: 53 (original emphasis).

25 On Xenophanes' contribution to moral philosophy, see Lesher 1992: 47–77.

26 For the *testimonia* see Graham 2010: 98–101. 27 Lesher 1992: 3.

28 The source is Plut. *Quaest. conv.* (746b), citing a habitual remark his teacher Ammonius made when concluding his own presentation of ideas and awaiting his students' expression of their own opinions. Cited by Lesher 1992: 169 and discussed by Bryan 2012: 8–9 and 20–23.

"Speak your mind!" *Parrhēsia* and democratic knowledge

If attempts to model the mind were "an early and recurrent feature of Greek thought" because they "contrast the perfect knowledge possessed by the gods with the blinkered limitations of the human perspective," then perhaps "modeling the mind" is an alternate way to describe knowledge production as a joint action by a group agency.[29] But how could *ta eikota*, an apparent but possibly specious sort of knowledge, empower ordinary citizens to become speaking subjects pursuing a semblance of truth as a group agent in a joint action of knowledge production? Athenian democratic institutions opened several paths that by the last half of the fifth century enabled speakers to achieve this. Ober's discussion about aggregation examines social networks and teams in the Assembly, Council of 500, and on boards of magistrates.[30] But Gagarin's and Tordoff's contributions to this volume indicate another path opening in forensic and deliberative rhetoric, where *eikos* arguments based on what was believed to resemble the truth (or to be probably, plausibly, or fittingly true) appear in the 420s BCE, if not sooner.

However, in addition to *eikos* arguments there was another type of "eikotologia" that at this same time promoted knowledge aggregation. It developed from the practice of *isēgoria*, which Athenians cherished as the freedom to speak publicly – or, more precisely, having an equal right to speak publicly. By the last third of the fifth century, a new coinage, *parrhēsia*, means the freedom to speak one's mind, to say everything one wishes, to speak the whole truth, or to have an "unrestricted right to express [one's] opinion."[31] Its most powerful manifestation was obviously in the political arena, where elite citizens displayed it in Assembly or courtroom deliberation, but the term also characterizes the theatrical speech of comic characters and the concerns of some characters in Euripides. In private life *parrhēsia* might characterize relations between friends and family members as well as interlocutors in philosophical dialogue. Apparently even menial folk, foreigners, and slaves in Athens spoke in this way.[32]

29 Nightingale and Sedley 2010: 3. 30 Ober 2008: 118–67.

31 For the relation between *isēgoria* and *parrhēsia*, see Raaflaub 2004: 222–23 and Carter 2004. Whereas the first term could be synonymous with Athens' democratic *politeia*, the second term's emphasis on free speech became "almost synonymous with citizenship" (Raaflaub 2004: 223). Carter argues that *parrhēsia* was not a "right" in our modern sense or even in the limited sense that applies to Athens; rather it is an "attribute" of citizenship (2004: 198, 205). Wallace 1996 compares Athenian *parrhēsia* to the modern American concept of citizen rights. (The word first appears in 428 BCE in Eur. *Hipp*. 422.)

32 See Wallace 2004: 222 on the scope of the term's use. On non-citizen and servile usage, see, e.g., Dem. 9.3; cf. ps.-Xen. *Ath. Pol.* 1.2 and 12, where the word *isēgoria* is used. On *parrhēsia* in comedy

There may be multiple explanations why Athenians invented a new term to describe a certain way of speaking publicly and privately.[33] I suggest that this particular way of speaking reflects changes in: (1) the agency of individual citizenship, particularly in the influence an individual's character (*ēthos*) and opinion (*doxa*) had on his performance of citizenship; and (2) a civic need to accommodate these increasingly individualistic performances of citizenship into a more efficient aggregation of collective political knowledge. If we follow List and Pettit on agency, we might look for reasons why speaking one's mind through *parrhēsia* enabled individual Athenian citizens to form markedly subjective representations, motivations, and intentions about their world. But, following Ober, we might also ask how it enabled Athenians as a group agent to broaden their epistemic horizon and integrate their personal representations, motivations, and intentions into a production of democratic knowledge. Either way, the years after *c.* 430 BCE redefine problems of *noos*, *to noein*, *doxa*, and *ta eikota*, asking how individual citizens can not only *open* their minds to civic deliberation but also feel free to *speak* their minds about what appears to them to be true – but might well turn out to be false.[34]

To answer these questions we need to keep one eye on the protocols (rules for correct conduct) that governed public speaking in Athens, in particular the do's and don'ts for exercising *parrhēsia*. But we need to keep another eye on what Ober calls the "incentives" individuals had for "information sharing" in "an economy of esteem" as well as the sanctions they faced for proposing false or misleading knowledge. *Parrhēsia* functioned in this regard as "an epistemic sorting device" intended to "weed out false or irrelevant information" or "inexpert knowledge."[35]

To speak one's mind candidly and fully implicated an individual in a dilemma that was ethical as well as political. As Carter argues, Athenians had "ambiguous attitudes" toward this privilege, for while every citizen had an equal right to speak in public (*isēgoria*), individuals might go too far when they exercised this right through *parrhēsia*.[36] On the one hand, a *parrhēsiastēs* was praised as a patriot who led his listeners to the truth, or to

see Henderson 1998, where women should enjoy its use at Ar. *Wasps* 469–99, 1388–1414 and *Lys.* 379 and 507–28; the word appears in Euripides at *Bacch.* 668, *El.* 1049, *Hipp.* 422, *Or.* 905, *Ion* 672 and 675, *Phoen.* 391, and fr. 737N. Monoson 1994 compares its democratic use to Plato's philosophical "appropriation" of it.

33 Raaflaub offers suggestions concerning the political climate in post-Periclean Athens during the early years of the Peloponnesian War (2004: 224).

34 In Roisman's formulation *parrhēsia* pitted an Assembly speaker and his audience in a struggle not only over his right to speak publicly but over "his right to speak his mind" (2004: 268).

35 Ober 2008: 118–20.

36 Carter 2004: 218.

more cogent arguments, especially when in the Assembly he offered novel or contradictory arguments.[37] In this way he encouraged that plurality of individual contributions for which Aristotle praised the many-handed, many-footed, multi-perceptive Super-citizen.[38] Fourth-century orators in particular praised the individual courage and manhood of those who (like themselves) spoke with *parrhēsia*, in effect transferring the traditional honor for military valor to ordinary citizens.[39] But, on the other hand, this need for courage speaks to the fact that when a citizen dared to speak his mind he put himself at risk. Ethically he had to be a citizen in good standing and one whose opinions – unlike a true gadfly's – complemented rather than antagonized the demos. Politically and legally he had to be seen as promoting state welfare rather than personal gain, thereby opening himself to charges that could cost him his citizenship, his fortune, or even his life.[40] Self-confidence was therefore a requirement, along with a thick skin to fend off the physical abuse when the crowd itself expressed its own version of *parrhēsia* in the *thorubos* (uproar) that humiliated and silenced him through heckling, shouts, and insults.[41]

In effect the protocol on speaking with *parrhēsia* amounted to a pact between a speaker and his listeners.[42] Each time a *parrhēsiastēs* had the courage to share his sincere if uncertain thought, each time his utterance invoked the illocutionary force of *moi dokei* ("it seems a good idea to me . . . ") or *moi eoike* ("it seems reasonable (or plausible or appropriate) to me . . . "), he took on the role of what List and Pettit call an "activist" group member.[43] Like Solon, this permitted him to criticize, even berate, group members for lapsing from group principles or values (e.g. Dem. 3.32, 23.204); he may also have, like Solon, argued to make the group more "responsibilizable" for its moral choices.[44] At the same time his activism left him vulnerable to what Ober calls "communication costs" that put his

37 On the link between *parrhēsia* and speaking the truth, see Dem. 3.3; 10.54, 76; 60.26.

38 Balot 2004: 239–40, expanding on Ober 1989: 163–64, discusses Aristotle's "summation argument" along with earlier praise of the "wisdom of the masses," including Thuc. 6.39; Dem. 3.15, 23.45–46; cf. Isoc. 13.8, where the expression of many *doxai* (opinions) more readily leads to agreement and can be superior to genuine knowledge (*epistēmē*).

39 On the courage of the *parrhēsiastēs* see Balot 2004 and Roisman 2004.

40 On the risk of speaking in the Assembly with too much *parrhēsia*, see Carter 2004: 218–19, citing Dem. 3.3, 4.51, 8.32, 9.3, 10.54, and 15.1. On ethical risks see Wallace 2004: 223. On political and legal risks linked to the need for courage, see Balot 2004: 243, citing Dem. 1.16, 3.11–13, 3.32, 4.51, 8.71–72, 10.70–72, and 23.5. On risks see also Monoson 1994: 177–78 and Foucault 2009: 12–13.

41 On self-confidence see Carter 2004: 214; on the *thorubos*, Wallace 2004: 223–25. Roisman 2004: 264 cites Aeschin. 3.80 on the crowd's expression of *parrhēsia*.

42 See Foucault 2009: 13–14 on this "pact" or *jeu* between *parrhēsiastēs* and listeners, and Monoson 1994: 183 on "collaboration."

43 List and Pettit 2011: 191–93.

44 List and Pettit 2011: 193.

moral character, social standing, and legal status up for public scrutiny. In Sara Monoson's words, "One's critical and bold speech was taken to express – perhaps to expose – something of who one is, what one cares about, how one has chosen to live. One's convictions were on display when one spoke with parrhesia."[45] To lessen those costs, however, the *parrhēsiastēs* operated in the cognitive zone of *to eikos*, and so he was not held to a standard of absolute truth; the "flawlessness" of his *logos* (speech, argument) was not at issue.[46]

Conclusion

I hope I have demonstrated that citizen agency in Greece underwent a sequence of epistemic shifts in the 250 years that separate Solonian Athens (590s BCE) from the post-Periclean age (420s) and fourth century BCE. In discussing Solon, I have tried to highlight how civic and intellectual leaders "modeled the mind" at least in part to overcome the traditional epistemic denigration of the personal knowledge (*doxa*) that individuals acquired through their own use of *noos* and *to noein*. The challenge, I have argued, was to find ways to convert this flawed understanding into forms of the collective knowledge that Greeks relied upon when they deliberated the fate of their political communities. A pivotal contribution to this epistemological transformation was Xenophanes' claim that human knowledge had a precarious, flawed relation to the truth – and that this very defect in *doxa* opened "eikotologic" possibilities for knowledge acquisition from which a political community could benefit. As one manifestation of these possibilities, I focused on the Athenian practice of *parrhēsia* as a potent expression of possibly specious subjective knowledge. My discussion utilized not only a theory of agency (individual and collective) but also a hypothesis about the production of democratic knowledge to explain how (to give one final example) an activist leader like Demosthenes could assure the Assembly that by speaking the truth with *parrhēsia* he was revealing his goodwill (*eunoia*) and lack of intent to flatter, harm, or deceive (10.76), that he was expressing "openly" or "simply" (*haplōs*) "everything I know" (4.51), and that this transformation of his interior states of mind into the bricks and mortar of a public policy under construction would make his fellow citizens "of one mind" (*mia gnōmē*, 19.298).

45 Monoson 1994: 184. 46 Monoson 1994: 175.

CHAPTER 6

Counterfactual history and Thucydides

Robert Tordoff

διὸ δεῖ σπεύδειν καὶ μηδεμίαν ποιεῖσθαι διατριβήν, ἵνα μὴ πάθωμεν ὅπερ οἱ πατέρες ἡμῶν. ἐκεῖνοι γὰρ ὑστερίσαντες τῶν βαρβάρων καὶ προέμενοί τινας τῶν συμμάχων ἠναγκάσθησαν ὀλίγοι πρὸς πολλοὺς κινδυνεύειν, ἐξὸν αὐτοῖς προτέροις διαβᾶσιν εἰς τὴν ἤπειρον μετὰ πάσης τῆς τῶν Ἑλλήνων δυνάμεως ἐν μέρει τῶν ἐθνῶν ἕκαστον χειροῦσθαι.

Therefore we must hurry and make no delay so that we do not suffer what our fathers did. For because they were slower to act than the barbarians and lost some of their allies, they were forced to run the risks of being a few against many, although if they had crossed over to Asia first with the entire strength of the Greeks, they could have subdued each of the peoples there one by one.

(Isocrates, *Panegyricus* 164)[1]

For a fleeting moment in the *Panegyricus*, in a plea for a collective Hellenic effort against the barbarian foe, Isocrates imagines what might have happened if the Persian Wars of the early fifth century had been fought on Persian soil. Counterfactual historical claims, usually in the form of protasis with the particle *ei* and a historic tense of the indicative and an apodosis with the particle *an* and another historic tense of the indicative,[2] can be found scattered throughout ancient Greek literature. In this chapter I examine a small number of counterfactual statements found in documents from early fourth-century Athens relating to the events of the Peloponnesian War and its aftermath: Athens' defeat, the rise of the Spartan-backed oligarchy,

I would like to thank Victoria Wohl for the opportunity to contribute to this volume and for suggesting numerous improvements to the text; participants at the *Eikos* workshop held at the University of Toronto in May 2010 for stimulating discussion; and Roger Brock for his help and advice on an earlier draft of this chapter. I would also like to thank the audiences at the American Philological Society meeting in Chicago in 2008 and at Université Laval and Cambridge University in 2010, where I presented versions of this chapter.

[1] All translations are my own unless otherwise indicated.

[2] There are other syntactical possibilities yielding very similar sense, for example those using accusative and genitive absolute constructions. Isoc. 4.164 (quoted above) is one example of the former.

the civil war fought between Athenian democrats and oligarchs, and the restoration of democracy in 403.

The period is one of enduring fascination, not least because of the surprising way in which the polis sought to reconcile its citizens with oaths to forget the events of the oligarchy and civil war. How Athenians remembered and commemorated the events of the last years of the fifth century following this institutional act of forgetting has been a subject of numerous studies.[3] The emphasis of this one is different: I am not concerned primarily with what Athenians remembered and how they remembered it, still less with what they swore to forget, but with how they thought about how things might have turned out differently. This is a subject that has generally eluded attention in discussion of the reconciliation agreement and the politics of early fourth-century democratic Athens. Yet, as I hope to persuade the reader, the discourse of counterfactual thinking in Athenian political life was extremely prominent in the first two decades following the Peloponnesian War. It appears in law-court speeches, in epideictic oratory like the passage of Isocrates quoted above and even in the deliberative processes of the democratic Assembly.

Counterfactual thinking also appears in the fundamental document in my inquiry, Thucydides' *History*. I am not the first to collect and study counterfactual passages in Thucydides, but the emphasis I place on them is new in that it has its roots in the wider social and historical context in which they are found. Rather than treating them as purely narratological features of Thucydides' contribution to the nascent genre of historiography or a biographical effect of his regret at his part in the struggle against the Peloponnesians, I see them as productively engaged with other contemporary currents of thought about how recent history might have happened differently. This is not to say that I deny the importance of biographical information or narratological analysis, merely that I believe that to understand the counterfactual statements made in Thucydides we must do three things that scholarship has not yet done: first, establish a clear view of what historical counterfactual claims are and how they work; second, distinguish the counterfactual statements plainly made in

[3] Most influential have been Nicole Loraux's studies of the politics of memory and forgetting in Athens post 403; Loraux 2002 usefully collects and translates her essays. More recent contributions by James Quillin and Andrew Wolpert, building on Thomas Loening's study of the reconciliation, have explored the implementation and the (apparent) success of the Athenian agreement not to remember past wrongs: Loening 1987; Quillin 2002; Wolpert 2002a, 2002b. For discussion of the legal scope and chronology of the amnesty see Carawan 2002, 2006; Joyce 2008. Recently Shear 2011 (esp. 166–312) has radically advanced our understanding of the memory of and response to oligarchy in the years following the 403 restoration of democracy, showing how archaeology and topography situate the documentary evidence of oratory and epigraphy in wider public discourses of remembering and forgetting.

Thucydides' authorial voice; and, third, place Thucydides' counterfactuals in their full historical context, which I take to be Athens in the period down to the King's Peace in 386 BCE.[4]

Among professional historians 'counterfactual history' is sufficiently notorious to warrant an immediate apology: this is especially the case in the field of ancient history, where the difficulties of establishing what actually did happen are so formidable that speculation about what did not happen but might have done appears a singularly idle and unprofitable activity.[5] Any historian who believes in some overarching determinism governing vast impersonal forces and the march of history is likely to find it a distraction, if not a betrayal of the historian's vocation. Descriptions of it from such quarters range from E. H. Carr's politely disparaging 'parlour game' to E. P. Thompson's refreshingly abusive neologism 'Geschichtenscheissenschlopff'.[6] But in practice, in everyday life, people do learn and act on the basis that if they had done something differently in the past, the present really would have been different.[7] Furthermore, for a historian to reject counterfactual thinking is not without serious methodological consequences; for whenever we wish to specify a cause – for example it was Spartan fear of the growth of Athenian power that ultimately caused the Peloponnesian War – we must accept the counterfactual claim that, *ceteris paribus*, but for Sparta's misapprehensions at the escalation of Athenian imperial ambition, the Peloponnesian War, at least as we know it, would not have happened.[8] In the words of Martin Bunzl, even 'narrative of the most innocent-seeming sort is replete with locutions that

[4] It is usually assumed that Thucydides had his work *in manibus* until his death, the date of which is unknown. He shows clear knowledge of the outcome of the Ionian war and the destruction of the Athenian Empire (5.26.1, 4–5; cf. 2.65.12, 6.15.3), and 404 was long held to be the earliest possible *terminus post quem*, while the testimony of Pausanias (1.23.9) has been taken to show that Thucydides died in or soon after 404/3. However, evidence internal and external supports the view of Thucydides writing in the first half of the 390s: see Pouilloux and Salviat 1983, 1985 (though on the Thasos archon list, see Cartledge 1984). Munn 2000: 315–23 suggests that the *History* was written in the form in which we have it in 396/5 as a deliberative brief (*sungraphē*) for the Athenian *Boulē* on the eve of the Corinthian War. For the view, based on 3.116 and Diod. Sic. 14.59.3, that Thucydides seems not to have lived to see the eruption of Mount Etna in 396, see Hornblower 1987: 143–44; contra: Munn 2000: 318 n. 61; note that the arguments lead to a date only for that particular passage, not to a firm date for Thucydides' death. For general discussion of the problems of the chronology of Thucydides' composition of the *History*, see esp. de Romilly 1963: 3–10; cf. Dover 1973: 14–20; Rawlings 1981: 250–54.

[5] Nevertheless, over the last generation classical scholars have taken an increasing interest in historical counterfactuals: see, for example, Demandt 1986, Flory 1988, Suerbaum 1997, Will 2000, Morello 2002. More recently, see Brock 2013, Grethlein 2010. I am very grateful to Dr Brock for sharing his unpublished work. Regrettably, Powell 2013 (*non vidi*) appeared after this chapter went to press.

[6] Carr 1962: 97, E. P. Thompson 1978: 300.

[7] Cf. D. K. Lewis 1973: 84.

[8] My choice of the legal phrase 'but for' is deliberate. English law, for example, has constant recourse to counterfactual analysis in matters such as determining responsibility and assessing damages. For discussion, see Weber 2009.

carry lurking counterfactual commitments'.[9] Since counterfactual claims are implicit in every non-determinist analysis of historical causation, it seems unlikely that any kind of historiography could emerge unmarked by the traces of counterfactual thinking, except perhaps the driest chronicling of 'one damn thing after another'[10] or the teleological faith of the most defiantly doctrinaire Calvinism or Marxism.[11]

It will be objected that there is a great difference between unconscious assent to counterfactual claims and actually making an explicit counterfactual assertion. Nevertheless, historians do the latter quite frequently to isolate and underline a cause or causes. Yet only rarely do academics write counterfactual narrative historiography of the kind found in Niall Ferguson's collection of essays, *Virtual History*, where what might have happened but did not is explored over many pages: for example, what if there had been no American War of Independence or if Hitler's plans for the invasion of Great Britain had succeeded?[12] I think historians are not greatly perturbed by a causal claim in the form 'If A had not happened, B would not have happened', but are very hesitant about answering questions in the form 'If A had not occurred, what would have happened next?' History at this point seems to metamorphose into fiction. Perhaps surprisingly to some admirers of the father of 'scientific history',[13] Thucydides in his own voice asks and even answers quite a few of these questions.

Much of the notoriety of counterfactual history is explained by the humorous or ironic mode in which the 'Cleopatra's nose' school of history has characteristically written.[14] Typically, essays in this vein attribute absurdly large counterfactual ramifications to minute, inconsequential or trivial imagined changes in the course of history. The insincerity here is both the effect and ongoing cause of the failure to develop an adequately robust methodology for asking counterfactual historical questions. If the

9 Bunzl 2004: 846.

10 The phrase, adapted from the well-known witticism attributed to Mark Twain 'Life is just one damn thing after another', is from Toynbee 1987: vol. II, 267 where he imputes this view of history to H. A. L. Fisher.

11 For discussion of counterfactual analysis and historical method, see Fearon 1991, Hawthorn 1991, Ferguson 1997b, Bunzl 2004, Lebow 2010.

12 Ferguson 1997a. Further collections include Cowley 1999, Tetlock *et al.* 2006.

13 The classic statement of Thucydides as 'scientific historian', identifying his methods with those of the medical writers of the Hippocratic Corpus, is Cochrane 1929. For recent discussion, see Crane 1996: 51–65.

14 The well-worn reference is to Pascal's *Pensée* 162: 'Le nez de Cléopâtre: s'il eût été plus court, toute la face de la terre aurait changé.' This example of historical contingency is famously discussed by Bury 1930: 60–69.

simplest way to begin counterfactual investigation is to imagine the consequences of a real historical event A, B or C not happening, is there any way to decide that ¬ A was any more probable than ¬ B or ¬ C and is therefore a more worthwhile historical question? At an elementary level, it seems easy to distinguish the sober questions from the frivolous. Asking whether the face of history would have been greatly altered had Cleopatra undergone rhinoplasty is clearly an entirely different order of question from asking how Athens would have fared in the Peloponnesian War had Pericles not perished in the plague in 429. But there is a vast array of potential counterfactual questions that may be asked about the Peloponnesian War without such obvious offence against common sense as the question just asked about Cleopatra. Is it any more important to ask how the war might have been different had Pericles survived than it is to ask what would have happened if Cleon had not died at Amphipolis in 422 – and if so, why?

The most pressing methodological difficulty is that any number of counterfactual questions may be raised, and there seems to be no systematic way to distinguish between more and less probable counterfactual scenarios. Here, the work of Geoffrey Hawthorn has pioneered a way forward. In *Plausible Worlds* Hawthorn suggests that one means of isolating a manageably small set of counterfactual questions is to base them upon what people at the time thought might happen in the future that *they* faced: in short, the possible futures envisioned by historical agents as described in documentary evidence constitute the field within which counterfactual questions may be asked. This approach has the virtue of nailing down a finite number of counterfactual questions based on hard historical data and at the same time alerting historians to the openness and fluidity of the historical situation *as it was for people at the time*.[15]

It is, of course, difficult to perform such operations with classical Greek history. Lacking the richness of documentation available for twentieth-century political history, for example, the approach is necessarily impoverished. No surviving sources independent of Thucydides and written during the Peloponnesian War present a sincere and private statement of how the author thought any given event of the conflict, or the war as a whole, might

[15] Hawthorn 1991; cf. Ferguson 1997a: 86–87: 'if all history is the history of (recorded) thought, surely we must attach equal significance to all the outcomes thought about. The historian who allows his knowledge as to which of these outcomes subsequently happened to obliterate the other outcomes people regarded as plausible cannot hope to recapture the past "as it actually was". For, in considering only the possibility which was actually realised, he commits the most elementary teleological error. To understand how it actually was, we therefore need to understand how it actually wasn't – but how, to contemporaries, it might have been.'

turn out.[16] However, there are many passages in Thucydides in which his characters or agents tell us what they think is likely to happen in the future, and insofar as we take Thucydides at his word in general they furnish us with a wonderfully vivid picture of the openness of historical experience as it happened.[17] There are instances in which a character or other agent expresses an intention, desire or threat, an expectation, anticipation or hope, or a fear of what they believe is to come – and when these go unfulfilled, the reader is provoked to imagine what might have been.[18] Similarly, when Thucydides tells us that something was tried or not done when it might have been, or that something nearly happened, or was avoided, or expectations were disappointed, or that an event had a single major cause (especially if the cause appears highly contingent or attributable to chance), the reader may readily wonder if history might have worked out differently.[19] There are many such passages in Thucydides, and even a number in which characters or agents make counterfactual assertions,[20] but there are only a few places where the historian endorses a historical counterfactual with his authorial voice.[21] The first major task of this chapter will be to isolate these, but, before that, a description of the *status quaestionis* is essential.

Counterfactuals in Thucydides have attracted a certain, limited amount of attention. In the main they have been studied as examples of Thucydides' narrative technique. Kenneth Dover, in a paper originally published in 1981, mentions but does not catalogue in full 'twenty occasions' on which Thucydides makes a counterfactual judgment; among these he refers to 'the half dozen most substantial' and later identifies six as 'substantial'.[22] Stewart Flory (1988) isolates and explores what he calls 'Thucydides'

[16] All sorts of inferences may be drawn from drama (for example Ar. *Frogs* 703–705, 734–37), but the public context fails to satisfy the criteria of private sincerity.

[17] Ancient recognition of Thucydides' extraordinary vividness: Plut. *De Glor. Ath.* 346f–347c.

[18] For example, intention: 2.80.1; desire: 2.59.2 (cf. 4.21.1); threat 1.71.4–5; expectation: 4.76.5; anticipation: 5.57.1; hope: 2.20.2; fear: 3.79.1.

[19] For example, failed attempt: 5.3.3; failure to act: 3.30.1; intention forestalled: 1.101.2; a near miss: 7.2.4; avoidance: 3.102.4–5; disappointment of expectation: 2.79.2; a singular cause: 4.81.2; contingency: 4.29.2–3.

[20] Thucydides' agents make historical counterfactual claims: 1.69.1, 73.4, 74.4, 75.4, 76.1; 2.18.4, 70.4; 3.11.1, 11.3, 40.5, 57.3; 4.27.5, 73.3, 104.2 (presumably the Amphipolitans); 5.7.5; 6.61.2; 7.42.3, 68.2; 8.2.1.

[21] Discussing Thucydides' 'authorial interventions', Gribble 1998: 47–49, 64–65 lists a number of counterfactual passages. However, his interest in 'authorial interventions' is in places where the author's voice markedly obtrudes into the otherwise 'objective' third-person narrative. My definition of authorial in this context is more capacious, extending to all statements in the text from which Thucydides does not explicitly withhold his full endorsement or which he does not attribute to someone else. Thus my list of 'authorial counterfactuals' is considerably longer.

[22] Dover 1987: 74, 75, 81. The 'half dozen most substantial' are: 1.11.1–2; 2.94.1; 3.113.6; 7.42.3; 8.87.4, 96.4.

hypotheses about the Peloponnesian War', a list of nineteen passages. Both Dover and Flory see the use of historical counterfactuals as an important aspect of Thucydides' style and they find his 'what-might-have-been thinking' to be influenced by his experience at Amphipolis. Since Flory focuses on counterfactual statements to do with the Peloponnesian War, he is not greatly interested in a passage like 1.11, where Thucydides makes a counterfactual claim about the Trojan War. In different ways Dover's and Flory's studies are eclectic and neither presents a full list of all the counterfactual claims made in Thucydides, nor even, more surprisingly, a catalogue of those in Thucydides' authorial voice.

Subsequently, Hornblower and Rood have elaborated on counterfactuals as a special feature of Thucydides' narrative technique.[23] Opinion seems to be converging on the view that 'full-dress counterfactuality' is 'a mode of rhetorical utterance, an emphatic focusing device, designed to draw special attention to a hinge-moment in history'.[24] In other words, Thucydides' counterfactuals alert the reader to the singular importance of certain links in a chain of historical events. It is hard not to agree that a counterfactual claim in historiography should drastically alter the contours of a kind of narrative that consists, for the most part, of facts,[25] and that (as Hornblower's term 'hinge-moment' implies) it will help to identify and isolate the relative importance of historical causes. However, unrealized possibilities in fiction have similar formal properties but analysis of their functions shows that they do more than underline causes, as Gary Saul Morson's study of 'side-shadowing' so illuminatingly demonstrates.[26] Therefore, from the viewpoint of narratology as well as that of the historian, it seems important to recognize that thinking about what might have been

[23] Hornblower 1994: 158–59; Rood 1998: 278–80. [24] Hornblower 2011: 8.

[25] I am aware that treating historical narrative as the presentation of a series of causally linked facts will strike some readers as a naïvely optimistic view of what history is. My purpose is merely to invoke the common-sense distinction (made by Arist. *Poet.* 1451b) between a genre that purports to convey (some of) what actually or probably happened in the past (but not what did not or could not have happened) and other genres not bound by these constraints. For thought-provoking discussion of the (literary) practices of writing history: Veyne 1984; White 1978, 1987. To put it another way, writers of history and fiction use propositions referring to entities and predicate actions, states, and processes of them; the difference is that only with history is the reader expected to believe that propositions refer to entities that once existed and that the predicates were at one time true. In the terminology of J. L. Austin 1971, historical propositions are constative (they are subject to truth valuation), whereas fictional propositions are performative (it makes no sense to ask whether Shakespeare is truthful when he has Gertrude die by poisoning in the world of *Hamlet*: there is nothing to be known). However, propositions *about* fictional worlds are subject to truth valuation: they are 'bet sensitive' as described by Woods 1974: 13–14; I will lose my money if I bet that Sherlock Holmes lived at 222a Baker Street.

[26] For the formal similarities of 'alternate history' fiction and counterfactual history, cf. Hellekson 2001: esp. 29. For 'side-shadowing' see Morson 1994: 117–72. I avoid terming these unrealized possibilities 'counterfactuals' to avoid confusion.

has more to offer than merely a more emphatic account of why history happened as it did.

Among Thucydidean scholars there is a stubborn, unspoken insistence on restricting the analysis of counterfactuals to the realm of the literary and its texture.[27] Rood's treatment of them is illustrative. He dismisses Dover's attempt (Dover 1987: 74–75, 81) at distinguishing between 'substantial' (e.g. 8.87.4, 96.4), 'rhetorical' (e.g. 8.86.5) and 'trivial' counterfactuals (e.g. 2.77.5) as 'arbitrary and unhelpful'.[28] Then his discussion of the counterfactuals in Thucydides Book 8 seeks to integrate these disturbingly questioning passages into a 'unitarian' text of Thucydides. He notes that in Book 8 the historian evokes three times the idea that the Peloponnesian War might have ended earlier than it did (if the Peloponnesians had exploited the revolt of Euboea more energetically)[29] and proceeds to suggest that these speculations explain 'the unexpected continuation of the war after the Sicilian disaster',[30] citing 2.65.12 and Pericles' grand strategy. Thus Athens' astonishing endurance after defeat in Sicily justifies Thucydides' presentation of Pericles' strategic vision of the defence of the Athenian Empire in Book 2. In short, the counterfactuals in Book 8 are held to support the case for the coherence and continuity of Thucydides' narrative.[31]

I do not doubt the general coherence of 2.65.12 and Book 8.[32] Indeed, Wolfgang Will (2000: 35) has argued that the entire *History* may be read as a vast counterfactual meditation on what would have happened if Pericles had not died in 429. But it seems to me that examining Thucydides'

[27] This is sustained by the comparison of historiographical counterfactuals with counterfactuals found in Homer. For Homeric counterfactuals (the majority from the *Iliad*), see de Jong 1987: 68–81; Lang 1989; Nesselrath 1992: 5–38. But as Flory 1988: 49 shows, Thucydides does not simply adopt a stylistic turn from Homer. Some of Thucydides' counterfactuals are genuine reflection on how events might have turned out differently, as opposed to brief 'cliff-hangers' of the form 'then X would have happened, if A had not prevented it'.

[28] Rood 1998: 279 n. 84. [29] Thuc. 8.86.4–5, 8.87.4 and 8.96.1, 4–5. [30] Rood 1998: 279.

[31] In what looks to be a generational pendulum swing against analyst views, for which see Andrewes 1961, the case for the unity of Thucydides is now increasingly enthusiastically embraced. To me the unfinished state of the text seems an insuperable objection. For criticisms of the analyst approach, see Rood 1998: 16–17, 154, 251–52, who rightly employs the term 'coherence' (see next note). For a sensitive assessment of what is at stake, see Pelling 2000: 93.

[32] I use the term 'coherence' in preference to 'unity' advisedly. By it I mean a text consisting of cohesive sections (i.e. a unit of usually more than one sentence, whose constituent parts are syntactically linked by grammatical items such as conjunctions and adversatives and by such features as internal references, substitutions, ellipses, recapitulations and lexical collocations) that develop in sequence logically, thematically and narratively. On cohesion, see Halliday and Hasan 1976. I doubt that any text of considerable length will on close inspection meet the absolute standard implied by the term 'unity', but I find it especially strange to speak of a text left unfinished by its author as 'unified'. It may reveal, as Thucydides' *History* does, a fairly high degree of coherence, but that is importantly different.

text under the microscope, without looking at other near-contemporary sources, has blurred our vision of Thucydides on the level of political discourse. More promising for a discursive analysis is Flory's study. He looks beyond Thucydides to Herodotus and Xenophon and draws two important conclusions. He finds, first, that Thucydides makes counterfactual assertions far more frequently than his forerunner Herodotus and his follower Xenophon in the *Hellenica*;[33] and second, that Thucydides' counterfactual claims show a marked tendency towards illustrating how events of the Peloponnesian War might have turned out worse than they in fact did. Flory explains this biographically, as an effect of Thucydides' changed perceptions after the death of Pericles and his exile in 424.[34]

I agree with Flory that the counterfactual in 4.106.4 was probably written with a deep sense of personal regret; but I think that Thucydides was not the only Athenian to regret the past, that his text is symptomatic of this and is constructed, in part, to interrogate the political ramifications of this sense that the past might have turned out differently. Thus, while I accept the general thrust of Flory's argument and most of the details of his analysis, my views differ from his on the question of a 'counterfactual discourse' in Athens post 404. I am not persuaded that 'there evidently did not prevail at Athens during and after the war a pervasive counterfactual "if only" feeling'.[35] A number of oratorical passages from the twenty years immediately following the war suggest quite the opposite, as does a unique epigraphic record of an Athenian Assembly decree dating to 386 BCE. Thucydides' authorial counterfactual statements in the *History* must be read in the context of these public speeches and documents. The clustering of counterfactuals in Thucydides, Lysias, and contemporary evidence is no accident but a historical phenomenon to be directly attributed to the experience of the defeat of Athens and the ensuing civil war.

The idea of Athenians indulging in 'the favourite consolation of the defeated' does not seem improbable.[36] The literature on counterfactual

33 Flory 1988: 47–48. Flory finds that Xenophon makes no use of counterfactuals in *Hellenica* 1.1.1–2.4.23 (the continuation of Thucydides down to 404). I suggest the absence of this feature underlines the significance of Thucydides' use of counterfactuals for an Athenian audience post 403, despite the obvious objection that the earlier part of the *Hellenica* may be very close in date to Thucydides' *History*. That Xenophon does not involve himself in counterfactual speculation perhaps reflects his absence from Athens. The chronology of composition of the *Hellenica* is a vexed question. Most accept that the first two books are early, but how early is impossible to say. Certain passages in Books 3–7 can be securely dated to the 370s and 350s. See Dillery 1995: 13–15 with further bibliography.

34 Flory 1988: 53; similarly, Dover 1987: 81–82.

35 Flory 1988: 48 n. 13.

36 The phrase is Carr's (1962: 99 n. 2).

thinking in experimental psychology makes three pertinent observations.[37] First, counterfactual thinking is most easily inspired by strong, negative stimuli. The experience of Athens' defeat surely meets those criteria.[38] Second, when people are moved to think counterfactually, they generally think about how things might have turned out better ('upward counterfactuals' in the parlance of experimental psychology).[39] This means that Thucydides' insistence on how events of the Peloponnesian War might have been worse ('downward counterfactuals') is curious and requires explanation. Third, counterfactual thinking is subject to considerable bias: subjects quickly forget the views of the future they held before an event happened and readily come to think of past events, however unlikely they seemed at the time, as having been fairly probable.[40] Thus Thucydides' suggestion that Athens very nearly suffered an *étrange défaite* as early as 411 looks extraordinarily provocative.

In Thucydides there are twelve passages ([a] 1.11.2, [b] 2.77.5, [c] 2.94.1, [d] 3.75.4, [e] 3.113.6, [f] 4.54.3, [g] 4.106.4, [h] 5.5.3, [i] 5.73.1, [j] 8.86.4, [k] 8.87.4, [l] 8.96.4–5) containing historical counterfactuals that may be said to be clearly in the author's voice. They appear in a number of syntactical arrangements and have various functions (as do the counterfactuals that are not clearly endorsed by our author), but one type of counterfactual in particular, whose form is directly traceable to Homer, has bedevilled the investigation of the entire subject. We shall examine it first.

Homer's 'unrealized possibilities' are especially numerous in the *Iliad*.[41] Typically they occur in the thick of battle, at moments when we are told that something would have happened if something else had not prevented it.[42] For example, at *Iliad* 11.310–12 Odysseus averts what would have been a disastrous rout by exhorting Diomedes to stand beside him in the battle line. As Flory remarks, passages like this create 'a moment of artificial suspense in an otherwise predictable narrative' (Flory 1988: 49). Whereas in Homer the dominant syntactic patterning of unrealized possibilities is what Flory calls 'protasis negative' with the 'counterfactual' apodosis placed before the negating protasis – raising a possibility only to foreclose it immediately – in Thucydides the syntactic arrangement of counterfactuals privileges a counterfactual protasis followed by a counterfactual apodosis. Although

37 For discussion, see Roese and Olson 1995, Roese and Morrison 2009.

38 Xen. *Hell.* 2.2.3; Plut. *Lys.* 13.1–15.5.

39 Roese and Morrison 2009: 16–17.

40 The classic study is Fischhoff 1975; cf. Hawkins and Hastie 1990.

41 See Flory 1988: 48–49. For the term 'unrealized possibility' applied to a fictional narrative, see above n. 26.

42 Flory lists these instances: Flory 1988: 48 n. 14.

Thucydides occasionally uses the Homeric arrangement, in general his counterfactuals are syntactically constructed to be more 'open-ended' in their implications than Homer's unrealized possibilities, or to have a higher 'degree of speculation' (Flory 1988: 50).

Hornblower refers to this Homeric type of counterfactual as the '"if... not" device' and urges that in his use of it 'Thucydides surely differs not at all from the poets and novelists' (Hornblower 2011: 90). But far from all of Thucydides' counterfactuals adopt this form, even in the restricted group in Thucydides' authorial voice. Therefore, our task is to construct a general typology of historical counterfactuals; we may then be able to understand better how Thucydides' historiography differs from the practices of poets and novelists.

In the following discussion the series A, B, C specifies a causal chain of actual historical events; P is a hypothetical event; X, Y, Z describe a chain of causally linked counterfactual events. I distinguish (on nonarbitrary grounds) between three kinds of developed counterfactual statements in Thucydides: rhetorical, analytical and inferential. A 'developed counterfactual' means that Thucydides has constructed a counterfactual conditional statement, with a counterfactual protasis, even if it is grammatically suppressed, and a counterfactual apodosis, and he has linked them by causal reasoning (i.e. X→Y).

By 'rhetorical' I mean counterfactuals following the syntax of the Homeric model, which make a momentary interruption into the narrative without the author exploring the consequences of a postulated counterfactual event; this does not mean that the reader is not provoked to reflect on how history might have been different, but the syntactical form of the counterfactual does not extend maximal encouragement to such a response. These passages are tantamount in historical sense to the mere acknowledgement of an alternative historical possibility without any exploration of its consequences; an example of the latter is 1.101.2: the Spartans were intending to invade Attica at the time of the rebellion of Thasos but an earthquake prevented them. Although this may be paraphrased as 'If there had not been an earthquake, the Spartans would have invaded Attica', that is not the form that Thucydides has given it and I exclude it as a developed counterfactual claim because it does not exhibit the syntax 'if X *had happened*... Y *would have happened*' in the form of two conditional clauses (of which the protasis may be suppressed) linked by causal reasoning.

'Analytical' counterfactuals are used to reason about the necessity or contingency of historical events in a causal chain. The functions of analytical counterfactuals are to construct hypotheses about events that presumably happened but for which evidence is absent, to identify a cause as necessary for a given consequence or to specify the degree of an event's contingency.

Lastly, and most importantly, 'inferential' counterfactuals pursue the consequences of a counterfactual event in a causal chain. These may be simple (X→Y) or complex (X→Y→Z . . .). If at 1.101.2 Thucydides had written εἰ γὰρ μὴ ἐγένετο τότε δὴ σεισμός, ἐς τὴν Ἀττικὴν ἂν ἐσέβαλον οἱ Λακεδαιμόνιοι ('If an earthquake had not occurred at that time, the Lacedaemonians would have invaded Attica'), the passage would be a simple inferential counterfactual. If he had added '. . . and then Z would have happened as a result of the invasion', the passage would present a complex inferential counterfactual. In other words, a simple inferential counterfactual is a piece of counterfactual causal reasoning depending on a counterfactual event. It becomes complex if one piece of counterfactual reasoning itself depends on another piece of counterfactual reasoning.

The rhetorical counterfactual is rare in Thucydides, as Flory has shown.[43] The great majority of Thucydides' counterfactuals do not follow the syntactical structure of the Homeric model:

(1) X would have happened, if A had not happened.

For an example of a passage that does this in Thucydides' authorial voice, we may turn to [i] 5.73.1:

> καὶ μάλιστ' ἂν τοῦ στρατεύματος ἐταλαιπώρησαν, εἰ μὴ οἱ ἱππῆς παρόντες αὐτοῖς ὠφέλιμοι ἦσαν.
>
> And [the Athenians] would have suffered more heavily than any other part of the army, if their cavalry had not been there in support.

The counterfactual scenario is conjured up in the reader's imagination only to be dispelled immediately by the delayed negative protasis. To the rhetorical type belong [c], [d] and [i], only a quarter of the counterfactuals in Thucydides' authorial voice. In [c] and [d] the protasis and apodosis are interwoven but the apodosis is complete before the protasis; in [i] the apodosis is presented first and followed by the negating protasis exactly as in the typical Homeric counterfactual.

Analytical counterfactuals may be divided into hypothetical and causative. In the first case the counterfactual allows for a hypothesized event that either probably or necessarily happened for some known event to have occurred.

[43] Cf. Flory 1988: 49–50.

(2) If P had not occurred, A would not have happened.

There is no example of this precise construction in Thucydides' authorial voice with a protasis introduced by the particle *ei*, but one passage yields relevantly similar sense. At 3.89.5 Thucydides mentions the occurrence of a series of earthquakes, some of which he believes caused the sea to retreat and then return in a terrible inundation:

> ἄνευ δὲ σεισμοῦ οὐκ ἄν μοι δοκεῖ τὸ τοιοῦτο ξυμβῆναι γενέσθαι.
>
> Without an earthquake I do not see how such an event could happen.

Similarly Thucydides uses the hypothetical to reason about events in the distant past for which there is no direct evidence, as in 1.11.1, where he conjectures that the Greeks landing at Troy must have fought and won a battle, otherwise they could not have fortified their camp.

Two further kinds of analytical counterfactual are used to identify the necessity or over-determination of a cause. First, there is the identification of a necessary cause:

(3) If A had not happened, B would not have happened.

The only strictly analytical counterfactual in Thucydides' own voice is [h], where Thucydides explains that if the Locrians had not been embroiled in other wars at the time (422 BCE) they would not have made peace with Athens. The passage illustrates the necessity of the 'other wars' for explaining why the Locrians chose to make peace with Athens when they had earlier (in 427) refused to do so.

Second, there is the postulation of a counterfactual event that would not have changed some causal chain of events, showing that the event was over-determined:

(4) Even if X had happened, A would still have happened.

Here the counterfactual is used to illustrate the extent to which an actual event was not obviously contingent or was over-determined. There is no example of this construction among the clearly authorial counterfactual statements, but in 7.42.3[44] it is held that a counterfactual event would not have disrupted a chain of other counterfactual events:

44 There has been much discussion of whether this passage represents Thucydides' authorial view as well as the point of view of his character Demosthenes: see Donini 1964, Dover 1987: 75–81. Despite Dover's meticulous arguments, I find the obstacles to accepting this counterfactual as clearly and uncomplicatedly Thucydides' authorial view too great to include it with the others identified above.

> . . . ὥστε μηδ᾽ εἰ μετέπεμψαν ἔτι ὁμοίως ἂν αὐτοὺς ὠφελεῖν . . .
>
> . . . even if they [the Syracusans] had sent for help then, it would not have benefited them . . .

This brings us to the inferential counterfactual constructions.

(5) If X had happened, Y would have happened.

A good example of this appears during the siege of Plataea in Book 2, at the point where Thucydides informs us that if the wind had risen, the Plataeans would not have escaped the resulting inferno (2.77.5):

> . . . πνεῦμά τε εἰ ἐπεγένετο αὐτῇ ἐπίφορον, ὅπερ καὶ ἤλπιζον οἱ ἐναντίοι, οὐκ ἂν διέφυγον.
>
> . . . if the wind had risen [and blown upon the flames], which was exactly what the enemy were hoping would happen, the Plataeans would not have got away.

On the face of it (5) is logically equivalent to (3) since positing the non-instantiation of an actual event (A) is by definition a counterfactual scenario 'if X . . . ' and the consequent non-instantiation of a real-world result (B) is equivalent to positing a counterfactual result (Y). Or, to put it another way, $\neg A \rightarrow \neg B$ looks equivalent to $X \rightarrow Y$. But this leads into a trap in which all historical counterfactuals are seen as examples of the 'If. . . not' device. This is a profound mistake because although all negations of a fact constitute a counterfactual, and though all counterfactual claims must (at least implicitly) negate some fact or set of facts, not all counterfactuals stipulate precisely what fact or set of facts they negate.

An acceptable historical counterfactual with a negative protasis must negate a fact:

If Pericles had not died of the plague . . .

Where a counterfactual proposition about the past like this is introduced as 'If $\neg$ A', the consequence (Y) is syntactically and logically severely devalued in importance even before the reader or listener encounters it (hence the inversion of word order in Homer's favourite 'unrealized possibility' construction 'then X would have happened, if A had not prevented it'). The statement emerges as a 'discounted' or negligible possibility. This makes it less likely that further links in the counterfactual causal chain will be presented, and, where they are, the demands on the reader's imaginative suspension of historical sense will be commensurately greater. But counterfactual claims do not always begin with the explicit negation of a fact.

Where they are presented in the 'positive' form 'If X . . . ,' they are offered as something more than a negligible historical possibility.

Inferential counterfactuals in Thucydides' authorial voice are found in [a], [b], [e], [f], [g], [j], [k] and [l]. Among these all but the last may be said to be simple (i.e. they involve causal reasoning only of the extent X→Y). This may be illustrated by passage [a]:

> περιουσίαν δὲ εἰ ἦλθον ἔχοντες τροφῆς καὶ ὄντες ἀθρόοι ἄνευ λῃστείας καὶ γεωργίας ξυνεχῶς τὸν πόλεμον διέφερον, ῥᾳδίως ἂν μάχῃ κρατοῦντες εἷλον . . . πολιορκίᾳ δ᾽ ἂν προσκαθεζόμενοι ἐν ἐλάσσονί τε χρόνῳ καὶ ἀπονώτερον τὴν Τροίαν εἷλον.
>
> If, however, Agamemnon had had plenty of supplies with him when he arrived, and if they had used their whole force in making war continuously, without breaking off for plundering expeditions and for cultivating the land, they would have won easily . . . If, therefore, they had settled down to a siege at once, they would have taken Troy in a shorter time and with less trouble. (1.11.2)

Passage [a] claims that under a set of counterfactual conditions (X) – the Greeks arriving at Troy with abundant resources and conducting a protracted siege without expending their energies on plundering or agriculture – Troy would have fallen sooner than it did (Y). The passage is lengthy but defining it as substantial misses the point that it is only a simple inference. As the first inferential counterfactual in the *History*, it seems to have programmatic force: wars may be forced to a resolution more quickly if the appropriate strategy is used.

Passage [l] is a complex inference: it presents a counterfactual chain X→Y→Z.

> . . . εἰ τολμηρότεροι ἦσαν, ῥᾳδίως ἂν ἐποίησαν . . . καὶ ἢ διέστησαν ἂν ἔτι μᾶλλον τὴν πόλιν ἐφορμοῦντες ἤ, εἰ ἐπολιόρκουν μένοντες, καὶ τὰς ἀπ᾽ Ἰωνίας ναῦς ἠνάγκασαν ἂν καίπερ πολεμίας οὔσας τῇ ὀλιγαρχίᾳ τοῖς σφετέροις οἰκείοις καὶ τῇ ξυμπάσῃ πόλει βοηθῆσαι· καὶ ἐν τούτῳ Ἑλλήσποντός τε ἂν ἦν αὐτοῖς καὶ Ἰωνία καὶ αἱ νῆσοι καὶ τὰ μέχρι Εὐβοίας καὶ ὡς εἰπεῖν ἡ Ἀθηναίων ἀρχὴ πᾶσα.
>
> . . . if the Peloponnesians had been more daring, they could easily have done this [sailed to Piraeus]. They would then have produced, simply by anchoring off the city, still greater dissension inside, or, if they stayed there and undertook siege operations, they would have forced the fleet in Ionia, however hostile it might be to the oligarchy, to come to the help of their own people and of the city itself; and meanwhile the Hellespont and Ionia would have fallen into their hands, together with the islands and everything as far as Euboea – the whole Athenian Empire in fact. (8.96.4–5)

The series of judgments depends on the counterfactual (X) 'if the Peloponnesians had sailed to Piraeus'. It proposes that there would have been a set of effects (Y): first, greater threat of revolution in Athens (the consequences of which are not explored); second, the necessity of the Ionian fleet returning to Athens. And if the latter had happened, the Hellespont, Ionia, the Aegean islands and Euboea would have fallen into the hands of the Peloponnesians (Z). In short, Thucydides only once constructs a complex inferential counterfactual (and it is in the positive form 'if X') and in it he imagines the collapse of the Athenian Empire in 411. He clearly means it to strike the reader as significant.[45]

There is every reason to think that questions like Herodotus' 'What would have happened if the Athenians had fled or Medized in 480?' (7.139) were a subject of the greatest interest to Greeks for many years after the final Persian defeat at Plataea in 479.[46] It seems hardly probable that counterfactual questions about the Peloponnesian War were not widely asked after the fall of the Athenian Empire in 404. If such was the case, what might these questions have been? The evidence indicates three in particular. First, could Athens have concluded a more favourable peace with Sparta in 404? Second, what would have happened if the democratic counter-revolution in 403 had failed? And third, could the Athenian Empire have survived the Peloponnesian War?

Lysias 14.16 illustrates the use to which counterfactual claims were put in Athenian forensic discourse in the years after 403. Here, the speaker asserts that if Alcibiades had been put to death at the age his son is now, Athens would have been spared terrible misfortunes:

> ... <ὃν> εἰ τηλικοῦτον ὄντα ἀπεκτείνατε, ὅτε πρῶτον εἰς ὑμᾶς ἐλάβετε ἐξαμαρτάνοντα, οὐκ ἂν ἐγένοντο συμφοραὶ τοσαῦται τῇ πόλει.
>
> If you had executed [Alcibiades] at this man's age, the first time you caught him harming you, the city would not have experienced such great disasters.

There are a number of parallel passages in Lysias. We shall consider two examples. The first is from a speech delivered in 382 in an attempt to block an Athenian called Evander from becoming archon (26.18). The speaker claims that more than twenty years ago Evander was among the Athenian citizens who supported the oligarchic terror:

45 There have been doubts whether Thucydides' judgment here is persuasive. Lazenby 2004: 193 argues on political and military grounds that Thucydides is misguided.

46 Cf. Thucydides 1.73.4–74.4.

> καὶ ἡγοῦνταί γε διὰ μὲν τοὺς τοιούτους συλληφθέντας τότε ἀποθανεῖν, διὰ δὲ τοὺς ἄλλους ἐκφυγεῖν, ἐπεὶ εἴ γε τὴν αὐτὴν γνώμην ἅπαντες ἔσχον, οὔτ᾽ ἂν φυγὴ οὔτ᾽ ἂν κάθοδος οὔτ᾽ ἄλλο οὐδὲν ἂν τῶν γεγενημένων τῇ πόλει συνέβη.
>
> The people believe it was owing to men like these that they suffered arrests and executions during that period, and owing to the others that they escaped, because if everyone had shared the same attitude, neither the exile nor the counter-revolution nor any of the other events would have happened to the city.

In the period, the sentiment 'if we had remained united and every man had done as I did, the oligarchy and civil war would not have happened' became a cliché of Athenian forensic rhetoric.[47]

Conversely, at 12.98 Lysias imagines the counterfactual scenario in which the democrats lost the battle for the city after the return from Phyle.[48]

> εἰ δὲ ἐδυστυχήσατε καὶ τούτων ἡμάρτετε, αὐτοὶ μὲν ἂν δείσαντες ἐφεύγετε μὴ πάθητε τοιαῦτα οἷα καὶ πρότερον, καὶ οὔτ᾽ ἂν ἱερὰ οὔτε βωμοὶ ὑμᾶς ἀδικουμένους διὰ τοὺς τούτων τρόπους ὠφέλησαν, ἃ καὶ τοῖς ἀδικοῦσι σωτήρια γίγνεται· οἱ δὲ παῖδες ὑμῶν, ὅσοι μὲν ἐνθάδε ἦσαν, ὑπὸ τούτων ἂν ὑβρίζοντο, οἱ δ᾽ ἐπὶ ξένης μικρῶν ἂν ἕνεκα συμβολαίων ἐδούλευον ἐρημίᾳ τῶν ἐπικουρησόντων.
>
> If you had been unlucky and had failed in your aim, you yourselves would have gone into exile fearing you would suffer as you had done before, and because of the ways of these men no temples or altars, which are a place of safety even for wrongdoers, would have protected you from being wronged. However many of your children were here would have been treated with no respect by these men, while those abroad would have been enslaved for petty debts and bereft of help.

The rhetorical strategy that became well worn in the courts found its way into other kinds of rhetoric. Isocrates in the *Panegyricus* (142) invokes a counterfactual to attribute victory at the battle of Cnidus in 394 to the leadership of Conon: had it not been for Conon, the Persian King would not have had the initiative to stop his fleet being blockaded by the Spartans and the soldiers would have deserted, being unpaid for fifteen months.

Athens' attempt (from the mid 390s into the early 380s) to resurrect its Aegean hegemony failed when Antalcidas recaptured the Hellespont in 387. In an inscription dating to 386 BCE counterfactual history appears in the proceedings of the Athenian assembly.

[47] Cf. Lys. 25.15; Andoc. 4.41; Isoc. 21.11.

[48] Lys. 12.92 also contemplates the survival of the oligarchy.

Κέφαλος εἶπε· τὰ μὲν ἄλλα καθάπερ [τ]-
ῆι βολεῖ· ἀναγράψαι δὲ Φανόκριτο[ν]-
τὸν Παριανὸν πρόξενον καὶ εὐεργ[έ]-
την αὐτὸν καὶ τὸς ἐκγόνος ἐν στήλ[ε]-
ι λιθίνει καὶ στῆσαι ἐν ἀκροπόλε[ι]
[τ]ὸγ γραμματέα τῆς βολῆς, ἐπειδὴ π[α]-
[ρ]ήγγελε τοῖς στρατηγοῖς περὶ [τῶν]
[ν]εῶν τõ παράπλο, καὶ εἰ οἱ στρατ[ηγο]-
[ὶ] ἐπίθοντο ἑάλωσαν ἂν α[ἱ] τρ[ι]ήρε[ι]ς
αἱ πολεμίαι· ἀντὶ τούτων ἔναι [κ]αὶ τ-
ὴν προξενίαν καὶ τὴν εὐεργεσί[αν.

Cephalus proposed: In other respects in accordance with the council; but Phanocritus of Parium shall be written up as a *proxenos* and benefactor, himself and his descendants, on a stone *stēlē* and it shall be placed on the Acropolis, by the secretary of the council, because he passed over to the generals a message about the passage of the ships, and if the generals had believed him the enemy triremes would have been captured: it is in return for this that he is to receive the status of *proxenos* and benefactor.[49]

As far as I am aware, this is an unparalleled example of counterfactual thinking in the epigraphic record of the deliberative processes of the Athenian Assembly;[50] in this case, the issue was voting civic honours for a certain Phanocritus of Parium. Phanocritus had passed information to the Athenian commanders that they had an opportunity to capture some Spartan triremes. Unfortunately, they failed to act upon the information. This may have reminded some Athenians of the incident in 405 just before the fateful battle of Aegospotami (Xen. *Hell.* 2.1.26), when Alcibiades advised the Athenian commanders to move their ships to a safer anchorage at Sestos; they ignored him and a large part of the fleet was captured by Lysander, leading swiftly to the blockade of Athens and the end of the war. In this case, Phanocritus is awarded civic privileges since the Assembly accepted that if his information had been trusted, the enemy triremes *would have been captured.* The counterfactual reasoning for awarding these honours to Phanocritus is an amendment to the original grant. Cephalus, a supporter of war against Sparta in the 390s (cf. *Hell. Oxy.* 7.2), amended the grant with the historical counterfactual explaining the reason for awarding

49 *IG* II^2 29 = *RO* 19, trans. Rhodes and Osborne 2003.

50 The only other example of a historical counterfactual in Athenian epigraphy to my knowledge is a plausible restoration in a funerary epitaph: *IG* II^2 9052. A present counterfactual conditional is found in another funerary inscription: *IG* II^2 11594.

the honour and this will then have had to be approved by the Assembly. Thus we find historical counterfactuals at the very heart of the democratic deliberation.

Two texts illustrate the two central counterfactual questions that Athenians asked in the years immediately following the Peloponnesian War. First we find the counterfactual claim that the Athenians could have concluded a less destructive peace with Sparta if the supporters of oligarchy had not prevented it with the prosecutions brought by Agoratus (Lysias 13.16):

> . . . βουλόμενοι βελτίω ταύτης εἰρήνην τῷ δήμῳ τῷ Ἀθηναίων ποιήσασθαι. ἐνόμιζον δὲ δυνήσεσθαι, καὶ ἔπραξαν ἂν ταῦτα, εἰ μὴ ὑπ' Ἀγοράτου τουτουὶ ἀπώλοντο.
>
> . . . they wanted to make a better peace than this [i.e. the peace of Theramenes] for the Athenian people. They thought it was feasible, and they would have done it, if they had not perished because of Agoratus here.

In this context it is worth mentioning that the 'Theramenes Papyrus' (P. Mich. 5982), if it is indeed an early fourth-century defence of Theramenes and his peace negotiations, is indicative of bitter wrangling in these years over this very issue.[51]

Second, in Lysias' *Funeral Oration* (2.65) we find the counterfactual historical scenario of the period that haunted Athenians more than any other. The valour of the democrats returning from Phyle was such that surely, if Athens had not been divided by oligarchy and civil war, the Athenians united could have defeated the Peloponnesians?

> εἰ γὰρ στασιάσαντες πρὸς ἀλλήλους βίᾳ παρόντων Πελοποννησίων καὶ τῶν ἄλλων ἐχθρῶν εἰς τὴν αὑτῶν οἷοί τε ἐγένοντο κατελθεῖν, δῆλον ὅτι ῥᾳδίως ἂν ὁμονοοῦντες πολεμεῖν αὐτοῖς ἐδύναντο.
>
> If they managed to return to their city while fighting against other citizens and facing the Peloponnesians and other enemy forces, it is clear that if they had been united, they could easily have carried on the war.[52]

The funeral oration is undoubtedly the most ideologically fraught locus of political rhetoric at Athens. In that regard, perhaps Plato's *Menexenus* is designedly set in a counterfactual Athens in which Socrates is still alive after the King's Peace. The irony of that scenario would then be an interrogation

[51] Text of P. Mich. 5982: Youtie and Merkelbach 1968; interpretation: Andrewes 1970.

[52] Stephen Todd's (2000) translation of this passage ('it is clear that united they would easily have defeated them') is, I think, overly optimistic about Athenian strategic capabilities even in a context like the funeral oration. The construction with the present infinitive πολεμεῖν is perhaps deliberately vague, but better rendered as 'carried on the war'.

of the funeral oration prompting the reader to ask what in recent Athenian history would have had to have been different for Socrates to participate in such a dialogue.[53]

By contrast with the aggrandizing counterfactual claims of Isocrates' *Panegyricus*, of Lysias' *Funeral Oration*, and the less grandiose but still perhaps inflated claims of *IG* II^2 29 and Lysias' prosecution speeches, Thucydides' counterfactuals fairly consistently point out that the Peloponnesian War could have been worse. But – and this point is crucial – over half of the counterfactuals in Thucydides' authorial voice show our historian focusing on how the war could have been worse *expressly* for Athens and the allies ([b], [c], [g], [i], [j], [k], [l]).[54]

For Thucydides, I suggest, there was a sense of horror that the Athenians were quickly setting themselves back on the road towards renewed maritime imperialism and conflict with the Peloponnesians. By 395 Athens was indeed at war, and such events as the massacre at Corinth in the winter of 393/2 (Xen. *Hell.* 4.4.1–6) were harbingers of a return to the internecine violence and destruction of the twenty-seven years from 431 to 404. Under the influence of discourses like the funeral oration, Thucydides saw that the democracy was quickly forgetting the horrors of the Peloponnesian War. According to the ideologically distorted history of Athens presented in such public contexts, the city had not been defeated by Sparta and her allies but ultimately by itself through internal division.[55] In opposition to this kind of history, which was evidently widespread and popular in the Athenian imagination, Thucydides not only emphasizes how terrible the war really was but underlines moments when it might have been worse for Athens and her citizens. Furthermore, he tackles the claim of the ideological history of the polis that Athens was defeated in 404 only by internal strife

[53] On the anachronistic difficulties of the dialogue, see Kahn 1963, M. M. Henderson 1975. Kahn identifies a number of problems in the interpretation of this text, one of which is blatant anachronism. Various explanations have been advanced, irony of one kind or another being the favourite. For example, Rosenstock 1994 argues that the ironic point is that the ideal Athenian funeral oration is presented by what must be the ghosts of Socrates and Aspasia. As far as I am aware, the idea that the *Menexenus* is intended to provoke counterfactual reflection – how would Athenian history have to be different for Socrates to be alive after 387 BCE? – has not yet been suggested or explored. A number of studies of the dialogue see Plato engaging directly with Thucydides' funeral oration: for example, Kahn 1963, Salkever 1993.

[54] In regard to [b], note that the Plataeans were granted Athenian citizenship ([Dem.] 59.104, Isoc. 12.94) and were thus part of the audiences of the kinds of public rhetoric discussed (for their continued presence in Athens in the period, see Lys. 23.5–7), at least until some point after 386 BCE when Plataea was refounded (Paus. 9.1.4).

[55] For the tendency of the Athenians after 404 to blame the defeat on betrayal and bad luck, see Lévy 1976: 29–55.

when he suggests with counterfactuals in Book 8 that Athens could have been forced to surrender much earlier, most provocatively of all at 8.86.4, where he writes that in one crisis the city was saved by Alcibiades – of all people! To hard-line democrats the idea of defeat at that point was surely unbearable, if not unthinkable – primarily because it would then have come before the disaster at Arginusae in 406, on which much could be blamed. But I wonder if Thucydides might not have thought that it would actually have been better if Athens had been forced to surrender then, for if the Peloponnesian War had ended seven years earlier than it did, Arginusae and Aegospotami would not have happened and thousands of lives might have been spared.

I am painfully aware that arguing for a highly politicized post-403 Thucydides places strain on the traditional dating of his work and argues against the author's insistence that he wrote it as a possession for all time not a show piece for the present moment (1.22.4), but perhaps another passage may stand as an alternative valedictory: καὶ γὰρ ὁ μὴ ῥηθεὶς λόγος τοῖς ὧδ' ἔχουσιν αἰτίαν ἂν παράσχοι ὡς, εἰ ἐλέχθη, σωτήριος ἂν ἦν ('To those in a situation such as ours, words left unsaid would provide a reason for thinking that if they had been spoken, they would have saved us', 3.53.3).

CHAPTER 7

Homer's Achaean wall and the hypothetical past

Karen Bassi

According to John Camp, walls are the "most enduring evidence of antiquity surviving in the landscape today."[1] As features in the built environment, the remains of walls delimit the lives of ancient humans in both time and space; they give substance to the past. Not surprisingly, the prominence of walls as archaeological features is shared by their prominence as narrative plot devices; they stand out (as it were) in both contexts. It might even be said that walls occupy a unique position as both signifiers of cultural production and structures within narrative. Between their existence as physical remains and their role in the representation of events (in both poetry and history writing), they also traverse the distinction between fact and fiction or, in the phrases made famous by Aristotle in the *Poetics*, between "things that have happened" (*ta genomena*) and "the sorts of things that might be expected to happen and are possible according to what is likely or necessary" (*hoia an genoito kai dunata kata to eikos ē to anankaion*), 1451a36–38).[2] In short, walls stand for and perhaps even epitomize the temporal and ontological categories that bring "the actual" into contact with "the possible" in what Wolfgang Ernst has called the "fragmented landscape" of the past.[3]

The basis for this premise is Homer's account of the Achaean wall in the *Iliad*. In this chapter I make two related arguments. First, that the ontological disposition of the wall, framed by the question of whether or not it actually existed, offers a unique – even formative – opportunity for investigating the prehistory of the criteria that Aristotle introduces to define *historia* in contrast to *poiēsis* in the *Poetics*. And, second, that the fate of the wall is linked with the ethical dimension of the poem, measured in the epic hero's achievement of *kleos aphthiton* (immortal glory). The link

[1] Camp 2000: 41. More recently, see Morris and Papadopoulos 2005: 155. Hauser 2013: 1 makes the more general claim that archaeological features and artifacts are "tangible evidence for the past" (haptisches Zeugnis der Vergangenheit).

[2] The translation is that of Else 1957, with some modifications.

[3] Ernst 1999: 53.

between these two arguments is the account of the wall's destruction in a hypothetical past, a past that, from the point of view of the poem, has not yet happened. This temporal anomaly, in which the *kleos* of the wall is at stake, demonstrates by analogy that ethical action is contingent upon the possibility that the past can be known. The purpose of the chapter is to examine how and why the Achaean wall fulfills this function in the *Iliad* and in the history of its reception.

As is well known, the history of scholarship on the *Iliad* equates the historicity of the poem itself with the fate of the wall; in a somewhat circular fashion, the poem's credibility and consistency – what is likely or necessary in Aristotle's terms – rise and fall on the question of whether or not the wall actually existed. In James I. Porter's succinct summation:

> [A]t stake in the wall . . . and underlying all the debates around it, is its basic status as a fictional object, and by extension its exemplary role vis-à-vis fictional objects generally in the Homeric poems. The Achaean wall cannot help but have this claim to interest by virtue of being an object that once so magnificently and palpably and uniquely – but also, so curiously and suddenly – *was* and then so utterly *is no more*.[4]

With reference to Eustathius' somewhat enigmatic comments on the fate of the wall (690.63–64 = 2.498.15–17), Porter concludes that it is "an emblem for poetic fiction." And he makes the further observation, following Roman Jakobson, that "ontological ambiguity . . . appears to be a generic trait of fictionality worldwide."[5] Given the history that Porter outlines, the wall's "exemplary role vis-à-vis fictional objects generally in the Homeric poems" is clear. At the same time, however, its "ontological ambiguity" is more than a trans-historical or universal truism; for, as the history of scholarship on the wall demonstrates, this ambiguity is initiated first of all in the fact that, as narrative plot devices, material objects and architectural features have a particular referential potential: they can refer to a phenomenal world that is prior to and outside the text.

This potential is implicit in the earliest attested commentary on the Achaean wall in the *Iliad* (Strabo, *Geographica* 13.1.36 = Rose fr. 162):

> Καὶ μὴν τό γε ναύσταθμον τὸ νῦν ἔτι λεγόμενον πλησίον οὕτως ἐστὶ τῆς νῦν πόλεως, ὥστε θαυμάζειν εἰκότως ἄν τινα τῶν μὲν τῆς ἀπονοίας τῶν δὲ τοὐναντίον τῆς ἀψυχίας· ἀπονοίας μέν, εἰ τοσοῦτον χρόνον ἀτείχιστον αὐτὸ εἶχον, πλησίον οὔσης τῆς πόλεως καὶ τοσούτου πλήθους τοῦ τ'

4 Porter 2011: 5 (original emphasis). Porter offers a detailed and comprehensive account of ancient and modern scholarship on the wall. I summarize this history in Bassi 2005: 21–23.

5 Porter 2011: 17 with reference to Jakobson 1960: 371.

> ἐν αὐτῇ καὶ τοῦ ἐπικουρικοῦ· νεωστὶ γὰρ γεγονέναι φησὶ τὸ τεῖχος (ἢ οὐδ᾽ ἐγένετο, ὁ δὲ πλάσας ποιητὴς ἠφάνισεν, ὡς Ἀριστοτέλης φησίν)· ἀψυχίας δέ, εἰ γενομένου τοῦ τείχους ἐτειχομάχουν καὶ εἰσέπεσον εἰς αὐτὸ τὸ ναύσταθμον καὶ προσεμάχοντο ταῖς ναυσίν, ἀτείχιστον δὲ ἔχοντες οὐκ ἐθάρρουν προσιόντες πολιορκεῖν μικροῦ τοῦ διαστήματος ὄντος·
>
> In fact, that which is now still called the naval station is so near the present city that it is likely (*eikotōs*) that someone would be amazed at [the Greeks'] lack of sense and [the Trojans'] lack of courage; lack of sense, if [the Greeks] kept the naval station un-walled for so long a period of time, when they were near to the city and to so great a multitude, both in the city and among the allies. For [Homer] says that the wall was recently built, or it was not: having fashioned it, the poet made it disappear, as Aristotle says. And lack of courage, if [the Trojans], after the wall was built, besieged it and broke into the naval station itself and attacked the ships but did not dare to march up and besiege the station when it was un-walled and with only a small interval in-between.

Here Strabo marshals Aristotle as an authority in a discussion of the location of the ancient city of Troy (*to palaion Ilion*, 13.1.35.2) relative to the present-day village (*hē tōn Ilieōn kōmē*, 13.1.35.1). More specifically, Aristotle's statement introduces and, it seems, endorses Strabo's skepticism about the tactical utility of a wall built so late in the war, in the tenth year: why would the Greeks leave the ships unprotected for so long in the face of such a multitude of enemies? And why did the Trojans attack the ships after the wall was built but not before? Aristotle's comment seems to respond to the inconsistencies that Strabo enumerates by inferring the non-existence of the wall outside the poem's fictional universe, where the inconsistencies themselves arise. In short, Homer made up the wall (*ho de plasas poiētēs*) and then made it disappear (*ēphanisen*) because it did not really exist.[6]

As a way of introducing this argument, Strabo begins with the criterion of what is likely (*eikotōs*), namely that an indefinite "someone" would be amazed at the combatants' failure to act. These failures are judged in relation to spatial and temporal contingencies: the nearness of the ships to the current city, on the one hand, and the time before and after the building of the wall, on the other. Embedded in a "natural" result clause, what is likely starts from a statement of fact (that the naval station is near the present city) and ends in a series of hypothetical or counterfactual claims.

[6] On Strabo's debts to and assessment of Homer as a source, see Dueck 2000: 31–40. Strabo refers to Homer as the "founder (*arkhēgetēs*) of geographical science" (1.1.2). But he also famously remarks that "Homer leaves us to guess about most things. And it is necessary to arbitrate between his statements and those of the others" (13.1.1; see also 7.3.6). His account of the Achaean wall finds Strabo negotiating between these two views of Homer.

Amazement is the likely or expected response because, given the wall's existence in the poem, the Greeks did show a lack of sense (*aponoia*) and the Trojans did show a lack of courage (*apsukhia*)! But is this the conclusion Strabo endorses? Coming between the two counterfactual statements, the quotation from Aristotle exonerates the Greeks and Trojans on the grounds that, although Homer made the wall, it did not really exist. In other words, the hypothetical conclusions to which the wall's construction within the poem are likely to lead are mitigated by the fact that it did not exist in reality. These conclusions are tactical but have ethical consequences: *aponoia* and *apsukhia* denote the combatants' failure to take the right action at the right time.

Thus, although the statement attributed to Aristotle can be interpreted as a simple statement of fact, he is also our earliest source for a rationalizing discourse in which the Achaean wall signifies a competition between historical (and archaeological) fact and poetic fiction in accounting for the past, a competition based in part on what is likely to happen.[7] As suggested above, this rationalizing discourse extends to the proposition that Homer was trying to cover up the fact that there were no physical remains of the wall in his (i.e. Homer's) own time.[8] Strabo's contribution to this discourse includes the observation that no trace of the ancient city of Troy survives in *his* own time (*Geographica* 13.1.38; cf. 13.1.41):

> Οὐδὲν δ' ἴχνος σώζεται τῆς ἀρχαίας πόλεως. εἰκότως· ἅτε γὰρ ἐκπεπορθημένων τῶν κύκλῳ πόλεων, οὐ τελέως δὲ κατεσπασμένων, ταύτης δ' ἐκ βάθρων ἀνατετραμμένης, οἱ λίθοι πάντες εἰς τὴν ἐκείνων ἀνάληψιν μετηνέχθησαν.
>
> No trace of the ancient city is preserved. This is to be expected (*eikotōs*). For while the cities all round it were sacked they were not completely dismantled. But because that city was demolished down to its foundations all the stones were carried from it for rebuilding those other cities.

In response to suggestions made by the Ilians to the contrary, Strabo enlists Homer as proof that the ancient city was in fact wiped out (*Geographica* 13.1.41):

> Ὅμηρος δὲ ῥητῶς τὸν ἀφανισμὸν τῆς πόλεως εἴρηκεν ἔσσεται ἦμαρ ὅταν ποτ' ὀλώλῃ Ἴλιος ἱρή.
>
> Homer explicitly speaks about the obliteration of the city, "There will come a day when sacred Ilios will perish."

[7] Cf. Porter 2011: 32. [8] Cf. Scodel 1982: 33 with n. 2.

Quoting from Hector's exchange with Andromache at *Iliad* 6.448, Strabo takes what is internal to the poem and, moreover, spoken by a character and not by "Homer," as proof for what the present topography of the site reveals to Strabo: because no trace (*ikhnos*) of the ancient city survives it must have been completely destroyed. In the same passage and as further support for this hypothesis, Strabo quotes *Iliad* 12.15 (*Geographica* 13.1.41): "The city of Priam was destroyed in the tenth year" (*pertheto de Priamoio polis dekatōi eniautōi*). But again the context of the line complicates Strabo's intention; for it occurs in what might be called the preamble to the detailed description of the total destruction of the Achaean wall (*Iliad* 12.17ff.). There is, in short, a contradictory logic at play in Strabo's text: in the case of the Achaean wall, Homer's account of its total destruction proves that it did not exist; in the case of Troy, Homer's testimony that the city was obliterated is based on the assumption that it did exist. More surprisingly, Strabo finds evidence for the latter in the context of the former.

To say that there is ontological ambiguity at work here seems an understatement. Strabo reads the topography of Troy together with the text of Homer to make propositions about structures that existed in fact and structures that existed in fiction. To reiterate, the source of this ambiguity is the referential potential inherent in architectural features designated variously as walls, stones, and the foundations of cities in Strabo's text. When he describes the destruction of the ancient city of Troy down to its foundations (*ek bathrōn*), moreover, he is clearly projecting Homer's description of what happened to the Achaean wall onto the city of Troy. Or, more accurately, he is reiterating Homer's projection of the destruction of the former onto the latter even as he (Strabo) seems to doubt the existence of the wall. As Porter says of its function in the *Iliad*, the Achaean wall is "like a virtual copy of Troy."[9] In Strabo's text this function is complicated by the conflicting ontological categories to which each is assigned.

At stake in this complex history is the truth about what actually happened in the distant past. Demonstrated in the negotiation between what the poet says and what the geographer sees, Strabo's truth claims are based on empirical evidence about the built environment in the present. When he states that "No trace of the ancient city is preserved. This is to be expected" (*ouden d'ikhnos sōzetai tēs arkhaias poleōs. Eikotōs*, 13.1.38), however, he again introduces the criterion of what is likely or expected into this history. On first glance, the adverb *eikotōs* simply anticipates and credits an explanation based on empirical fact; the city was dismantled down to its stones and these

[9] Porter 2011: 3.

were then used as building materials for other cities. As a result, no trace of the ancient city is preserved. The progression of thought and its reliance on concrete material data seem straightforward and convincing. Coming between the declarative statement and the details that support it, moreover, the rhetorical force of *eikotōs* registers an equivalence between what is likely or expected to happen and what Strabo maintains has happened in fact.

At the same time, however, and given the ambiguities discussed above, the passage also registers a certain ambivalence or hesitation about this equivalence. Or, perhaps more accurately, it betrays confidence in the power of what is likely or expected (*eikotōs*) to substantiate what cannot be proven in fact (that the ancient city existed even though there is no material evidence in the present to support this conclusion). On the other hand, Strabo's text demonstrates that what is likely or expected is not restricted to propositions about future events; somewhat paradoxically, it can also be invoked to confirm assertions about what has actually happened in the past.

As Porter notes, "A high proportion of the ancient discussions of the Achaean wall descend, or seem to descend, from Aristotle."[10] Thus, although there is nothing in the extant works of Aristotle to corroborate Strabo's quotation, the philosopher stands at the beginning of the complicated tradition about the wall's existence, a tradition in which the ancient city of Troy is also implicated. What I want to suggest is that the ontological ambiguities in Strabo's text are an expression of the temporal criteria that Aristotle assigns to *historia* and *poiēsis* – history and fiction – as discrete literary genres in the *Poetics*, namely, the distinction between what has happened (*ta genomena*) and what is likely to happen (*kata to eikos*, 1451a36–38).[11] In other words, Strabo's inconsistent claims about the past in the *Geographica*, claims for which both Homer and Aristotle are invoked as sources, are part of an extended discourse that finds its earliest theoretical expression in the *Poetics*.

Given what Stephen Halliwell has called the "unpolished state" of the *Poetics*, it is no surprise that Aristotle's criteria for defining *historia* and *poiēsis* often seem inconsistent.[12] The principal passage is 1451a36–38:

> Φανερὸν δὲ ἐκ τῶν εἰρημένων καὶ ὅτι οὐ τὸ τὰ γενόμενα λέγειν, τοῦτο ποιητοῦ ἔργον ἐστίν, ἀλλ' οἷα ἂν γένοιτο καὶ τὰ δυνατὰ κατὰ τὸ εἰκὸς ἢ τὸ ἀναγκαῖον.

[10] Porter 2011: 32.

[11] It may be relevant that Strabo claims to have written a historical work, which he calls *hupomnēmata historika*, prior to the *Geographica* (*Geographica* 1.1.23).

[12] Halliwell 1998: 28. On the reception history of the *Poetics*, see also Halliwell 1992.

> It is clear from what has been said that a poet's object is not to tell what has happened but the sorts of things that could have happened and are possible according to what is likely or necessary.

A few lines later (1451b4–5) Aristotle repeats the relevant phrases: history tells what has happened (*ta genomena*) and poetry tells what might happen (*hoia an genoito*).[13] As periphrases with *gignomai*, these criteria are both temporal and ontological; they make a distinction between actual and hypothetical events. As a means of defining *historia* and *poiēsis* as distinct genres, however, they tend to collapse in the *Poetics*. This is perhaps best illustrated when Aristotle maintains that the tragic poets "adhere to the existing names" (*epi dē tēs tragōidias tōn genomenōn onomatōn antekhontai*, 1451b15–16).[14] The problem has to do with *tōn genomenōn*: does it refer to the names of actual persons or to the traditional names of fictional characters? In his discussion of this passage, Lucas concludes that:

> The easiest solution is to suppose that A[ristotle], like Thucydides, believed that Greek myth, or much of it, was basically historical, or at least that names like Heracles or Achilles belonged to the class of γενόμενοι, real people, but that he distinguished between legends such as those of Troy or Thebes, and history of recent events like the Persian Wars.[15]

This easy and often repeated solution relies on the premise that narratives about historical actors do not necessarily constitute historical events; or, alternatively, that "real people" can be the subjects of fictional narratives. At the level of the content of tragic *muthoi* (stories, plots) these premises are not contradictory. But Lucas' hesitation, expressed in the phrase "at least," also testifies to the uneasy distinction between actual and hypothetical events as temporal as opposed to generic categories in the *Poetics*. This becomes more apparent when Aristotle returns to these criteria to explain the reason or "cause" (*aition*) behind the tragic poets' use of these "existing" names (1451b16–19):

13 For the translation of *hoia an genoito* in this passage, see Lucas 1968. Halliwell 1998: 99–106 provides a detailed discussion of necessity and probability in Aristotle's works in general and in the *Poetics* in particular; see also Halliwell 1987: 105–12. Moles 1993: 107–9 discusses Aristotle's distinction in the *Poetics* between *poiēsis* and *historia* as it is anticipated in the works of Herodotus and Thucydides.

14 In the *Poetics* both Alcibiades, the exemplary figure of *historia*, and the heroes of tragedy are described as those who have "suffered or done terrible things" (*pathein deina ē poiēsai*, 1453a22; cf. 1451b11, *epraxen ē ti epathen*). In other words, the line between the historical and the fictional (or the actual and the possible) is blurred here in terms both of past actors and of past events. On the choice of Alcibiades, see Else 1957: 304 n. 9; Hornblower 1987: 10, 38; de Ste. Croix 1992: 24–29; and Wohl in this volume.

15 Lucas 1968 ad loc. So also, Else 1957: 303–4; Finley 1975: 14; and Halliwell 1987: 105, with n. 1.

> αἴτιον δ' ὅτι πιθανόν ἐστι τὸ δυνατόν· τὰ μὲν οὖν μὴ γενόμενα οὔπω πιστεύομεν εἶναι δυνατά, τὰ δὲ γενόμενα φανερὸν ὅτι δυνατά·
>
> The reason for this is that what is possible is credible. Therefore, we do not believe that things that have not happened are possible, but it is clear that what has happened is possible. For, it would not have happened, if it were impossible.

Although the logic is somewhat circular, this passage clearly echoes the temporal criteria in the earlier passage: what is credible is judged by the criterion of what is possible, on the one hand, and what has or has not happened, on the other. And while what is possible (*to dunaton*) is not further qualified by "what is likely or necessary" it covers a similar semantic range. These criteria are then reassigned to their respective genres a few lines later when Aristotle brings the poet and, implicitly, the historian, back into the picture (1451b29–32):

> κἂν ἄρα συμβῇ γενόμενα ποιεῖν, οὐθὲν ἧττον ποιητής ἐστι· τῶν γὰρ γενομένων ἔνια οὐδὲν κωλύει τοιαῦτα εἶναι οἷα ἂν εἰκὸς γενέσθαι καὶ δυνατὰ γενέσθαι, καθ' ὃ ἐκεῖνος αὐτῶν ποιητής ἐστιν.
>
> Even if [the poet] happens to make things that have happened, he is none the less a poet; for there is nothing to prevent some actual occurrences being the sorts of things that are likely to happen and are possible, and it is in virtue of this that that man is the maker of these things.

Here Aristotle admits that the line between what has happened and what might be expected to happen, as the defining characteristics of historical events and poetic events respectively, can be crossed. As I have suggested above, this is the line that Strabo confronts when he brings the fates of the Achaean wall and the ancient city of Troy into evidence in the *Geographica*. Thus, even if we doubt the authenticity of the quotation Strabo attributes to Aristotle (or perhaps especially if we do doubt it), the statement that "the poet fashioned the wall and then made it disappear" concisely links the temporal and ontological variables that Aristotle assigns to genres in the *Poetics* with those that Strabo assigns to the built environment in the *Geographica*. In what follows, I will argue that the *Iliad* itself puts these variables into high relief.

Andrew Ford rightly observes that "The strangest object on the ground at Troy is the great defensive wall the Greeks erect in Book 7 of the *Iliad*, and which is finally destroyed without a trace when the Greeks go home, as we hear in Book 12."[16] To recall again the words attributed to

[16] Ford 1992: 147. See also Scodel 1982 and Maitland 1999: 3–4.

Aristotle by Strabo, this strangeness is due to the fact that the poet "made the wall disappear." More specifically, the wall's disappearance takes place in a hypothetical past, a past that, in the present moment of the poem, has not yet happened. As such, its fate not only tests the temporal and ontological limits of the epic, it also and perhaps most significantly tests its ethical limits, namely the achievement of immortal fame or *kleos*.[17] The ontological ambiguity expressed in the fate of the wall, in other words, finds its analogue in the fact of human mortality. This analogy is based in general terms on related conflicts about what is possible, that is, the ability to transcend death, on the one hand, and the ability to ensure something like "enduring evidence of antiquity," on the other. Emblematic of the contingencies associated with knowing what happened in the past, this temporal anomaly is the basis for the discussion of the wall that follows.

Following the indecisive single combat between Ajax and Hector in Book 7 of the *Iliad*, Nestor urges the Greek leaders to call for a cease-fire. This will allow them to gather the corpses for burning on a communal pyre so that "each man may take the bones home to their children, whenever we go back again to our fatherland" (7.334–35).[18] Immediately following this appeal to an uncertain return home, figured in the act of repatriating the bones of the dead, Nestor urges the Greeks to build a wall and ditch around the ships (7.336–43):

> τύμβον δ' ἀμφὶ πυρὴν ἕνα χεύομεν ἐξαγαγόντες
> ἄκριτον ἐκ πεδίου· ποτὶ δ' αὐτὸν δείμομεν ὦκα
> πύργους ὑψηλοὺς εἶλαρ νηῶν τε καὶ αὐτῶν.
> ἐν δ' αὐτοῖσι πύλας ποιήσομεν εὖ ἀραρυίας,
> ὄφρα δι' αὐτάων ἱππηλασίη ὁδὸς εἴη·
> ἔκτοσθεν δὲ βαθεῖαν ὀρύξομεν ἐγγύθι τάφρον,
> ἥ χ' ἵππον καὶ λαὸν ἐρυκάκοι ἀμφὶς ἐοῦσα,
> μή ποτ' ἐπιβρίσῃ πόλεμος Τρώων ἀγερώχων.

> And around the pyre let us heap up a single tomb
> building it up from the plain; and in front of it let us swiftly build
> high walls to be a defense for the ships and ourselves.
> And let us make well-fitted gates within them
> So that there can be a passageway for driving horses.
> And outside [the walls] let us dig a deep ditch nearby
> which, surrounding the walls, can obstruct both horse and men,
> so that the war of the noble Trojans does not press us down.

[17] Scodel 1982: 48 n. 1, notes that the wall's destruction involves the destruction of *kleos*. See also Ford 1992: 150.

[18] For the conclusion that *Iliad* 7.334–35 are an interpolation, based on a confusion of burial practices, see West 1969: 259 and Porter 2011: 7–8.

The ships, of course, are the only means by which the Greeks can return home so that Nestor's plan to build a wall around them seems to respond naturally to the uncertainty of that return. Situating the wall in the poem's temporal landscape is explicitly expressed here in the negative purpose clause (*mē pot' epibrisēi*), that is, in its function as a defense against a hypothetical – and significant – future event. But the proximity of wall, tomb, and ship, a proximity that refers to both their placement in the narrative and their location on the Trojan plain, constitutes a more complex temporal relationship; for, if building the wall is implicitly equated with heaping up the tomb, then the ships function as a mediating third term; they can be equated with either a tomb or a wall.[19] In other words, the Greeks are destined to die in Troy insofar as the destruction of the wall anticipates the destruction of the ships. Dividing the Trojan plain from the open sea, the wall stands at this point in the poem for the precarious dividing line between victory and defeat, homecoming and abandonment, and life and death; this is clarified in the statement that the wall is a defense both for the ships and for the men themselves (*eilar nēōn te kai autōn*, 7.338).

The wall's significance in the narrative at large becomes more apparent at the end of Book 7 where, following the Trojans' failed attempt to end the war with an offer to return the treasure that Paris had brought home with Helen (7.389–91), Poseidon compares the Achaean wall with the great wall of Troy that he and Apollo had built (7.443–53):[20]

> οἳ δὲ θεοὶ πὰρ Ζηνὶ καθήμενοι ἀστεροπητῇ
> θηεῦντο μέγα ἔργον Ἀχαιῶν χαλκοχιτώνων.
> τοῖσι δὲ μύθων ἦρχε Ποσειδάων ἐνοσίχθων·
> Ζεῦ πάτερ, ἦ ῥά τίς ἐστι βροτῶν ἐπ' ἀπείρονα γαῖαν
> ὅς τις ἔτ' ἀθανάτοισι νόον καὶ μῆτιν ἐνίψει;
> οὐχ ὁράᾳς ὅτι δ' αὖτε κάρη κομόωντες Ἀχαιοὶ
> τεῖχος ἐτειχίσσαντο νεῶν ὕπερ, ἀμφὶ δὲ τάφρον
> ἤλασαν, οὐδὲ θεοῖσι δόσαν κλειτὰς ἑκατόμβας;
> τοῦ δ' ἤτοι κλέος ἔσται ὅσον τ' ἐπικίδναται ἠώς·
> τοῦ δ' ἐπιλήσονται τὸ ἐγὼ καὶ Φοῖβος Ἀπόλλων
> ἥρῳ Λαομέδοντι πολίσσαμεν ἀθλήσαντε.

19 For a similar collocation of walls, ships, and tombs in Thucydides, see Bassi 2007: 190–96. Cf. Grethlein 2006: 148: "Als Medium des Ruhms gleicht die Mauer nicht nur den Gräbern, sondern auch dem Epos." Grethlein goes on to argue that the wall can be understood as a substitute (Ersatz) for Achilles (149). See also Goldhill 2007: 144–47.

20 Kirk 1990 ad loc. summarizes the arguments for taking these verses to be an interpolation; none is decisive.

> And the gods sitting beside Zeus, the Lightener,
> gazed at the great work of the bronze-clad Achaeans.
> And Poseidon the Earth-Shaker began to speak to them:
> "Father Zeus, is there any mortal on the wide earth
> who will declare his thought and counsel to the immortals?
> Don't you see that again the long-haired Achaeans have built
> a wall in front of their ships, and have drawn a trench
> around it, and that they have not given glorious hecatombs to the gods?
> For sure the fame (*kleos*) of this wall will spread like the dawn;
> and men will forget (*epilēsontai*) the [city wall][21] that Phoebus Apollo
> and I built with toil for the hero Laomedon.

This debate over the significance of the two walls rests on the question of which one will be remembered and which forgotten or, in other terms, which will have a past.[22] As we have seen, this is the debate that animates Strabo's intertwined accounts of the existence of the wall and the city of Troy in the *Geographica*. In the *Iliad* the debate takes the form of an unusual, if not unique, appeal to the wall's fame or *kleos*. In the epic poems *kleos* is almost exclusively used as an attribute of human agents.[23] In the one other example in which it is an attribute of a physical object (and this may be an interpolation) Nestor's shield is endowed with *kleos* that, says Hector, "reaches to heaven" (*aspida Nestoreēn, tēs nun kleos ouranon hikei*, 8.192).[24] Of course, the *kleos* of the wall may be understood as an example of hypallage in which the fame of the wall's creators, the Greeks, has been transferred to the wall itself. But for this very reason it has something of a unique status in the poem, one that not even the great wall of Troy can claim.[25]

As an object of *kleos*, the Achaean wall is also distinguished from other objects and architectural features in the *Iliad* which, when remarked upon,

[21] *Polizō* is found only one other time in Homer (*Il.* 20.217), where it means "to build a city," specifically to build the city of Troy. I take it to have this same sense here, where building a wall is equated with building a city.

[22] See Goldhill 2007 for a pertinent discussion of the symbolic significance of the walls of Troy and Thebes in Aeschylus and Euripides.

[23] For example, in the phrase *klea andrōn*, *Il.* 9.189 and 524; *Od.* 8.73. Cf. *Od.* 23.137, where *kleos* is attributed to the killing of the suitors (*kleos . . . phonou*). Edwards 1985: 71–93 offers a general overview of *kleos* and its sources in the Homeric poems. See also Nagy 1990: 26.

[24] On Nestor's shield, see Kirk 1990 ad loc. It can be noted that not even the arms made for Achilles by Hephaestus have *kleos*, including the famous shield. These are called *kluta teukhea* (19.10) and *aglaa dōra* (19.18), where the former formulaic expression is found ten times in the *Iliad* (and twice in the *Odyssey*).

[25] The Trojan walls are nowhere explicitly endowed with *kleos* in the *Iliad*, although they are called *kluta teikhea* once (21.295).

are frequently called *aphthiton* or *aphthiton aiei* "imperishable" or, hyperbolically, "imperishable forever."[26] What emerges from this usage is a discourse of physical objects and features defined by their tendency to decay and be forgotten and, as a compensatory gesture, by their resistance to destruction (*phthinō*). As suggested above, the building of the wall performs this compensatory function in the context of and as a response to the act of heaping the bodies of unidentified warriors on funeral pyres (*epinēneon*, 7.428 and 431). The *kleos* of the Achaean wall, in other words, implicitly compensates for the *kleos* that eludes these unknown warriors.[27] In this way, the question of which of the two walls will be remembered and which one will be forgotten – or which one will have a past – links the disposition of the poem's built environment to its ethical dimension, that is, to the possible attainment of *kleos* as a means of evaluating and preserving the heroic past.

The particular attributes of the wall's *kleos* are spatial and temporal extension, encompassed in the simile of the spreading dawn.[28] Thus, while the literal meaning of *kleos* is "an act of hearing," the simile figures it here as a visual phenomenon and, more specifically, one that constitutes the primary means of reckoning time in the epic. The formulaic appearance of the dawn structures the epic narrative both in qualitative terms (as a perpetual recurrence) and in quantitative terms (as a specific number of days).[29] As the only example in Homer in which the dawn is an element of comparison, the simile connects *kleos* with time that is both eternal and transient. In general terms, *kleos* oscillates between these two temporal registers in the epics; it refers primarily to the promise of the hero's transcendent reputation or "fame" which will be achieved in a single day (or dawn), on the day of the hero's death. Here we can recall Achilles' words to Lycaon regarding his own death in Book 21: "There will come a dawn, or early afternoon, or midday when some man will take my life in war" (*essetai ē ēōs ē deila ē*

[26] Agamemnon's scepter is *aphthiton* (2.46, 2.186); the palace of Poseidon is *aphthiton aiei* (13.21–22); the throne made by Hephaestus is *aphthiton aiei* (14.328); the house of Hephaestus is *aphthiton* (18.369–370); the wheel rim of the goddesses' war chariot is *aphthitos* (5.724). The remaining two examples of *aphthitos* in the *Iliad* refer to Zeus's "counsels" (24.88) and, of course, to Achilles' unique "fame" (9.413). On this phrase, including a discussion of the "typical" application of *aphthiton* to material objects, see Volk 2002: 66–67 and the sources cited there. Cf. Finkelberg 1986: 4 with n. 16.

[27] Cf. Sinos 1980: 47 on the warrior's tomb as the physical and visible manifestation of his *kleos* in the *Iliad*. Sinos is cited by Nagy 1990: 215.

[28] It can be objected that the construction with *hoson* is not a simile in a strict sense. But the comparison fulfills a similar function in the passage.

[29] In its formulaic expressions, the dawn "approaches" (*bainō*), "spreads" (*kidnamai*), "rises" (*ornumi*), and "comes" (*erkhomai*), etc., but most frequently it "appears" (*phainō*). The dawn is also the principal means of measuring a specific number of days, e.g. "on the twelfth dawn" (*Il.* 1.493).

meson hēmar..., 21.111–13). Poseidon's claim that "the fame (*kleos*) of this wall will spread like the dawn" operates within this same tempero-semantic range in which the wall's *kleos*, like that of the hero, depends on its physical destruction.

As Gregory Nagy has argued, the attainment of *kleos* in Homer means "transcending concerns about material wealth and security" through the vehicle of epic poetry; moreover, it is what the hero "earns" by dying.[30] What is most immediately transcended in this process, in other words, is the hero's mortal body.[31] Put another way, Achilles' *kleos* defies destruction: it is *aphthiton*, because his physical body is subject to death and decay.[32] Zeus's response to Poseidon's fear that the Achaean wall will rob the Trojan wall of its *kleos* should be understood in the context of this interplay between transcendence or eternal fame, on the one hand, and immanence or mortality, on the other (7.454–63):

> Τὸν δὲ μέγ' ὀχθήσας προσέφη νεφεληγερέτα Ζεύς·
> ὢ πόποι ἐννοσίγαι' εὐρυσθενές, οἷον ἔειπες.
> ἄλλός κέν τις τοῦτο θεῶν δείσειε νόημα,
> ὃς σέο πολλὸν ἀφαυρότερος χεῖράς τε μένος τε·
> σὸν δ' ἤτοι κλέος ἔσται ὅσον τ' ἐπικίδναται ἠώς.
> ἄγρει μὰν ὅτ' ἂν αὖτε κάρη κομόωντες Ἀχαιοὶ
> οἴχωνται σὺν νηυσὶ φίλην ἐς πατρίδα γαῖαν
> τεῖχος ἀναρρήξας τὸ μὲν εἰς ἅλα πᾶν καταχεῦαι,
> αὖτις δ' ἠϊόνα μεγάλην ψαμάθοισι καλύψαι,
> ὥς κέν τοι μέγα τεῖχος ἀμαλδύνηται Ἀχαιῶν.

> Then greatly disturbed Zeus the cloud-gatherer said,
> "Oh! Earth-Shaker, what have you said!
> Some other god might fear this thought,
> one far less powerful than you in both hands and strength;
> But your fame (*kleos*) will spread like the dawn.
> Now when the long-haired Achaeans return to their
> dear native land in their ships, take the wall
> and break it up and heap it all into the sea,
> and cover (*kalupsai*) the great beach again with sand
> so that the great wall of the Achaeans is destroyed."

30 Nagy 1990: 123 and 138 respectively. The *locus classicus* is the speech in which Achilles rejects Agamemnon's gifts (9.378–416). See Ready 2007: 19–21.

31 See the discussion of Nagy 1990: 85 on the disposition of the body after death in Indic and Greek contexts.

32 Cf. the compound *phthisimbrotos* ("mortal destroying") used of battle at *Il.* 13.339 and of Athena's aegis at *Od.* 22.297. The epithet is discussed by Russo *et al.* 1992: 271.

Here Zeus uses the same simile to refer to the *kleos* of Poseidon that Poseidon had used at 7.451 to refer to the *kleos* of the Achaean wall. In repeating the simile here and transferring it to the *kleos* of the god who built the Trojan wall, Zeus's words imply that the Achaean wall, destined for destruction, will have no *kleos*. But, as suggested above, this conclusion is clearly contradicted by the poem, in which the wall's destruction and absence are the constituent features of its *kleos*. As Porter notes:

> The monumental obliteration of the Achaean wall, rather than erasing its memory and fame, quite the contrary ensures that the same wall will go down in the annals of memory as one of the most unforgettable and most famous walls that was ever – or better yet *never* – constructed.[33]

A similar paradox holds for the *kleos* of the heroic warrior; it can be achieved only after his death or, in the language of the poem, after death has "covered" (*kaluptō*) him. The implied equation of the destruction of the wall with the death of the hero is supported by the fact that death dominates the semantic and formulaic ranges of *kaluptō* in the epic poems; it is used of the darkness that "covers" the warrior's eyes at the moment of death (*skotos osse kalupse*) and in the phrase "the end of death covered [him]" (*telos thanatoio kalupse*).[34] In these phrases *kaluptō* is euphemistic; it constitutes a rhetorical deferral in which death is a disappearance rather than non-existence. The "death" of the wall, figured in the act of being covered up, thus constitutes a euphemistic or compensatory reference to the death of the Greek and Trojan warriors who, once upon a time, fought around it. Within this complex configuration, the basis of *kleos* as an oral report is also linked with the wall's temporal and ontological transformation from visibility to invisibility. In short, the story of the wall's appearance and disappearance is related to the principal conflict in the poem, namely the conflict between what is possible for humans (*kleos*) and what is certain (death).

I have been arguing that the disposition of the Achaean wall, by virtue of its singular claim to *kleos*, is a key to understanding the ways in which the poem's temporal, ontological, and ethical dimensions are mutually productive. It remains to situate this argument within the narrative of the wall's complete and utter destruction (12.9–18):

33 Porter 2011: 19 (original emphasis).

34 "Darkness covered his eyes" (*skotos osse kalupse*): *Il.* 4.461, 6.11, 13.577, 14.519, 15.578, 16.316, 16.325, 20.393, 21.181, where *kalupsō* is in the same metrical position as *kalupsai* at 7.462 and *kalupse* at 12.31. "The end of death covered [him]" (*telos thanatoio kalupse*): *Il.* 5.553, 16.502, 16.855, 22.361. Other examples include *Il.* 12.116, where the death of Asius is described: "[A]n ill-named fate covered him up" (*min moira dusōnumos amphekalupsen*). Of *erōs*, *Il.* 3.442 and 14.294.

τὸ καὶ οὔ τι πολὺν χρόνον ἔμπεδον ἦεν.
ὄφρα μὲν Ἕκτωρ ζωὸς ἔην καὶ μήνι' Ἀχιλλεὺς
καὶ Πριάμοιο ἄνακτος ἀπόρθητος πόλις ἔπλεν,
τόφρα δὲ καὶ μέγα τεῖχος Ἀχαιῶν ἔμπεδον ἦεν.
αὐτὰρ ἐπεὶ κατὰ μὲν Τρώων θάνον ὅσσοι ἄριστοι,
πολλοὶ δ' Ἀργείων οἱ μὲν δάμεν, οἱ δὲ λίποντο,
πέρθετο δὲ Πριάμοιο πόλις δεκάτῳ ἐνιαυτῷ,
Ἀργεῖοι δ' ἐν νηυσὶ φίλην ἐς πατρίδ' ἔβησαν,
δὴ τότε μητιόωντο Ποσειδάων καὶ Ἀπόλλων
τεῖχος ἀμαλδῦναι ποταμῶν μένος εἰσαγαγόντες.

The wall could not endure (*empedon*) for long.
While Hector was living and Achilles raged
and the city of king Priam was still not sacked,
the great wall of the Greeks stood firm (*empedon*).
But when those who were the best Trojans had died
and of the many Greeks some had fallen and some were left,
and the city of Priam had fallen in the tenth year
and the Greeks had gone back to their dear native land,
at that time Poseidon and Apollo planned to destroy
the wall by bringing together the strength of the rivers.

Oscillating between temporal and ontological categories, from future to past and from monument to ruin, the wall will disappear without a trace as Apollo and Poseidon cover it with water and bury it in sand along with the "Many bull-hide shields and helmets . . . [of] a generation of men who were half-divine" (12.22–23; cf. 12.105).[35] At the level of the plot, the fate of the wall is described while the battle is being waged around it and up until the point when the Trojans, led by Hector, will finally breach it (12.435–41).

But its destruction is recounted from the point of view of a projected future, that is, as a hypothetical past event, before returning to the present moment of the narrative (12.27–37):

αὐτὸς δ' ἐννοσίγαιος ἔχων χείρεσσι τρίαιναν
ἡγεῖτ', ἐκ δ' ἄρα πάντα θεμείλια κύμασι πέμπε
φιτρῶν καὶ λάων, τὰ θέσαν μογέοντες Ἀχαιοί,
λεῖα δ' ἐποίησεν παρ' ἀγάρροον Ἑλλήσποντον,
αὖτις δ' ἠϊόνα μεγάλην ψαμάθοισι κάλυψε
τεῖχος ἀμαλδύνας· ποταμοὺς δ' ἔτρεψε νέεσθαι
κὰρ ῥόον, ᾗ περ πρόσθεν ἵεν καλλίρροον ὕδωρ.
Ὣς ἄρ' ἔμελλον ὄπισθε Ποσειδάων καὶ Ἀπόλλων
θησέμεναι· τότε δ' ἀμφὶ μάχη ἐνοπή τε δεδήει

[35] Cf. Ford 1992: 149 on *hēmitheoi* as "a word one uses in speaking of heroes from a distance, contemplating the heroic age retrospectively as something apart, utterly remote in time." See also Taplin 1992: 139–40 and Porter 2011: 31–32.

τεῖχος ἐΰδμητον, κανάχιζε δὲ δούρατα πύργων
βαλλόμεν'·

And the Earth-shaker himself, holding
the trident in his hands, led the way
and sent into the waves all the foundations
of beams and stones the Greeks had set up with toil.
He made everything smooth along the strong-flowing Hellespont
and again covered (*kalupse*) the great shore with sand,
after he had swept away the wall; then he turned the rivers to flow back
into the channel, where they had sent their beautiful-flowing water before.
This Poseidon and Apollo were going to do in time to come (*opisthe*).
But then (*tote d'*) the battle raged on both sides
of the well-built wall (*teikhos eudmēton*), and the beams of the towers rang
as they were struck.

As I have suggested above, the wall's absolute absence, described in terms of being covered or hidden, also evokes the death of the warriors whom the *Iliad* would immortalize, with Achilles first among them. This particular cross-reference is decisively illustrated in the description of Achilles' battle with the river Scamander when, following the death of Patroclus, he re-enters the war and, in a killing rage, clogs the river's streams with the corpses of Trojans. Calling for help from his brother, the river Simois, Scamander describes the effect of their joint effort in curbing Achilles' onslaught (21.308–23):[36]

φίλε κασίγνητε σθένος ἀνέρος ἀμφότεροί περ
σχῶμεν, ἐπεὶ τάχα ἄστυ μέγα Πριάμοιο ἄνακτος
ἐκπέρσει, Τρῶες δὲ κατὰ μόθον οὐ μενέουσιν.
ἀλλ' ἐπάμυνε τάχιστα, καὶ ἐμπίπληθι ῥέεθρα
ὕδατος ἐκ πηγέων, πάντας δ' ὀρόθυνον ἐναύλους,
ἵστη δὲ μέγα κῦμα, πολὺν δ' ὀρυμαγδὸν ὄρινε
φιτρῶν καὶ λάων, ἵνα παύσομεν ἄγριον ἄνδρα
ὃς δὴ νῦν κρατέει, μέμονεν δ' ὅ γε ἶσα θεοῖσι.
φημὶ γὰρ οὔτε βίην χραισμησέμεν οὔτέ τι εἶδος
οὔτε τὰ τεύχεα καλά, τά που μάλα νειόθι λίμνης
κείσεθ' ὑπ' ἰλύος κεκαλυμμένα· κὰδ δέ μιν αὐτὸν
εἰλύσω ψαμάθοισιν ἅλις χέραδος περιχεύας
μυρίον, οὐδέ οἱ ὀστέ' ἐπιστήσονται Ἀχαιοὶ
ἀλλέξαι· τόσσην οἱ ἄσιν καθύπερθε καλύψω.
αὐτοῦ οἱ καὶ σῆμα τετεύξεται, οὐδέ τί μιν χρεὼ
ἔσται τυμβοχόης, ὅτε μιν θάπτωσιν Ἀχαιοί.

Dear brother, let us both hold back the strength
of this man, since he will soon destroy the great city of

[36] On this battle, see Fenno 2005: 498–503.

king Priam, and the Trojans will not curb him in battle.
Bring aid as quickly as you can, and fill your streams
with water from their sources, and rouse up all your torrents,
raise up a great wave, and stir up a great cacophony
of logs and stones, so that we can stop this savage man,
who now is victorious and thinks that he is equal to the gods.
For I say that neither his might will let him prevail, nor his looks
nor his beautiful armor, which will lie at the bottom of a marsh,
covered (*kekalummena*) by slime; and I will enfold him
in heaps of sand after I've spread gravel beyond counting over him,
and the Achaeans will not know how to care for his bones;
such is the silt that I will cover him under (*kalupsō*).
And his tomb (*sēma*) will have been built here, and there will be no need
of heaping up a tomb, whenever the Achaeans bury him.

The similarities between this passage and the description of the destruction of the Achaean wall in Book 12 are evident, beginning with the fact that the Trojan rivers are the immediate sources of destruction in both cases and including a more subtle correlation between the projected absence of Achilles' tomb or *sēma* and the projected absence of the Achaean wall.[37] Here, the river will cover up the dead hero just as it will cover up the wall. And the fact that "a great heap of silt will be Akhilleus' burial-mound," as Kirk summarizes, goes again to the heart of the epic ethos, namely to the achievement (or loss) of *kleos* as exemplified in the choice that Achilles must make.[38] Covered (*kaluptō*) by sand and gravel, in other words, Achilles' non-tomb in Scamander's vision is part of the same discursive formation as the Achaean wall; both refer to a hypothetical past characterized by what is *not* visually available or, with specific reference to the wall, what has disappeared from view.

To the extent that heroic *kleos* signifies a struggle against the fact of human mortality, this relationship is also part of the compensatory logic referred to above. And if tombs are the most obvious physical manifestations of this struggle (see *Il.* 7.84–91), the Achaean wall and the walls of Troy constitute an analogous category in the *Iliad.* In temporal terms, the wall is to the collective past as the tomb is to the past of the individual hero.[39] This collective past, expressed in terms of a collective mortality, is invoked again in the midst of Achilles' battle with the river, when Poseidon reminds Apollo of the tasks the two of them had performed for the

[37] Cf. Ford 1992: 153–54 on the "antifunereal" aspects of this passage. Also, the brief remarks of Grethlein 2006: 150, with n. 345.

[38] The quotation is from Kirk 1993 ad loc. Cf. Nagy 1990: 215–17 on the *sēma* (tomb) as a "reminder of *kleos*" (217). Also Purves 2010: 80–84.

[39] Cf. Ready 2007: 19.

treacherous Laomedon. Specifically, he tells Apollo that he had "built a wide and very beautiful wall around the city of the Trojans, so that the city would be unbroken" (ἤτοι ἐγὼ Τρώεσσι πόλιν πέρι τεῖχος ἔδειμα εὐρύ τε καὶ μάλα καλόν, ἵν' ἄρρηκτος πόλις εἴη, 21.446–47). As we have seen, this event had already been described in Book 7 where Poseidon had put the Trojan wall into competition with the wall of the Achaeans. In Book 21 the ill treatment the two gods had received from the Trojan king after building the wall fuels Poseidon's argument against giving aid to the Trojans in the present moment of the poem. In accepting Poseidon's advice, Apollo responds with a brief but remarkable digression on the ephemerality of human life (21.462–67):

> Τὸν δ' αὖτε προσέειπεν ἄναξ ἑκάεργος Ἀπόλλων·
> ἐννοσίγαι' οὐκ ἄν με σαόφρονα μυθήσαιο
> ἔμμεναι, εἰ δὴ σοί γε βροτῶν ἕνεκα πτολεμίξω
> δειλῶν, οἳ φύλλοισιν ἐοικότες ἄλλοτε μέν τε
> ζαφλεγέες τελέθουσιν ἀρούρης καρπὸν ἔδοντες,
> ἄλλοτε δὲ φθινύθουσιν ἀκήριοι. ἀλλὰ τάχιστα
> παυώμεσθα μάχης· οἳ δ' αὐτοὶ δηριαάσθων.
>
> And then in turn king Apollo, who works from far-off, spoke:
> "Earth-Shaker, you would not think that I was of sound mind
> if I should do battle with you for the sake of mortals,
> those wretched creatures, who at one time like leaves
> thrive in the fullness of life [of fire?], eating the produce of the field,
> but then waste away and become lifeless. But as quickly as possible
> let us cease from the fight, and let them fight by themselves."

The point of Apollo's response seems obvious: why should he fight against a fellow god on behalf of mortals who, no matter what aid the gods give them, will die anyway? But, given the mutual self-definition of humans and gods in the epic, the response also lacks force. Humans die, but while they live they motivate the intentions and actions of the gods, with whom they even share genealogies. In the given context, moreover, the fact that humans "waste away and become lifeless" (*akērioi*, 466) is logically if contrastively correlated with the notion that Troy will be "unbroken" (*arrēktos*, 447).[40] I am not suggesting that these two terms are etymologically related, of course; nor does their proximity in the poem emphasize a phonetic resemblance.[41] But insofar as the wall built by the gods guarantees that the city will endure, it is implicitly contrasted with the ephemerality of the human beings for

[40] Leaf 1971 ad loc. believes that these lines are a "deliberate parody" of the famous simile of the leaves at *Il.* 6.146–49. On the leaves simile in the *Iliad* as an expression of *Schicksalkontingenz*, see Grethlein 2006: 84–97.

[41] On contiguity in poetry, see Jakobson 1960.

whom it was built; we can note here too that in transferring the adjective "unbroken" from the wall to the city the poet has effectively equated their vulnerability.

For, like Strabo, we know that Troy is destined to fall or "break," a fact alluded to when Apollo subsequently enters the city out of concern that the Greeks will destroy the city wall before its allotted time (21.515–17):

> αὐτὰρ Ἀπόλλων Φοῖβος ἐδύσετο Ἴλιον ἱρήν·
> μέμβλετο γάρ οἱ τεῖχος ἐϋδμήτοιο πόληος
> μὴ Δαναοὶ πέρσειαν ὑπὲρ μόρον ἤματι κείνῳ.
>
> But Phoebus Apollo entered sacred Ilion,
> for he was concerned that on that day the Danaans
> would destroy the wall of the well-built city beyond what was ordained.

Here, in the only instance in the epic poems where the epithet "well-built" (*eudmēton*) is used of a city, it has arguably been transferred from the city wall (*teikhos*); in other words, we should expect "the well-built wall of the city" rather than "the wall of the well-built city." This is perhaps less convincing than the example discussed above, in which I have suggested that the adjective "unbroken" constitutes a transference of this sort. But outside this one example *eudmēton* applies only to specific architectural features in both the *Iliad* and the *Odyssey*, namely to a wall (*teikhos*), a tower (*purgos*), or an altar (*bōmos*).[42] And in the two instances where it describes a wall (*teikhos*) in the *Iliad* it refers to the Achaean wall (*teikhos eudmēton*, 12.36 and 12.137), to the wall that is destined to be utterly destroyed. It should be noted here too that Apollo is concerned that the Greeks will destroy the *wall* of Troy, as a metonym for the city and its people.

Thus, while I realize that any conclusion based on such a small sample is speculative, taken together these examples suggest that Troy is well built in the same sense that the Achaean wall is well built; the adjective amounts to a euphemism that only anticipates their shared future destruction.[43] Understood in the context of Priam's response to Achilles' killing frenzy, in which the king calls for the Trojans to come within the walls of the city for protection and then orders the gates to be closed after them for fear that

[42] *Il.* 1.448, 12.36, 12.137, 12.154, 16.700, 21.516, 22.195; *Od.* 7.100, 20.302, 22.24, 22.126. Cf. Porter 2011: 5 with n. 9.

[43] Porter 2011: 30–31 comments on the "different styles" attributed to the Achaean wall in Homer, i.e., that it is both flimsy and stately. I attribute this inconsistency to the point I am making here, namely, that the wall's oscillation between being well built and utterly destroyed is a key to its function in the poem's temporal and ethical landscape, in which the walls of Troy are necessary comparanda.

Achilles will "leap inside the wall" (*es teikhos alētai*, 21.536), the progression of events constitutes an exercise in delayed inevitability and, in the process, focuses attention on the question of what is likely to happen to the city. It is in the context of this misplaced or counterfactual faith in the endurance of the built environment that Apollo's simile about human mortality is both true in an absolute sense and prescient in a contingent one; it exemplifies Proust's observation that "What seems eternal . . . contains within itself the impulse of its own destruction."[44]

Buried in a hypothetical past in which it can no longer be seen, the Achaean wall in the *Iliad* introduces the principal challenge faced by history and archaeology as disciplines in search of visible or enduring proof about what happened in the past. Speaking of the shared objectives of ancient epic and history writing, François Hartog vividly captures this aspect of the wall's fate:

> If all human speech deals with death, if men tell stories because they know themselves to be mortal, then the epic and history – set on the boundary between the visible and invisible to call to mind those who are no longer – together play the role of taming death while socializing it.[45]

"Set on the boundary between the visible and invisible," the wall shows us that what has happened in Aristotle's terms is never fully available to us or, in the language of the poem, it is covered up. At the same time, the fate of the wall suggests that the invisibility of the past is the very condition out of which the desire to transcend death arises, as Hartog implies. As such, the history of the Achaean wall – from Homer to Aristotle (?) to Strabo – is one in which ontological ambiguity, specified in the remains of the built environment, is tied to the ethical dimension of what is possible. It demonstrates that the achievement of *kleos* is a negotiation between what is likely to happen, viewed from within the poem's fictional universe, and what has happened, as a consequence of the poem's production.

[44] Quoted by Adorno 1983: 178. [45] Hartog 1992: 85.

CHAPTER 8

Play of the improbable
Euripides' unlikely Helen

Victoria Wohl

> ἔστιν δὲ τοῦτο καὶ εἰκὸς ὥσπερ Ἀγάθων λέγει, εἰκὸς γὰρ γίνεσθαι πολλὰ καὶ παρὰ τὸ εἰκός.
>
> And this is also probable, as Agathon says, for it is probable that many things happen against probability.
>
> (Aristotle, *Poetics* 1456a23–25)

Tragedy is the genre par excellence of *eikos*. Aristotle defines tragedy as the realm of the probable, as opposed to history, which is the narration of the actual. But that opposition proves hard to maintain even in Aristotle's own terms: *eikos* refuses to stay put on one side of the history–poetry divide and thus dismantles, even as it defines, the distinction between fiction and reality. Euripides' *Helen* puts further pressure on that dichotomy. This play, as critics have long recognized, is uncommonly self-aware about its own status as fiction: its constant play on illusion and reality warns us of its fictionality even as the action-packed plot compels us to mistake that fiction for reality. Moreover, since that fictional plot is ostentatiously improbable – Helen was in Egypt the whole time, and chaste! – the play presents a sophisticated meditation on the force of probability within fiction and life that anticipates Aristotle's own. And if within *Helen*'s hypothetical world improbable fictions and probable history merge and blend, beyond it – in the realm of contemporary Athenian reality – history takes on a decidedly fictional cast.

In section 9 of the *Poetics* (1451a36–b5), Aristotle defines *poiēsis* (fiction) in contradistinction to *historia* (history):

> Φανερὸν δὲ ἐκ τῶν εἰρημένων καὶ ὅτι οὐ τὸ τὰ γενόμενα λέγειν, τοῦτο ποιητοῦ ἔργον ἐστίν, ἀλλ' οἷα ἂν γένοιτο καὶ τὰ δυνατὰ κατὰ τὸ εἰκὸς ἢ τὸ ἀναγκαῖον. ὁ γὰρ ἱστορικὸς καὶ ὁ ποιητὴς οὐ τῷ ἢ ἔμμετρα λέγειν ἢ ἄμετρα διαφέρουσιν (εἴη γὰρ ἂν τὰ Ἡροδότου εἰς μέτρα τεθῆναι καὶ οὐδὲν ἧττον ἂν εἴη ἱστορία τις μετὰ μέτρου ἢ ἄνευ μέτρων)· ἀλλὰ τούτῳ διαφέρει, τῷ τὸν μὲν τὰ γενόμενα λέγειν, τὸν δὲ οἷα ἂν γένοιτο.

> It is clear from what has been said [about the nature of tragic unity] that it is the poet's task to speak not of things that have happened (*ta genomena*) but of the sort of things that might happen (*hoia an genoito*) and that are possible (*dunata*) according to probability (*to eikos*) or necessity. For the historian and the poet differ not in that one writes in meter and the other does not (for you could translate Herodotus into verse and it would still be a history no less in verse than in prose), but they differ in this respect: that one speaks about what has happened, the other about what might happen.

This dichotomy between "things that have happened" (*ta genomena*) and "the sort of things that might happen" (*hoia an genoito*) carves out a distinct realm for *poiēsis* as a mimesis of probable not actual events, and for tragedy in particular as the paradigmatic form of *poiēsis*.[1] The criterion of the probable is Aristotle's answer to Plato's characterization of poetry as the enemy of truth. In the *Phaedrus* Socrates opposes the probable to the true: orators often speak not about what has actually been done (*ta prakhthenta*) but about probabilities (*ta eikota*) and "bid the truth farewell" (272d7–e5).[2] Like orators, poets deal in mere probabilities, deceptive illusions, and thus stand apart from and against the truth.

For Aristotle, by contrast, *to eikos* provides an independent criterion for aesthetic evaluation, one not bound by the Platonic standard of veracity.[3] Fiction is distinguished, for Aristotle as for Plato, from reality, "what actually happened." History as the discourse of *ta genomena* can be judged by its fidelity to reality; fiction, on the other hand, should be assessed not by its literal truth-value but by its plausibility "according to probability or necessity" (*kata to eikos ē to anankaion*).[4] *Eikos* thus allows one to judge the work of art on its own terms: to determine the proper length of the plot and its unity of action (1451a12–15), evaluate the appropriateness of characterization (1454a33–36), assess the reversal (1451a12–15, 1452a14–24) and recognition (1455a7, a18, b10). This criterion is not purely formal or internal, though. While Aristotle does not define *eikos* in the *Poetics*, nor specify whether its probabilities are logical, natural, social, or ethico-psychological, by any of these definitions the patterns of likelihood and expectation that

1 *Poiēsis* for Aristotle means not just poetry but all imaginative mimetic creation, that is, fiction broadly conceived. See Halliwell 1987: 71–72; 2002: 166; cf. Else 1957: 1–9. I focus here on the literary side of this polarity but it is obviously open to critique from the side of history, too: see, e.g., White 1978: 97–99, and Bassi in this volume.

2 For a reinterpretation of this passage, see Bryan in this volume.

3 Halliwell 1987: 70–74, 99–101, 105–12, 177–80; 2002: 154–55, 166–67. Woodruff 1992 disputes the Platonic influence on *Poetics*.

4 Cf. *Poet.* § 25 (1460b6–1461b25), which explicitly holds poetry to a different standard of accuracy (*orthotēs*) than "politics or other arts" (1460b14–15) and admits even impossibilities (*adunata*) if they contribute to the work's emotional impact (1460b23–26).

determine tragic probability have their roots in the real world outside the play.[5] The play's internal world refers to external reality for its logic but does not merely or passively reflect it. In this way, *eikos* defines fictional *poiēsis* as a semi-autonomous domain, with its own "probable" verities.[6]

Aristotle's first distinction between history's *genomena* and poetry's *dunata kata to eikos* is then supplemented by a second. *Poiēsis* deals with generalities (*ta katholou*) and *historia* with particularities (*ta kath' hekaston*). This, Aristotle says, is why fiction is "more philosophical and more serious" than history (*philosophōteron kai spoudaioteron*, 1451b5–6). If the distinction between *ta genomena* and *ta dunata* defends fiction against a charge of falseness, this one defends it against a charge of triviality. Again *eikos* is key, for *eikos* is what allows fiction to translate real life's random particularities into meaningful universalities. Reality in itself lacks order or purpose, as does history as the affirmative representation of reality. As Aristotle says later in the *Poetics*, historical composition, unlike poetic, does not need to adhere to rules of unity since the events it describes are related to one another not by logic or probability but by mere happenstance (*hōs etukhen*, 1459a24). For instance, the battle of Salamis happened at the same time as the battle of the Carthaginians in Sicily, but these two events do not share a *telos* (aim, purpose) and the fact that one happened to follow the other is insignificant (1459a17–29).

Fiction adds to history's contingency (*tukhē*) a unifying purpose (*telos*), and it does so by means of *eikos*. This is because *eikos* means not just the probable (as opposed to the actual) but also the usual (as opposed to the exceptional). In the *Rhetoric* (1357a34; cf. 1402b14–16, 1402b21, 1403a1), *eikos* is defined as "what generally happens" (*epi to polu gignomenon*): it is a general pattern that stands between the random events of real life and the iron necessity of natural laws. A similar phrase appears in the *Poetics* in Aristotle's first discussion of tragic plot structure: the play's end is naturally connected to its beginning either by necessity (*ex anankēs*) or "as generally happens" (*hōs epi to polu*, 1450b29–30). This phrase seems to be a first attempt to articulate a concept of regularity for which Aristotle later settles on the set phrase *kata to eikos ē to anankaion*.[7] The apparent substitution of *to eikos* for *epi to polu* and the way in which these terms

[5] Logical: Halliwell 1987: 99–100 and 1998: 100–101; ethical/psychological: Else 1957: 304–6 and Frede 1992; natural: Belfiore 1992: 111–19; social: Gastaldi 1989: 88–89; cf. Pl. *Phdr.* 273b1, where *to eikos* is glossed as *to tōi plēthei dokoun* ("what the majority think"). See further Scodel 1990, C. Gill 1993: 74–77.

[6] Halliwell 1987: 108–10; Woodruff 1992. On probability and the distinctiveness of fiction, see Pavel 1986, Newsom 1988, Gallop 1991, Scodel 1999, Gallagher 2006.

[7] See Frede 1992 on the substitution.

are simultaneously linked to and differentiated from *anankē* suggests that Aristotle understands *to eikos* as a likelihood frequent enough to fall into a predictable general pattern but without becoming inevitable. As Aristotle specifies in the *Rhetoric*, "*to eikos* is that which happens for the most part, but not without qualification, as some define it, but in respect to that which could turn out differently" (*Rh.* 1357a34–36; cf. 1402b20–1403a1). *Eikos* thus stands between *tukhē* and *anankē*, sharing the contingency of the former and the regularity of the latter; in this way it allows poetry to generalize history's chance particularities, creating a more meaningful – philosophical and serious – whole.[8]

So for Aristotle, *eikos* defines fiction in its opposition to both history and reality. But that seemingly clear distinction and the role of *eikos* in it quickly become blurred. First of all, poetry's *dunata* are contrasted to history's *genomena*, but what exactly are these? Aristotle seems to imagine *historia* as the direct transcription of raw historical facts. As an example of history's focus on the particular (*ta kath' hekaston*), Aristotle offers "what Alcibiades did or what he suffered" (*ti Alkibiadēs epraxen ē ti epathen*, 1451b10–12). But not only is that formulation uncomfortably close to Aristotle's understanding of tragic plot (with its emphasis on *praxis* and *pathos*), but the specific example is one where historical facts were especially subject to historiographic interpretation: "what Alcibiades did or what he suffered" was the object of competing accounts even at the time and liable from the start to a good deal of partisan fictionalization. Indeed, for Thucydides – the historian Aristotle probably had in mind when citing this example[9] – the paradoxical probabilities of "what Alcibiades did or suffered" before the launching of the Sicilian expedition precipitate a historiographical crisis, sending him to seek historical certainties in an even more hypothetical past, the quasi-mythical tyrannicide (6.53–60).[10] In illustrating history's particularity this exemplum calls into question its actuality, as the improbable (and often downright implausible) *praxeis* of Alcibiades destabilize the positivist history against which Aristotle defines the probable world of fiction.[11]

Moreover, as section 9 progresses, historical *genomena* and fiction's *dunata* seem to cross and merge (1451b11–32). Aristotle begins the section,

[8] Cf. Halliwell 2002: 200. [9] De Ste. Croix 1992.

[10] So I argue in Wohl 2002: 152–58, 210–12. On the probabilities and improbabilities of Alcibiades' life, see Gribble 1999, Wohl 2002: 124–70.

[11] Likewise, one might question his claim in § 23 that the battles of Salamis and Himera were related only by contingent contemporaneity. One could easily construct a historical narrative in which the two are united by the shared *telos*, e.g., of Greek defense against foreign imperial control. The repudiation of narrative in history discussed by Ricoeur 1984: 95–120 can be traced back to Aristotle's distinction between *poiēsis* and *historia*.

as we have seen, by contrasting the particularities of history to drama's universality. Comedy, he goes on, starts with universal situations (he is probably thinking of Middle Comedy)[12] and then adds invented characters (*ta tukhonta onomata*), instead of starting with particular individuals as invective does. Tragedy, by contrast, starts with traditional characters (*tōn genomenōn onomatōn*), because "what is possible is plausible (*pithanon esti to dunaton*) and while we don't yet believe that what hasn't happened can happen, what has happened is obviously possible because if it weren't it wouldn't have happened" (1451b16–19). Some tragedians do make up the names of their characters, he concedes, and that is all right because the familiar plots give pleasure even if people do not know them. This proves his central thesis that it is the *muthos*, the plot, that counts in drama and that drama is the mimesis of action. He concludes: "for nothing prevents some real events (*tōn genomenōn*) from being the sort of thing that would happen according to *eikos* (*hoia an eikos genesthai*), and it is in accordance with *eikos* that one is a poet when he writes about them" (1451b30–32).

In this compressed passage, *ta genomena* seem to shift meaning from historical events to mythic plots. The distinction drawn at the beginning of the section between historical *genomena* and poetic *dunata* is muddied by the discussion of traditional and invented characters, which Aristotle introduces to illustrate the generality of tragedy. The rationale he then offers for tragedy's use of existing characters, that what has happened (*ta genomena*) seems possible (*dunata*) because it happened, seems to refer not to historical *genomena* – "what Alcibiades did or suffered" – but to the traditional *genomena* of Greek myth: what Oedipus did or suffered. In this section Aristotle slides from real individuals to mythic individuals to invented individuals and finally back to real individuals: in the concluding sentence *ta genomena* seems to mean "historical reality" again, but a reality now characterized by probability: at least in some cases, what has happened (*ta genomena*) is what might happen (*hoia an genoito*), and historical events are an appropriate subject of poetic mimesis.

What Aristotle is trying to say seems clear enough: just because an event really happened does not mean that it cannot have the kind of unity and universality that make for a good tragic *muthos*. But the admission of *eikos* into the realm of *historia*, along with the slide between historical and mythical *genomena*, obscures the clear distinction established at the start of the section: *eikos* now stands on the side both of *ta genomena* and of *hoia an genoito* and the opposition between *historia* and *poiēsis* seems to break down.

[12] O'Sullivan 1995: 59–63; cf. Else 1957: 309–12.

Section 9 then goes on to discuss the importance of *eikos* in development of the plot, condemning episodic plots where events follow one another without *eikos* or *anankē*. Such plots, Aristotle says, diminish the emotional impact of a play "since it should be a mimesis not only of a complete action but also of things which cause fear and pity, and these things occur most when events arise out of one another logically but unexpectedly; this creates more wonder (*to thaumaston*) than if they happen of their own accord or by chance" (*apo tou automatou kai tēs tukhēs*, 1452a1–6). Wonder, *thauma*, is created not by random events but by those that seem to have a purpose: for example, when the statue of Mitys fell on Mitys' murderer and killed him (1452a7–11). Through *eikos*, the logical ordering of events, the poet transforms the random accidents (*tukhai*) of history into purposeful probable narratives that generate theatrical *thauma*. *Eikos* thus contributes not only to the universality that makes tragedy serious and philosophical, but also to the emotional impact that makes it pleasurable.[13]

There is a danger, however, that if events become *too* probable the *thauma* will be lost: the plot will become predictable and will lose the element of surprise. Aristotle returns to the relation between *thauma* and *eikos* in section 18. Poets use *thauma* to achieve their desired effects and this, Aristotle notes, is *tragikon* and *philanthrōpon*, that is, it produces the tragic emotions of pity and fear by creating identification on the human level (1456a20–21).[14] It does this through a patterned reversal of probabilities: the wise but bad man is deceived, the brave but unjust man is beaten (1456a21–23).[15] Such results, though possible, seem unlikely – not least to the tragic protagonist himself, who recognizes only after the fact that this reversal is not only the kind of thing that might happen but that it in fact has happened in his own case: *hoia an genoito* become, tragically, *ta genomena*. And this unlikely conversion of the possible into the actual happens within the world of the play because it happens in the real world, as Aristotle explains: "And this is also probable, as Agathon says, for it is probable that many things happen against probability" (ἔστιν δὲ τοῦτο καὶ εἰκὸς ὥσπερ Ἀγάθων λέγει, εἰκὸς γὰρ γίνεσθαι πολλὰ καὶ παρὰ τὸ εἰκός, 1456a23–25 = Agathon fr. 9 Snell).

13 O'Sullivan 1995 stresses this aspect of *eikos* in the *Poetics*. On the ambiguous position of *thauma* in Aristotle's theory, see Halliwell 1987: 174–76; Gastaldi 1989: 93–100.

14 Halliwell 1987: 52.

15 These examples function like the reverse *eikos* argument attributed to Corax: a strong man argues that it is less probable that he started the fight because it is more probable that he would be suspected of having started the fight. See Antiphon, *First Tetralogy* 2.3, 2.9; Arist. *Rh*. 1402a17–20; Gagarin in this volume.

The section seems to acknowledge that if tragic plots become too probable, they lose the ability to produce *thauma*, and with it they lose both the pity and fear (*tragikon*) that make the action meaningful and the human identification (*philanthrōpon*) that makes it plausible. The solution to this problem is an infusion of the improbability, the paradox and reversal of expectation, that characterize real life. Agathon's aphorism is alluded to again at section 25, telegraphically and without attribution, in the course of a discussion of the irrational (*to alogon*) and the impossible (*to adunaton*) in drama. These are to be avoided in general but are acceptable if they conform to probability (1460a26–b2) and augment the work's emotional impact (1460b23–26). A poet can defend himself if he is cricitized for writing about *to alogon* by arguing that it is not really *alogon*, since "it is probable that the improbable also happens" (εἰκὸς γὰρ καὶ παρὰ τὸ εἰκὸς γίνεσθαι, 1461b15). Here as in section 9 the tragedian's maxim is a way of incorporating the unexpected into drama without compromising its probability and in fact increasing its verisimilitude.[16]

Agathon's paradox – it is probable that many improbable things happen – situates tragic probability in a delicate balance between the random, episodic *tukhē* of reality and a plot so predictable and contrived that its necessity chokes off all wonder. But, again, this paradox seems to collapse the distinction between fiction and reality. *Eikos*, which started off as a defining feature of fictional mimesis, now becomes a feature of real life, which is both more probable than fiction (it is, after all, *ta genomena* not merely *hoia an genoito*) and less (for in actuality it is probable that many improbable things happen). Meanwhile, tragedy borrows its *thauma* from history and is most realistic where it seems most improbable. Agathon's paradox thus marks the spot on the Moebius strip where improbable history and probable fiction (or is it the other way around?) cross and invert. *Eikos* confuses the very distinction it was introduced to elucidate, and Aristotle's theory of probability starts to look itself rather improbable.

Euripides' *Helen* takes this Moebius strip and gives it a couple of extra twists. The word *eikos* does not actually occur in the play, but questions of probability are insistently raised by the plot, which revolves around the image (*eikōn*, 73, 77) of Helen in Egypt. This plot is self-consciously improbable and, indeed, may have struck the audience as such: the play seems to exhibit a degree of free invention uncommon in tragedy, taking

[16] Cf. *Rh.* 1402a11–12, where Aristotle quotes the full dystich; he contemplates (and rejects) the syllogistic conclusion of Agathon's dictum: that the probable is, in general, improbable.

material from Homer, Herodotus, and Stesichorus and turning it into something new and – from the perspective of the traditional myth – ostentatiously fictional.[17] The play's theorization of its own practice as fictional mimesis poses a sophisticated critique *avant la lettre* of Aristotle's distinction between fiction and history.

The play stages a counterfactual history, as Matthew Wright has argued.[18] The "fact" that it counters is the Trojan War. Here, as in much of tragedy, the Trojan War represents the brute actuality of history. Associated with the finality of death, it is what has happened and cannot un-happen – *ta genomena*, in Aristotle's terms. Helen herself, again here as elsewhere in tragedy, is a way of thinking about an alternative or counterfactual history, *hoia an genoito*. This tradition begins already with Homer, where we find Helen wishing that death had been pleasing to her before she followed Paris to Troy (*Il.* 3.173–75), or that a whirlwind had whisked her away on the day her mother bore her (*Il.* 6.345–48).[19] These counterfactuals imagine Helen as a switch-point between *ta genomena* and *hoia an genoito*: she goes with Paris and the Trojan War happens; she does not go and it does not. The same idea is expressed countless times in tragedy, where one chorus after another wishes that Helen had never sailed to Troy, never met Paris, never been born.[20] This tragic topos imagines Helen's counterfactual history as a simple negation of the history of the Trojan War. A similar binary logic informs the remaining fragment of Stesichorus' Palinode: the story is either true or false; either Helen went to Troy or she did not, alternatives as stark and mutually exclusive as that between vision and blindness.[21] Euripides' play does something rather more complicated. It turns this either-or into a both-and: Helen both went to Troy and did

[17] Meltzer 1994; Dunn 1996: 152–53; Wright 2005: 57, 155. Scodel 1990: 78 notes Euripides' "positive fascination with problems of verisimilitude." As Mastronarde observes (2010: 3, 64), Euripides has been condemned since antiquity for not adhering closely enough to the Aristotelian principle of probability.

[18] Wright 2005: 58–59 *et passim*.

[19] On the ambiguous nature of the Homeric and post-Homeric Helen see Suzuki 1989, N. Austin 1994, Worman 1997. Gorgias, in his defense of Helen, does not use her for posing a counterfactuality to the Trojan War, but he does associate her with probability, as he explores at length "the probable reasons (*di' has eikos ēn*) for her voyage to Troy" (5). See Worman 1997: 171–80.

[20] In the *agōn* of Euripides' *Trojan Women*, for instance, Hecuba blames Helen for running off with Paris, while Helen blames Hecuba for bearing him in the first place.

[21] Pl. *Phdr.* 243a5–b3. Pl. *Resp.* 586c4–5 attributes to Stesichorus the story that Helen was replaced in Troy by a phantom, but there is no indication that her ambiguity infects the reality of Troy for him, as it does in Euripides' play. It is no mere coincidence that our fullest fragment of Stesichorus' Palinode is preserved in the same Platonic dialogue in which Socrates opposes *eikos* to truth: the Palinode suggests that the truth about Helen is the product of the poetic play of probabilities (243a1–2). On Stesichorus' Helen, see Bassi 1993, and on Stesichorus' influence on Euripides, Allan 2008: 18–22 (contra Wright 2005: 86–110).

not go to Troy. The counterfactual does not erase the factual but instead fictionalizes it: *ta genomena* and *hoia an genoito* coexist and the latter's optative, its potentiality or probability, comes to infect the former.

As in Aristotle, the line between history (*ta genomena* of the Trojan War) and fiction (Helen in Egypt) becomes blurred to such an extent that it becomes hard to tell which is which. The Trojan War, fought for an illusory Helen, becomes a mere illusion itself (115–22): all those labors and all those lives were wasted "for a cloud" (707), the *eidōlon* (image, phantom) of Helen that Hera fabricated out of thin air and sent to Paris in revenge for his verdict in the divine beauty contest. The reality of the war and the truths it was thought to represent are challenged throughout the play, not least by the representation of Menelaus as a bedraggled shipwrecked beggar; as the Old Porteress says, shooing him away from the palace door, "you may have been a mighty man somewhere but you aren't one here" (454). Here Menelaus' heroism is just a story and, the Porteress implies, a rather improbable one. This fictionalization of history comes to a head when the Trojan War is restaged at the end of the play as part of the elaborate make-believe of the escape plot. "Waging a great contest for Helen's bed" (843–44), Menelaus will both redeem and reaffirm his "Trojan *kleos*," as he says, and prove himself once again willing to risk his life to win Helen (843–50). "Where is the glory of Troy!" Helen exclaims, as she urges the sailors to fight off the invading barbarians and hasten their escape (1603; cf. 1560, 1593): the escape becomes a second Trojan War, but this time fought for the real Helen. In this sense, this overtly fictional plot – a fiction within a fiction – is more real than the reality of the events at Troy, which become, as Helen puts it, *erg' anerga* (363), deeds that are no deeds. The *mimēsis* undoes the *praxis*.[22]

Ta genomena become fictionalized. Likewise, *hoia an genoito*, the counterfactual history of Helen in Egypt, is presented as reality by means of the play's own mimetic art. The prologue establishes this tension between the "historical reality" of Troy and the improbable scenario being presented as reality on stage. Indeed, Helen's opening speech insists simultaneously on its improbability and on its reality. The very form of the prologue, a first-person account delivered more or less directly to the audience by the one person who might be expected to know what "really" happened, establishes this surprising new scenario as *ta genomena* within the world of the play. Helen tells us where we are, who she is, what has happened. And yet, even as she asserts the reality of this version of events, she also expresses her

[22] Pucci 1997 shows how reality is an effect of *eidōla* in this play.

doubt about its veracity. "They say Zeus flew to my mother in the shape of a swan . . . if this account is true" (18–21; cf. 256–60).[23] Stesichorus' two alternatives – either Helen went to Troy or she did not – seem holographically superimposed in the prologue, as Helen presents *genomena* that she does not seem to believe are *dunata*, much less *eikota*: this version of events might be real, but it is not probable. The ensuing conversation with a very bewildered Teucer underscores this and extends it to Troy, where he saw Menelaus drag off Helen with his own eyes, he tells this Helen, "no less than I see you now" (116–18). In this scene's troping of seeing and seeming, Helen of Troy and Helen of Egypt are equally probable, which is to say equally improbable. As in Aristotle, the distinction between *ta genomena* and *ta dunata* collapses around *eikos*, the *eikos* in this case of an *eikōn* and of the real woman of whom it is an improbable imitation (72–77).[24]

The rest of the *muthos* unfolds from this collapsing of fiction and reality, as the play demands that the audience accept as real a scenario that it explicitly marks as improbable. *Helen* insistently advocates skepticism: "there is nothing more useful for mortals than wise disbelief" (*sōphronos apistias*, 1617–18), says the messenger in conclusion to his account of the protagonists' daring escape. Likewise, the old servant who reports the miraculous disappearance of Helen's phantom seems barely to credit his own story. In the midst of his account he sees the "real" Helen: "So this is where you've been," he says sarcastically. "I was just reporting how you had gone up to the stars, little realizing that your body had wings" (616–19). Amidst relentless variations on the theme of illusion and reality, we are emphatically warned to question everything, trust no one, not to believe our eyes. It is no surprise, then, that many scholars read the play as a meditation on the impossibility of human knowledge.[25] Even its messengers do not believe their own messages.

And yet, for the theatrical experience to work – and presumably the playwright did want it to work – the audience must check its skepticism and believe, at least provisionally, in the reality of the fiction it is watching. It must believe, that is, that the play's *dunata* are *genomena*. The play is thus structured by a tension, in Aristotle's terms, between its *dianoia*

[23] Allan 2008: 148 *ad* 17: "H.'s scepticism provokes reflection upon the nature and truth value of mythology itself... the audience are encouraged to reflect upon (and enjoy) the fictiveness of Eur.'s own poetic 'reality'." Mastronarde 2010: 175 sees this metathetatrical distancing at work in Euripidean prologue-monologues in general.

[24] Downing 1990 offers a smart reading of the prologue in light of the play's themes of fiction and deception.

[25] See, e.g., Segal 1971; Meltzer 1994; Wright 2005: 278–337; Allan 2008: 46–49.

or "thought" (the repeated warnings not to believe what we see) and its *muthos*, which not only requires belief but works hard to produce it.[26]

This tension is particularly acute in the *anagnōrisis*. Menelaus is justly sceptical that the woman in front of him is his wife: he knows (or thinks he knows) that he has brought Helen from Troy and left her in a cave; what would she be doing here at Proteus' tomb? But the audience recognizes this Helen as the "real" Helen: she told us that she was in the prologue and we have been given no reason to doubt the veracity of her story, improbable as it seems. Menelaus is perplexed by the news that Spartan Helen is in Egypt but explains it with a rationalization akin to Agathon's paradox (489–99):

> Διὸς δ' ἔλεξε παῖδά νιν πεφυκέναι·
> ἀλλ' ἦ τις ἔστι Ζηνὸς ὄνομ' ἔχων ἀνὴρ
> Νείλου παρ' ὄχθας; εἷς γὰρ ὅ γε κατ' οὐρανόν.
> Σπάρτη δὲ ποῦ γῆς ἐστι πλὴν ἵνα ῥοαὶ
> τοῦ καλλιδόνακός εἰσιν Εὐρώτα μόνον;
> διπλοῦν δὲ Τυνδάρειον ὄνομα κλῄζεται,
> Λακεδαίμονος δὲ γαῖά τις ξυνώνυμος
> Τροίας τ'; ἐγὼ μὲν οὐκ ἔχω τί χρὴ λέγειν.
> πολλοὶ γάρ, ὡς εἴξασιν, ἐν πολλῇ χθονὶ
> ὀνόματα ταὔτ' ἔχουσι καὶ πόλις πόλει
> γυνή γυναικί τ'· οὐδὲν οὖν θαυμαστέον.
>
> [The Porteress] said she was the child of Zeus.
> Is there really a man named Zeus on the banks
> of the Nile? For there is only one Zeus in Heaven.
> And where on earth is Sparta, except only where
> the beautiful reedy streams of Eurotas are?
> Is the famous name of Tyndareus double?
> Is there a land with the same name as Lacedaemon
> or Troy? I don't know what to say.
> I guess in this wide world there are many people
> with the same names, cities like cities and women
> like women. There's nothing so amazing about that, then.

The improbability of real life, which in Aristotle is imported into drama to augment its *thauma*, here explains it away. It is probable that many improbable things happen, among them a woman named Helen daughter

[26] Compare the interesting observations of Gallagher 2006: 346 on disbelief as the condition of fictionality in the eighteen-century British novel: "Novels promoted a disposition of ironic credulity enabled by optimistic incredulity; one is dissuaded from believing the literal truth of a representation so that one can instead admire its likelihood."

of Tyndareus living in a town called Sparta-on-the-Nile: "there's nothing so amazing about that" (*ouden oun thaumasteon*).[27]

Menelaus would seem to practice the hermeneutic skepticism the play advocates. But his skepticism nearly ruins the play. The *anagnōrisis* begins with an extended stichomythic pas de deux. Helen and Menelaus meet face-to-face and are both instantly dumbstruck by the other's resemblance to their lost spouse (557–64). But whereas Helen's hopes are immediately confirmed – the stranger *is* her long-lost husband – Menelaus refuses to believe that she is his wife: he knows that he has left Helen in the cave and that he is not "the single husband of double wives" (571). Her insistence that she is his one and only wife produces a cognitive double vision. Believe your eyes, she tells him. My eyes must be diseased, he responds (575, 581). Convinced of the reality of Hera's nebulous fiction and of his heroic labors for her, he will not credit the improbable reality of the Helen in front of him. The recognition thus fails and, as Menelaus rejects his flesh-and-blood wife for a phantom's "empty bed" (590), the *muthos* seems to have stalled out. The play threatens to end in failure at line 596.[28]

But of course it does not end there. The cure for Menelaus' "eye disease," his refusal to believe the implausible, is something even more improbable, if not impossible. Just at that moment the old man comes on to report the miraculous disappearance of the phantom Helen. "It is a *thauma*," he begins, "less amazing in the telling than in the event" (601). This *thauma* is the most fictional of theatrical fictions: we never see this *eidōlon* – or *mimēma*, as she is later called (875) – with our own eyes; to us she was always just a shadowy illusion, and even the old man who witnessed her disappearance does not really believe it (616–21).[29] Aristotle proposes that plausible impossibilities are preferable to implausible possibilities (1460a26–27, 1461b9–12). This phantom seems neither possible nor plausible. And yet, it is this improbable fiction that precipitates the *anagnōrisis*. "This is it!" exclaims Menelaus. "Her stories have turned out true! Oh longed-for day, which has given me you to hold in my arms!" (622–24). The *anagnōrisis*, the joyful reunion of husband and wife, the emotional *peripeteia* of the plot and satisfying *lusis* of the play's dramatic *desis*: all this

[27] As Hose 2000: 22–24 notes, *eikos* arguments within tragedy tend to lead to erroneous conclusions, as here.

[28] Burnett 1971: 84.

[29] Allan 2008: 244 ad loc. notes that Euripides is the only poet to use *mimēma*, "a word used frequently by Plato of various (counterfeit) images, including artistic and musical representations." The fictionality of this phantom (which delivers a divine pronouncement as it flies away) is underlined by its resemblance to a *deus ex machina*, the most artificial of theatrical devices. Barone 1989 examines the *thaumaston* in *Helen* from the perspective of theatrical effect.

is accomplished only when Menelaus abandons his skepticism, when he comes to accept – as we have from the beginning – this improbable fiction for reality.

Of course, we do not really mistake these unlikely *dunata* for *genomena*, do we? *Helen* challenges us on this question throughout and especially in the escape plan that occupies the second half of the play. Overtly metatheatrical, this "contrivance of salvation" (*mēkhanē sōtērias*, 1034) is explicitly marked as a fiction.[30] The fantastic scheme is constructed before our eyes as a play of probabilities, *ta dunata kata to eikos ē to anankaion*. Helen and Menelaus conceive their plan after considering and rejecting a number of alternative hypothetical options as *adunata* (808, 1043); by contrast, this plan – they will tell the Egyptian king Theoclymenus that Menelaus is dead and stage a funeral for him at sea – is apparently *dunaton*.[31] They also work hard to make it *eikota*, overcoming Theoclymenus' warranted skepticism with detailed explanations of the made-up Greek burial ritual they claim to be enacting (1245–77). Part of the pleasure of this escape plot is the gap between our awareness of its fictionality and Theoclymenus' credence in its reality. The incessant ironies of the episode are a knowing wink at the audience over the barbarian's head. Who told you Menelaus was dead? asks Theoclymenus. This man here, Helen answers, pointing to the living Menelaus. What a miserable sight he is, remarks Theoclymenus. I think my husband must be in the same condition, Helen responds (1198–1205). And so on. Such heavy-handed ironies let us congratulate ourselves because we, unlike Theoclymenus, can tell the difference between fiction and reality. But, as we do, we forget that the reality we have subscribed to is itself a fiction. We see through the fiction of the escape plot only because we know the "truth" of the dramatic plot, the *muthos*. Does this make us cannier than the barbarian king or more deluded? We are both of us fooled into accepting *dunata* for *genomena*. Gorgias' riddling apothegm is apropos: "the deceiver is more just than the one who does not

[30] Burnett 1971: 84–85; Downing 1990: 11–12. Line 1056, in which Menelaus condemns the plan as "rather oldfashioned" is generally taken as a disparaging reference to Sophocles' *Electra*: see Dale 1967: 134 ad loc.; Scodel 1999: 100–101; Allan 2008: 260–61 ad loc. The delay of the first stasimon until 1107 and the concentration of all the choral odes in the last 500 lines adds to this sense of a play within a play, as if Euripides were lending his characters the theatrical resources of tragedy for use in their little drama.

[31] Allan 2008: 237 *ad* 802–14 remarks that "The technique of exploring and rejecting *adunata* ('impossible plans') before hitting upon the best scheme is typical of plotting scenes." This technique is structurally akin to that of the rhetorical *eikos* argument. See M. Lloyd 1994: 178 *ad* Eur. *Andr.* 192–204: "E.'s arguments from probability are always expressed through hypophora, a figure in which alternative answers to a problem are both proposed and rejected by means of a series of rhetorical questions." Cf. Scodel 1999: 99.

deceive, and the one who is deceived more wise than the one who is not deceived."[32]

The escape plot is extremely contrived, both at the diegetic and at the extradiegetic level. The characters may be uncertain it will achieve its *telos*, but the playwright deploys all the weapons in his arsenal to ensure its successful conclusion. He hints from the beginning that the escape is part of Zeus's plan (36–37), and that plan is proclaimed fulfilled at the end (1646–55).[33] The speech of Theonoe, Theoclymenus' sister and an infallible seer, aligns the escape with justice: she agrees to help them because it is pious and just (998–1032). The mythic paradigm of Persephone, which has been a subtle leitmotif throughout the play, also predicts the successful return of the still-virginal Helen.[34] Add to this the victory of a Greek hero over a barbarian tyrant and everything conspires to make the escape not only probable but inevitable: *eikos* verges on *anankē*. So, too, things that seemed random in the first half of the play, elements of mere happenstance, in the second half become a purposeful part of the escape plot.[35] Menelaus' rags, for instance, go from being the product of *tukhē* to props in the service of their escape. Indeed, as the escape plot re-enacts the epistemological confusions of the play's early episodes now as deliberate and self-conscious deceptions, the word *tukhē* – which recurs obsessively throughout the play – comes to refer primarily to the success of the plot and its *telos*.[36] Aristotle, as we saw, associates *tukhē* with *historia*: real life lacks a unified *telos* or purposeful order. It is the playwright's job, he says, to impose that *telos* and order, and he does so by means of *eikos*, the logical unfolding of one event from another. *Helen* stages that process, as all the improbabilities of the play's scenario – the double Helen, the shipwrecked

[32] Gorgias B 23 DK. Aristophanes anticipates this line of argument in his parody of *Helen* at *Thesmophoriazusai* 855–919, where Kritylla is tricked precisely because of her inability to enter into the parody's fiction.

[33] The *Dios boulē* (plan of Zeus) is a matter of both *tukhē* and *anankē*. Helen expresses faith in her father's care but also associates him with shape-shifting, deception, and incredible legend (20–21). The certainty of Zeus's plan is undermined by the shadowy role of Hera and Aphrodite, and of the divine council that Theonoe reports will meet that very day to determine whether Helen should return to Greece (878–91), a verdict that Theonoe herself ultimately seems to decide (887–91, 1005–1006).

[34] Foley 2001: 303–31. The death pact at 835–42 works similarly. If the escape does not work, the heroes vow to kill themselves right there on stage. This emotional blackmail forces the audience to invest in the escape plan.

[35] Downing 1990: 7–8.

[36] E.g. 738, 742, 1409, 1424, 1445, 1450. See esp. 1636: *tukhē* gave Helen to Theoclymenus; necessity (*to khreōn*) took her away. Note also that the first stasimon, which sings of the "contradictory unexpected *tukhai*" of the gods (1138–43), comes only after the escape plot is already under way. This play has the second most occurrences of *tukhē* of any tragedy (after Euripides' *Ion*). Cf. Dunn 1996: 147–51; Wright 2005: 373–80.

Menelaus – are redeployed to bolster the probability of the escape plot. The *tukhai* of the play's fictional reality are swept up by the *anankē* of its dramatic plotting and suborned to the *telos* of making the improbable seem real.

And this improbable fictional reality becomes, in the end, our own real reality. The play ends with a *deus ex machina* – *dei*, in fact, Helen's brothers Castor and Polydeuces. Francis Dunn comments on the "unusually artificial or ready-made quality" of *Helen*'s ending, what he calls its "avant-garde conventionality."[37] This ostentatiously fictional play concludes its ostentatiously fictional escape plot with an ostentatiously fictional device: the *deus ex machina* was mocked by Aristophanes and scorned by Aristotle for introducing irrational elements (*aloga*) external to the action (*Poet.* 1454b1–5).[38] These improbable gods reassert the necessity of the escape plot: it is fated, they tell us, for Helen to go home to Sparta with Menelaus (1646, 1660); it is necessary (1651, 1654); it is the gods' will (1648, 1661). The Dioscuri also connect this improbable narrative directly to our own extra-theatrical reality through the founding of a new (and in the fifth century presumably still existing) cult. Helen will be deified after her death and will share in cult offerings with her brothers by the command of Zeus (1667–69); the island off the coast of Attica where Hermes stopped with Helen en route from Sparta to Egypt will henceforth be called "Helen Island" (1670–75), as indeed it was in the fifth century. The *aitia* that close many Euripidean plays forge a link between the world of the play and the world of the audience, between *poiēsis* and *historia*. But here the connection is remarkably tenuous: the cult offerings to Helen are nowhere else attested and some scholars believe Euripides simply invented them; likewise Euripides' aetiology for the name of "Helen Island" conflicts with other fifth-century explanations.[39] It is distinctly possible that these *aitia*, the links between the fictional world of the play and the historical world

[37] Dunn 1996: 135, 142.

[38] The *dei* here are all the more fictional for their association with the illusions of the first half of the play. Teucer reported their deaths to Helen at 138: "they are dead and not dead; there are two stories." One story tells that they have become "similar to stars," the other that they have killed themselves from shame (139–42). According to Teucer the more improbable *logos*, the katasterization, is the "stronger" (139), and the final epiphany supports his claim. Wright 2005: 361; cf. Dunn 1996: 137–39.

[39] On "Helen Island" see Allan 2008: 343 *ad* 1670–75: Hecataeus says that Helen stopped there on her way back from Troy; Strabo, that the island was the first place where Helen and Paris made love on their way to Troy. Both of these aetiologies, as Allan notes, "are more plausible candidates than Eur.'s version, since they take as their basis the traditional story of H.'s abduction from Sparta and presence at Troy." On these *aitia* see further Wright 2005: 357–62; Allan 2008: 14–16, 339–40 *ad* 1642–79. Scullion 1999–2000 believes Euripides' closing *aitia* were in general largely invented.

of the audience, are entirely made up and that the audience would have recognized them as such.

Finally, the drama's improbability is projected as a theology that governs not only the fictional world of the play but our own world as well. *Helen* ends with a conventional choral closing, a formula so generic that it also ends four other Euripidean plays.[40]

πολλαὶ μορφαὶ τῶν δαιμονίων,
πολλὰ δ' ἀέλπτως κραίνουσι θεοί·
καὶ τὰ δοκηθέντ' οὐκ ἐτελέσθη,
τῶν δ' ἀδοκήτων πόρον ηὗρε θεός.
τοιόνδ' ἀπέβη τόδε πρᾶγμα.

Many are the shapes of the divinities;
many things the gods accomplish against all hope.
The expected is not accomplished,
for the unexpected the god finds a way.
That's how this affair has turned out.

(1688–92)

This final refrain recalls Agathon's maxim and establishes that paradox as the law of the play: it is probable that many improbable things happen. All the play's manifold improbabilities can now be seen as vehicles toward a higher *anankē*: if the escape plot seemed contrived, it *was* – not by the playwright but by the gods in service of their own inscrutible *telos*.[41] Mimetic *poiēsis* becomes a law of nature that extends beyond the bounds of the play. Formally, these lines function as a transition between the fictional world and the real: the chorus officially declares the play over and addresses the audience more or less directly to deliver the "take-home" message of what we have just seen. Thus the dogma of improbable probabilities becomes a universal law and the whole world is said to work like the unbelievable fictions inside the play.

Aristotle, as we saw, uses Agathon's paradox to inject a bit of life's contingency into the contrivance of the dramatic mimesis. For him tragedy, as a *mimēsis* of a *praxis*, relies on reality for its theatrical effect and

[40] *Alc.*, *Andr.*, *Bacch.*, and (with slight variation) *Med.* Dunn 1996: 25, 136 and Mastronarde 2010: 106 view this closing tag as empty of significance or particular relevance to the plays. Hose 2000: 30, however, notes its pertinence to the issues of *eikos* and *tukhē* in this play. D. Roberts 1987: 59–60 observes that these generic closing lines simplify the play's complexities, even as they universalize them.

[41] Cf. 711–13: the servant's response to the impossible/implausible disappearance of the phantom Helen is to chalk it up to the inscrutibility of the divine.

emotional impact: its *thauma* is the product of an artful arrangement of life's random surprises. But *Helen*'s final lines suggest an inverse relation between *poiēsis* and *historia*, in which the improbabilities of the drama are projected outward as the governing probabilities of the real world. If life is surprising, it is because it resembles this play and the gods work there as they have done here. Fiction is not a *mimēsis* of reality but its model: its improbable *eikones* are iconic of life's own improbability. Agathon was, after all, a tragedian and his paradox originally referred not to the real, extratheatrical world, as Aristotle's decontextualized quotation suggests, but to the fictional world of its play. The probability of the improbable that Aristotle discovers in real life was actually poetic *eikos* all along.[42]

Helen's apparent reversal of the relation between fiction and reality, between *poiēsis* and *historia*, raises the question of the connection between the play's *dunata* and the *genomena* of its historical context. This play is often seen as a direct mimesis of the Athenians' recent experience in Sicily, where, according to Thucydides, having set off with the highest expectations, they found they had been pursuing an *eidōlon* and discovered to their fear and astonishment (*phobos te kai kataplēxis*, Thuc. 8.1.2) that what they thought was reality was actually just an illusion. This reading subsumes the entire problematic of fictional probability within the firm realities of historical *genomena*: the play is improbable because life in Athens in 412 BCE was improbable and Agathon's paradox becomes straightforward historiographical reportage.[43]

But the end of *Helen* suggests that if in 412 it proved probable that many improbable things happen, that is not because fiction imitates life but because life imitates fiction. Far from the play's reporting on events in Sicily, Sicily becomes a real-life *mimēsis* of Euripides' dramatic *praxis*. *Helen* shows us how *poiēsis* can make the improbable seem probable – indeed, it can make the impossible seem necessary – and can make us mistake fiction

[42] We do not know which play the fragment is from, but the lines as fully quoted at Arist. *Rh.* 1402a11–12 are in iambic trimeter. Ricoeur 1991 dismantles the "apparently invincible exteriority between the inside of fiction and the outside of life" (151): on the one hand he posits the mimetic, narrative, even fictional prefiguration of real life; on the other hand, since writing is fulfilled in reading, he proposes "the transfiguration of the world of action in terms of fiction" (151).

[43] Dunn 1996: 156: "The action of *Helen* is firmly rooted in the Athenians' new experience of profound uncertainty . . . By closely observing his chaotic world, by honestly *reporting* how it resists the search for order, Euripides shows that chance can lead to success, as well as to failure, that an uncertain world can be both exciting and terrifying, and that the desire for happy endings may be entirely irrelevant even as it is realized" (emphasis added). For *Helen* as a reflection of current events, see, e.g., Meltzer 1994: 238, 254–55; Dunn 1996: 155–57; Steiner 2001: 292–93; Wright 2005: 47; Friedman 2007: 197–98, 208–9. Cf. the provocative comments of Arrowsmith 1973: 133–34 n. 5.

for reality even when we are told that it is fiction. If the Athenians did in fact mistake *ta dunata kata to eikos* for *ta genomena* in Sicily – if this is not just a figment of Thucydides' historiographical *poiēsis* – then they are merely living out the lessons of Euripides' play: the expected is not accomplished, for the unexpected the god finds a way. And that is how the affair turned out.

CHAPTER 9

Revision in Greek literary papyri

Sean Gurd

In a scene from his now lost comedy *Wine-Flask* (performed in 423 BCE), Cratinus staged a dialogue between himself and the personification of Comedy (who is, in this play, his wife, though scorned by his attention to a new mistress, the wine-flask of the title). During this dialogue Cratinus is writing a comedy, and Comedy is advising him on what to put in it. "You're talking nonsense!" she says; "put it into a single episode. Cleisthenes playing dice will be funny" (208 KA); then, "Rub out *Hyperbolus* and put in *in the market of oil lamps.*" Scenes of poetic tutelage are hardly rare in ancient culture: they appear on vase paintings and bas-reliefs, in Hesiod and Callimachus. But nowhere else, as far as I am aware, does the deity instruct a poet to *erase*, change, or restructure a text. These frustratingly meager snippets thus make a strong claim on our attention. No doubt, the scene was written and played for laughs – at the very least, having a poet and his wife/tutelary goddess debate what to put in a play gave Cratinus the opportunity to deliver two jokes instead of one (Comedy's "funnier" versions, and Cratinus' own). But the scene is also a radically meta-poetic moment. By showing *himself* as comic poet *changing* a text, Cratinus reminds us that this scene of revision may itself have been revised as well, that in addition to what we read there may have been other versions, other words, and other jokes that he explored but ultimately changed or rejected. I say that this is a *radically* meta-poetic moment because, while the dramatic texts of the fifth century can frequently be caught indulging in self-referential play (Old Comedy, thanks to the convention of the *parabasis*, does this almost systematically), they very rarely reflect on their own contingency, on the possibility that they were or might have been other than they are. In fact, the only other example I can think of is the unfinished and probably never performed *Clouds* of Aristophanes, which talks of itself as a revision – a revision made, certainly not accidentally, in response to the victory of Cratinus' play over Aristophanes' original *Clouds* in 423.

In showing a comedy being altered during the process of its genesis, Cratinus exposes his audience to the contingency of his text, inviting the reflection that this is only one possible version among others, and implicitly provoking us to ask why we have the version we do. The answer that the scene implies – "because the personified Comedy told me to write it this way" – is its most unavoidably counterfactual element: Comedy knows (or should know) what is right and what is wrong, and in following her Cratinus would be following the instructions of a quasi-muse. Because we know that this does not happen – goddesses do not instruct us on what to write (though I wish they would) – the scene is fantastic, compelling, and, ultimately, funny. But in enjoying it in this way we also admit that it came to be in a process that was much more difficult to explain and perhaps ultimately impossible to understand. What does it mean to contemplate the contingency of the form before us, to reflect as we observe that *what* we observe might have been something else? And how can we conceptualise a creative process in which what comes to be does so not thanks to inspiration or divine tutelage but thanks to a process better described as blind groping, hypothetical experimentation, or the indefinite exploration of possibilities?[1]

I want to respond to Cratinus' provocation, to use the creative process (and especially that part of it that has to do with revision) as a way in which to explore the themes of possibility, hypotheticals, and counterfactuality not as theoretical constructs or conceptual tools but as practices; and not as practices associable with an elaborated theory or *tekhnē* but as having no grounding in conceptuality, even with a close or intimate relation to non- or pre-conceptuality, to the un- or almost- or barely thought. My wager is that by looking at concrete instances of revision from Greek papyri we will get a glimpse into the role of the themes of this volume not as alternatives to actuality and the discourses which describe it, but as prior to it and essential to its constitution. Counterfactuals, possibility, and the hypothetical are part of the genesis of what there is, and what there is is what it is because it might have been otherwise. In the case of literature, a text is what it is because it *was* otherwise. And that, in turn, means we must look at it again, otherwise.

During the genesis of a literary text, authors and their collaborators do not know what they are doing. Revision happens because authors do not know what their texts should look like; when they do, they stop revising. From this perspective, revisions cannot be counterfactuals, though

[1] This chapter expands and gives additional detail to matters covered in the introduction to Gurd 2012. On the topic of revision and authorial practice, see also Dorandi 1991, 2000; Cavallo 2000. Papyrological abbreviations can be found at Oates *et al.* 2013.

they might represent possibles or hypotheticals. Revisions appear to be counterfactuals only when they are seen from a perspective which privileges a single version which lacks them. Once such a version exists, one might organize earlier versions in its light, so that some are called "right" and others "errors." Then, once it has been decided what will count as "fact," rejected versions of a work can seem counterfactual. Before this, the elaboration of a work in progress takes place by means of a field of possibilities and "as if" scenarios in which the work to come, as yet unknown, is explored and gradually but not inevitably formed.[2]

This picture is complicated by the fact that when an author engages in revisions, each new version seems to replace the old. Genetic critics refer to this fact by observing that every new draft is a final draft – until it is replaced. But that a revision or a deletion instantly recategorises a previous version as counterfactual does not mitigate the overall lack of teleology in a genetic process. The reason for this is simply that authors can and do revert to versions that they had changed or rejected in the course of rewriting. There are local teleologies in which hypothetical versions are converted into counterfactuals, but a global virtuality in which the field of revisions is only a field of possibles.

To illustrate these claims, I will describe three late-antique literary papyri which bear traces of authorial revision. My aim will be to show that while the chronology of revisions in a papyrus can usually be reconstructed, and in many cases it is also possible to conjecture the reason for a revision, neither chronology nor interpretation allows us to see the revision process as a linear one in which an author can be said to truly know what he is aiming for. Later versions are not necessarily preferred versions, even when their motivations can be discerned. When authors revised, they produced multiple variants from which they would choose only after the fact.

My arguments depend on close readings and genetic reconstructions of three literary papyri from the second to the fourth centuries CE. Literary papyri which can reasonably be thought to have been written by their

[2] Only two other ways to interpret revision are currently extant in scholarly literature. The first assumes a teleological progression from early and poor versions to a final and perfect version. This is a noble narrative, but it is not supported by the data – neither ancient nor modern writers work this way, as is clear from their actual drafts. Abundant documentation of the untenability of the teleological narrative is provided in the large body of work done by genetic critics. The second, very popular among genetic critics, posits a complete absence of teleology, so that the writing process is an open and rhizomorphic network of variants and versions, with published versions being very little more than accidental crystallizations of a much larger textual field. I sympathize with this perspective but would like to have micro-teleologies, in which authors momentarily work toward what they believe may be a better or final version, without macro-teleologies, in which the whole process is directed toward a known end. See Gurd 2012: 129 n. 1 for a starter bibliography on genetic criticism.

authors ("autographs") are not common.[3] Of these, not all contain evidence of revision; even fewer contain evidence of an author recopying a revised earlier draft.[4] The three literary texts presented here are the richest in detail and the most informative about genetic processes of which I am aware. P. Berol. 11632 is the only example of literary prose containing ample revision; PSI I 17 is a very well-preserved papyrus and offers evidence of a complex genetic process; P. Köln VI 245 is the most studied, having received a detailed commentary in 1991. That papyrus, however, has been cleaned and reassembled since 1991, and the restoration has made several new readings possible.[5] New texts of each of these papyri, based on my own inspections, begin each section, followed by commentary. I cannot give full reconstructions of the genetic process in any case, but I have endeavored to give examples of every kind of revision instantiated in the papyri.

The status of the texts I present repeats the argument I make. I am convinced that each of these papyri contains traces of revisions that can be called authorial (even in the case of P. Berol. 11632, where revision seems to have been the result of collating different historical accounts of the same event; there is nothing unusual in claiming that historiographers can revise on the basis of new information). Identifying a text as containing authorial revision involves a complex set of criteria, including the handwriting, the material on which the text is written, the extensiveness of the variants and their placement. Inevitably, the judgment falls far short of certainty: the farthest I am willing to go is to claim that these seem to me probably to contain traces of authorial revision. I was not the first to inspect any of these papyri, and what I "saw" was itself inevitably the result of a complex and ultimately circular series of constantly revised judgments based on previous

[3] I am not convinced that all of the "autographs" identified by previous scholars are indeed autographs; nor am I convinced that every papyrus containing authorial revisions was inscribed by its author. But I am also not convinced that the question of whether an author wrote or dictated is the most illuminating when it comes to understanding authorial process, though it certainly has important sociological implications. We need also to ask questions about the temporality of composition, the manner in which authors plan out, execute, and then reconsider their texts, the logical and intentional status of drafts and revisions. See Gurd 2012.

[4] Dorandi 2000: 53–60. See also Pasquali 1962, Dorandi 1991. On Dioscoros, see Fournet 1999.

[5] Literary papyri from before the fourth century CE that contain evidence of revision but are not discussed here include P. Köln III 128 (MP[3] 1763.4), P. Ross. Georg. I 11 (MP[3] 1861; see Sutton 1987, Furley 2007), P. Oxy. VII 1015 (MP[3] 1047), P. Berol. 10559A–B + P. Berol. 10558 (MP[3] 1851), PSI XV 1482 (MP[3] 1861.1). A selective list of documentary papyri with revisions would include: P. Petrie II.13 (two separate letters); P. Petrie II.4, from the Archive of Kleon (third century BCE), re-edited as 18a, 18b, and 9 by Bart Van Beek; P. Lond. I 24 verso, p. 26 (draft of a petition also found on P. Lond. I 35 verso, pp. 27–29 [161 BCE]); P. Tebt. III 775 (draft of a petition, second century BCE); P. Tebt. III 765 (draft of a letter, 153 BCE); P. Mich. inv. 1436 (a revised copy of P. Mich inv. 1436; see Hanson 1982, 1984); P. Mich. inv. 341 (second-/third-century CE letter; see Youtie 1979); *Chartae Latinae Antiquiores* III.202 (Latin petition, 341–42 CE). On petitions in the archive of Dioscoros of Aphrodite, see Fournet 2010.

editions, previous restorators' theories of the text, my own eye's ability to descry a letter within a recognizable word, my hand's ability to reproduce the movements of the stylus leading to the visible letter- and word-form, and (in more than a few cases) the assistance of digital technologies. I do not believe that I truly "saw" what I transcribe here, if by "see" one means an unmediated or unreflective judgment about things simply present to the visual sense. These texts are theories about very difficult documents, and they are the product of a process of visual, linguistic, and literary revision that includes myself and earlier scholars who have worked on the papyrus.[6] You read an argument about revision, probability, and hypotheticals that is itself intensely revised and no more than probable or hypothetical.[7]

I

P. Berol. 11632 (MP³ 2207)

Second century CE

Narrative of the Siege of Rhodes by Demetrios Poliorketes, 304 BCE.

First edition: Gaertringen 1918. See also Bilabel 1922: #8; Powell and Barber 1929: 66–68.

P. Berol. 11632 is a single piece of papyrus with the remains of the right half of one column and the remains of the left half of a second. The narrative, of Demetrius Poliorketes' siege of Rhodes in 304 BCE, resembles Diodorus Siculus' account of the same events at 20.93–94.

6 I have discussed the logical status of textual judgments of this sort in Gurd 2005.

7 The texts printed here conform to standard papyrological conventions, with two exceptions. I have represented superscript, subscript, and marginal additions in a smaller font and in a location that reflects their position on the papyri, and I have used ` ´ to surround corrections made in the line but directly on top of other writing. The normal convention of indicating insertions in-line proves unworkable here, since the exact spatial position of a revision is often central to determining how it is to be interpreted. I have accented according to the base text: where revisions inserted above the lines imply a change in accent, it is not represented.

In reconstructing the revision sequences for each papyrus, I followed the following guidelines. A similar methodology is followed by Andorlini and Maresch 2006: 45–46.

(1) I assumed that authors wrote on the front of a papyrus (i.e. with the fibers) before writing on the back, and that they wrote in lines and columns before writing above lines or in margins.
(2) Revisions imply versions: a single word, or even fragment of a word, inserted above, below, or beside a line needs to be integrated with the line itself. My genetic reconstructions thus display inferred versions as well as versions that are written out in full in the papyri.
(3) Although nothing prevents an author from writing nonsense, I avoid reconstructions of revision sequences in which a second version makes less sense or is less grammatical than the first version.

Text

Column 1

[   ] β̣αϲιλικοῖϲι γάρ ⟦ϲφι̣ν̣⟧ ἐοῦϲι χρέ-
[εσθαι οὐ] ϲ̣φί̣ϲιν ἦν. ⟦Ἀμύ⟧ κ[α]ὶ τὰ μὲν οὕ-
[τωϲ ἔπρη]ξ̣α̣ν. Ἀμύντηϲ δὲ πλέων ἐπὶ
κ̣αταπελ̣τ̣αφέτα̣ϲ̣ ἕνδεκα
[νήϲω]ν̣ ⟦ὄργανα⟧ πρὸϲ τὴν πολιορκ-
[ίην τὴν τ]ῶ̣ν̣ π̣ολίων Δημητρίῳ̣ φ̣ε-
[ο]υϲ   πολλούϲ τε
[ρόμεν] ⟦α̣ πολλὰ⟧ αἵρεε ⟦καὶ⟧ βελέων δη-
ούϲ
[μιουργ]ο̣ὺ̣ϲ̣ τ⟦ά⟧περ Δημητρίῳ̣ λυομ-
[ένῳ̣ λύτ]ρ̣ων τῶν τεταγμένων Ῥ̣[ό]-
/   /
[διοι οὐκ ἀ]π̣ε̣δ̣ί̣δοϲαν, χρεΐζειν αὐτῶν
/
[φάσκοντεϲ·] ὁ̣ δὲ τῶνδε ἀπαμαρτῷ̣ν̣
[οὐδ' αὐτὸ]ϲ̣ ϲφίϲιν ἔφη λύϲ̣ειν ἔτι, ἢν
[ἕλη̣. μετὰ] δ̣ὲ̣ τ̣αῦτα Δημητρίου τὰ Ῥο-
[δίων τ]είχεα ὑπορύϲϲοντοϲ ἐν-
[τόϲ τε τ]ῶν τειχέων ἤδη ἐόντων
[τῶν κατ]ὰ̣ γῆϲ ἐργ[αζ]ομένων, Ῥόδ̣ι-
[οί τε μη]ν̣υ̣ϲ̣ά̣ντων αὐτο̣μόλων τὸ
ο
[γεινόμ]ε̣ναν ⟦καὶ⟧ τόν τε χῶρον τὸν
[ὑπορυ]ϲ̣ϲόμενον ὑποδεξάντω(ν),
τ
[διορύξ]α̣ντεϲ ἐϲ τωὐτὸν ⟦κείν⟧οι-
[ϲι Μακεδ]ό̣νων μεταλλωρύχοιϲιν
[ὑπηντί]α̣ϲ̣αν· οἵ τε ὑπόνομοι πα-
ἐϲ τωὐτόν
[ρὰ μικρ]ὸ̣ν συνέμιξαν ἀλλήλοιϲι(ν).
ων
[ἔργου δ' ἑ]κάτερ⟦οι⟧ τοῦ ϲφετέρου φύλα-
[ἐπιστησά]ντων ἀλλήλοιϲί τε τῶν φυ̣λάκων
[καϲ ⟦ἐπι]ϲτήϲαντεϲ⟧, προδοϲίη
ἐϲ λόγουϲ ἀπὸ
τῶν προτε̣ι̣χ̣ι̣ϲ̣μ̣ά̣-
τ(ων)
ἀπικνε-
ομένων

Column 2

τιϲ Δημητρίῳ̣ δόξαϲα [ἀϲφαλὴϲ κάρτα]

μιν ἀπάτησεν. Ἀθηναγόρ̣η̣[ς γά]ρ̣
τις Μιλήσιος ἀ̣[ποσταλεὶς ἐξ Αἰγύ-]
πτου Ῥο[δίοισι, τὴν τοῦ τείχεος φυ-]
λακὴν ἔχω̣[ν, Δημητρίῳ ἐπὶ]
χρήμασιν̣ [ὡ]μ̣ο̣λ̣ό̣γ̣ησ̣ε̣ν̣ [παρεσ-]
ἀντίπαλον
ἄξειν στρατὸν ἐς τὸ ἄστυ· ὅ̣ρ̣κ̣ι̣α̣ δ̣ὲ̣ ἐ-
πὶ τοῖσ`ι´δε δούς τε καὶ λα[βών, εὐθὺς]
δὲ σημήνας τοσ`ι´σι Ῥο̣δ̣[ίων ἄρχου-]
παλιμπροδοσίην ἐργά̣[ζεται
σ̣ι̣ν, ν̣υ̣κτὸς τῆς ἐπιγεινο[μένης]
γὰρ
ἀγαγὼν - - - - - - - - - - - -[ ]
ἐ̣σ̣⟦ήγαγε παρ̣ [approx. 9 letters]⟧ Ἀλέξαν-]
δρον ἄνδρα Μακεδόνα [τῶν ἀμ-]
φὶ Δημήτριον̣ ἐόντω̣ν̣ [τὰ ἔν-]
δον πρὸ τοῦ πεμφθῆν̣α̣[ι μετὰ]
στρατοῦ κατοψόμενον Ῥο̣[δίοισί]
μιν ἐνεχείρισεν. Ῥόδιο[ι δὲ τὸν]
μὲν Ἀθηναγόρην̣ [ἐστεφάνωσαν]
χ̣[ρυσ]έ̣ῳ τε σ̣τ̣ε̣φ[άνῳ καὶ ἀργυ-]
ρίου ταλάντοισι πέντ[ε, ἅμα τοισί-]
δε τοὺς ἄλλους ξέ[νους πρὸς]
τὰ ὁμοῖα ἐκκαλεόμε[νοι· τὸν δὲ]
Ἀλέξανδρον κτείνειν.[........ἀ-]
πετράποντο κήρυκος̣ [παρὰ Δημη-]
κατὰ λύτρω[σιν]
τρίου εἴσω ἀπικομέν[ου]

Supplements are from the first edition (Gaertringen 1918).

6. *Superscript*: [στελλομένο]υς (Wilken's suggestion) seems too long for the space, but more natural in the revised context. 7. Possibly λυομε̣-|[νῳ̣]. 19–20. ⟦κείν⟧οι-|[σι Μακεδ]ο̣νων Reiter (personal communication). 21. ὕπονομοι in the papyrus. 30. Above the [ω], itself not legible, a horizontal line. 32. The second ι in τοισίδε is written above the δ. 45. μ̣[έλλοντες] Gaertringen; π̣[ειρώμενοι] Reiter (personal communication), without confidence.

Translation

Since ~~for them~~ they were appropriate for a king, it was not possible for them to use them. ~~Amy~~ And they did those things as I have said. But Amyntas, sailing towards the

eleven catapult-engineers brought
islands, seized ~~many machines~~ which were [~~brought~~ (?)] for Demetrius for the
and many makers which
siege of the cities, ~~and also~~ [ ] of weapons, ~~which~~ the Rhodians did not give back to Demetrius, although he was disposed to pay the established ransom, saying that they needed them. But deprived of these things, he said that he would not release to them anything he happened to capture. And after these things, when Demetrius was digging under the Rhodians' walls, and when those who were working under the earth were already inside the walls, the Rhodians, after deserters
g
had disclosed what was happenin and showed the place where they were digging, dug through to the same place and met with the Macedonian miners. And they
in the same place
almost clashed with each other under the ground.

when setting up
Then each side was ~~setting up~~ guards of their own work
and when the guards came to speak with each other at some distance from the advanced fortifications,

Demetrius was gravely deceived by a plan for a betrayal, which he thought was safe. For a certain Milesian, Athenagoras, who was sent from Egypt to the Rhodians and had watch over the wall, was bribed by Demetrius to lead the
right up
army to the town. Having sworn to do this, but immediately indicating the plan
he devised a counter-betrayal. leading
to the leaders of the Rhodians, the next night ~~he led~~ in [ ]
for
Alexander, a Macedonian from the inside circle of Demetrius, in order to see what was inside the city before having the army sent, he handed him over to the Rhodians. And the Rhodians crowned Athenagoras with a golden crown and gave him five talents of silver, summoning the other foreigners to engage in the same project of defense with them. But [intending (?)] to kill Alexander, they desisted
to talk about ransom
when a herald came from Demetrius

Comments

The author began this fragmentary text as follows:

[ ]β̣αcιλικοῖcι γὰρ cφ̣ι̣ν̣ —

Since for a king for them – (1)

He immediately stopped, however, and expunged cφιν by writing a line above it before going on to complete the sentence. The result was a new version:

> [ ] β̣ακιλικοῖcι γάρ ἐοῦcι χρέ[εσθαι οὐ] c̣φ̣ίcιν ἦν.

> Since they were appropriate for a king, it was not possible for them to use them. (1–2)

I have not reproduced the lineation in the second version because the revision does not have a textual existence: it is an inference from the bi-versional text. Version 1 is an inference as well, of course. The written text itself comes complete with variants written above and below lines and embodies multiple crossings of the text. A reader – either the editor or an author seeking to make a new version on a new sheet of papyrus – needs to read it less as a "text" than as an assemblage of possibilities, each one indicated only in part, according to the often extremely parsimonious logic of the annotation.

Only three words into the next sentence, our author stopped again.

> Ἀμυ —

> Amy – (2)

After striking through these letters, he continued:

> κ[α]ὶ τὰ μὲν οὕ[τωc ἔπρη]ξ̣α̣ν. Ἀμύντηc δὲ πλέων . . .

> And they did those things as I have said. But Amyntas, sailing . . . (2–3)

It is clear from the continuation why he crossed out Ἀμυ: that is the beginning of Ἀμύντηc, the name of the Ptolemaic general who assisted the Rhodians against Demetrius. Our author was about to begin a new episode, in which Amyntas captured some of Demetrius' military equipment (or personnel: more on this below). But his first transition was too abrupt, so he decided to close the previous episode with κ[α]ὶ τὰ μὲν οὕ[τωc ἔπρη]ξαν ("And they did those things as I have said").

The replacements for the deleted words cφιν and Ἀμύ[ντης] are written in the line rather than above it, a fact that indicates that they were deleted immediately after they were written down and before the author had written anything else. This is an example of revision taking place long before the draft was finished.

The sentence that began κ[α]ὶ τὰ μὲν οὕ[τωϲ ἔπρη]ξαν was long (2–10). The first version was written (apparently without interruption) as follows:

κ[α]ὶ τὰ μὲν οὕ-
[τωϲ ἔπρη]ξ̣α̣ν. Ἀμύντηϲ δὲ πλέων ἐπὶ
[νήϲω]ν̣ ὄργανα πρὸϲ τὴν πολιορκ-
[ίην τὴν τ]ῶ̣ν̣ π̣ολίων Δημητρίῳ φε-
[ρόμεν]α̣ πολλὰ αἵρεε καὶ βελέων δη-
[μιουργ]ο̣ὺ̣ϲ̣ τάπερ Δημητρίῳ λυομ-
[ένῳ λύτ]ρ̣ων τῶν τεταγμένων Ῥ̣[ό]-
[διοι οὐκ ἀ]π̣ε̣δ̣[̣ίδοϲαν, χρεΐζειν αὐτῶν
[φάϲκοντεϲ·]

And they did those things as I have said. But Amyntas, sailing towards the islands, seized many machines which were [brought (?)] for Demetrius for the siege of the cities, and also [makers] of weapons, which the Rhodians did not give back to Demetrius, although he was disposed to pay the established ransom, saying that they needed them.

Sometime later, however, our author made a series of alterations. The author now believed that Amyntas captured not siege machines (ὄργανα) but the men who ran or made them (καταπελταφέταϲ), and he believed that he knew their number. This change led to others. "Many" (πολλά) was now redundant (since the number eleven had been provided) and was therefore deleted. The neuter plural φερομένα needed to be changed to the masculine, which was done above the line. The deletion of "and" (καί) and the insertion of "and many" (πολλούϲ τε) in the same line can perhaps be explained by the fact that he was now writing about two groups of people rather than some machinery and some people, and so the strong conjunction καί was replaced with a weaker τε, and the absence of a number of missile forgers to balance "eleven" (ἕνδεκα) prompted the less specific "many" (πολλούϲ) – something was needed, in any case, to come before τε. Two masculine plural nouns now functioned as the antecedent of τά, which was accordingly changed to τούϲ; this was done by erasing only the vowel and adding the new ending above the line. This series of revisions was prompted by the author's discovery of new information: he thought that Amyntas had taken siege-engines but later found a precise number of engineers among the captured. This discovery was probably made after he wrote the sentence, and perhaps even after he wrote the whole draft.

The version he envisaged at this point can be reconstructed, though it is nowhere written out in full (I have underlined the revisions):

> κ[α]ὶ τὰ μὲν οὕ[τωϲ ἔπρη]ξ̣α̣ν. Ἀμύντηϲ δὲ πλέων ἐπὶ [νήϲω]ν̣ κ̣αταπελ̣τ̣αφέτα̣ϲ̣ ἕνδεκα πρὸϲ τὴν πολιορκ[ίην τὴν τ]ῶ̣ν̣ π̣ολίων Δημητρίῳ̣ φε[ρόμενο]υϲ αἵρεε πολλούϲ τε βελέων δη[μιουργ]ο̣ὺ̣ϲ̣ τούϲπερ Δημητρίῳ̣ λυομ[ένῳ̣ λύτ]ρων τῶν τεταγμένων Ῥ̣[όδιοι οὐκ ἀ]π̣ε̣δ̣ί̣δοϲαν, χρεΐζειν αὐτῶν [φάϲκοντεϲ·]
>
> And they did those things as I have said. But Amyntas, sailing toward the islands, seized eleven catapult-engineers which were [brought] for Demetrius for the siege of the cities, and many [makers] of weapons, which the Rhodians did not give back to Demetrius, although he was disposed to pay the established ransom, saying that they needed them.

A sequence of smaller corrections can be found in lines 18–21. "Happenin" ([γεινόμ]ε̣ναν, 18) was an error, so he made a small correction over the line (ο above α), implying the correct γεινόμενον ("happening" – the error in the Greek is not of spelling but of gender). Καί (18) he expunged, as he had ϲφιν in the first line of the column, by means of a line above it, then continued. The κείνοι in 19 is unnecessary, and the author wanted a dative plural article to agree with μεταλλωρύχοιϲιν (20). He accordingly crossed out κειν and wrote τ above it, presumably adding -ϲι above the next line (19–20).

In the next section of text, our author either lost track of his line of thought or changed his mind in the middle of the sentence. At first he wrote the following:

> οἵ τε ὑπόνομοι πα-
> [ρὰ μικρ]ὸ̣ν ϲυνέμιξαν ἀλλήλοιϲι(ν).
> [ἔργου δ' ἑ]κάτεροι τοῦ ϲφετέρου φύλα-
> [καϲ ἐπι]ϲτήϲαντεϲ, προδοϲίη
> τιϲ Δημητρίῳ̣ δόξαϲα [ἀϲφαλὴϲ κάρτα]
> μιν ἀπάτηϲεν.
>
> And they almost clashed with each other under the ground. Then each side setting up guards of their own work, Demetrius was gravely deceived by a plan for a betrayal, which he thought was safe. (21–26)

The nominative phrase "then each side setting up guards of their own work" (23–24) is syntactically disconnected with the rest of the sentence. At first, it might have been intended as the subject of a plural verb. But, as he wrote, προδοϲίη intruded as a new subject and left the participial phrase stranded. An extensive set of revisions solved the problem and expanded the sentence considerably. After adding a specification above "clashed" (συνέμιξαν) ("in the same place," ἐϲ τωὐτόν, is written above the line), he changed the nominatives ἑκατέροι . . . [ἐπι]στήϲαντεϲ into genitives and then added a whole new genitive phrase above the line and in the space at the bottom of column 1. The new version, inferable from the messy papyrus remains, reads as follows:

> οἵ τε ὑπόνομοι πα[ρὰ μικρ]ὸ̣ν ἐϲ τωὐτόν συνέμιξαν ἀλλήλοιϲι(ν). [ἔργου δ᾽ ἑ]κάτερων τοῦ ϲφετέρου φύλα[καϲ ἐπιστησά]ντων ἀλλήλοιϲί τε τῶν φυ̣λάκων ἐϲ λόγουϲ ἀπὸ τῶν προτε̣ι̣χ̣ι̣ϲ̣μ̣ά̣τ(ων) ἀπικνεομένων, προδοϲίη τιϲ Δημητρίῳ̣ δόξαϲα [ἀϲφαλὴϲ κάρτα] μιν ἀπάτηϲεν.
>
> And they almost clashed with each other under the ground. Then, when each side was setting up guards of their own work, and when the guards came to speak with each other at some distance from the advanced fortifications, Demetrius was gravely deceived by a plan for a betrayal, which he thought was safe.

There are several more revisions in the second column of this papyrus; the reader may consult the appended text. Only three of the revisions in this document are clearly corrections of errors; the rest project new versions which the author may or may not have ultimately preferred. It is in fact likely that he would have done so, since he almost always crossed out the original text when he made a change. But the erased text remains legible, and nothing prevented him from reverting to an older version or even creating a wholly new sentence in a subsequent copy (for examples of this, see below). So the micro-teleology implied by the revisions, in which an erasure shifts the previous version into counterfactuality, needs to be balanced against a more general virtuality, one in which we do not know (as the author did not) what would ultimately have been preferred.

II

P. Köln VI 245 (MP³ 1965.41)

Third century CE; *fragment of a draft of tragedy* (?), *from upper Egypt.*

First edition: P. Köln VI 245. Ed. Maior: Parca 1991.
See also Gianotti 2005.

This papyrus fragment contains the first forty-one lines of one column and the left and right edges of two more, but only the central column is intact enough to yield meaningful text. The text contains iambic trimeters on what seems to be the subject of Odysseus in disguise at Troy. This might not be a continuous draft. It might instead be a series of passages, separated by paragraphoi, which the poet had excerpted from a previous draft or intended to insert into a later one.

Text

Column 1
[*approximately 18 lines*]

Column 2
⟦ἢ πα̣ῖ̣δ̣ε̣ς̣ τ̣ὸ̣ν̣ [ἐ]χθρὸν π̣ο̣λ̣έ̣μ̣ιο̣ν̣ ω̣......ρ̣....[ ]⟧
ἢ παῖδας ἐχθροὺς Π̣ρ̣ι̣α̣μίδ̣ας ⟦λ̣⟧π̣⟦ι̣⟧ολ̣λ̣ο̣ὺ̣ς̣ μ̣ό̣ν̣[ος]
ἐκφεύξομαι; [.]...τ̣α̣ν̣....π̣ε̣ι̣ρ̣ω..ε̣.ι̣
—
ὀδμή τις η̣.[.]..ν π̣ι̣θάν.π̣ν̣ας̣τοθ..λ̣εα.
φωνὴ προσῆλθε προσφιλής
—
ὦ τεῖχος οὐκ ἄδοξον, ἐν θεοῖς μέγα,
στέφος
κρατὸς λόχ̣ευμα, Παλλάς, οὐρανοῦ θάλος,
ὁπλοφόρε
⟦γοργοφόνε⟧ καὶ λοιγωπὲ καὶ γοργοκτόνε,
φένγασπι, δρακοντόσ̣τ̣ηθε, δορατοδ̣έ̣ξιε,
ἀ̣μ̣[ή]τ̣ωρ θεῖε
εὔοπλε κα[ὶ ἀρ]ς̣ε̣ν̣ω̣πέ, κρηπειδόσφυρε,
ἄνυμφε (καὶ γὰ[ρ] ἄ̣νυμφον οἴκησας τόπο̣ν)
εεγλο
ἀπαρθε̣ν̣εύτο̣υ παρθένος λ̣ο̣χεύμ̣ατος
φέρεις, ⟨ς⟩θ̣ενεί̣[ς]τη πάντα, Τειτᾶνος βολάς̣
ἀκτεῖσι κεφα̣[λ]ῆς, εἶτα καὶ μήνης κ̣ύκλ̣α
θείαις παρε̣ι̣[αῖ]ς, κόσμον ὠλέναις διπλ̣α̣ῖς̣.
ς̣ὺ̣ κυρεῖς τ̣ὰ π[άν]τ̣α· διὰ σὲ δ᾽ εἰσορῶ φάος,
βυθῶν̣ π̣λ̣ά̣ν̣ος
σο̣ῦ μ̣οὶ [π]έ̣λα̣[ς] φύγοιμι καὶ θεῶν ὀργήν,
...
δ̣υνά̣μει δὲ τ̣ῇ σ̣ῇ καὶ τὸ νῦν θαρρ{ς}ῶ̣ν περῶ
γραφὰς κο̣μίζ[ω]ν̣ πρὸς λάκαιναν εἰς Φρύγ̣α̣ς̣.
—
ἅπ̣αντα δράσω̣· [κ]α̣θ᾽ ὁδὸν ἀλλάξω τύ̣πους,
κρύψω πέπ̣λ̣[ο]υς δὲ τοὺς ἐμοὺ⟦υ̣⟧ `θ´άμνω̣ τινί.
ς]ὲ̣ γὰρ ἔχω κἀκεῖ πάλιν
θαρρ̣ῶν ἄπε̣[ιμ]ι̣· σ̣ὺ δὲ θεὰ παρίστασο.
—

. . νι̣ δειν̣.[
. . δ̣ε̣ι̣ν̣ὰ καὶ̣[....]π̣ροcπλακέντ̣α̣ [πράγμ]α̣τ̣α̣.[
(δι)ἀ̣τῆ̣ν[
[..]..διπ̣λ̣ῆ̣ν[.....]. φιλονείκοιc̣ π̣ά̣λ̣ι̣ν̣
[..] ἔc̣χ̣α̣τ̣ο̣ν̣ [......].πον ἤ̣λυθεν τέλοc
[...].δο.[......]ε̣ὐ̣γενὴc Τρώων πόλιc
[3.8 cm].δε..τ̣ο̣ῦ̣τ̣ο̣ ἀ̣c̣π̣ά̣c̣ιον
λακ̣ωνί
[ποιο]ῦ̣cιν ἐχ[θρὸ]ν̣ εἵνε̣κε̣ν̣ τῆc̣ γ̣υ̣μ̣ν̣άδοc.
[....]δοc γὰρ εἶ̣c Π̣άρι̣ν̣ θελει.....cφα[γεῖν]
[....].ε τ̣όξοιc το̣ῖc Φ⟨ι⟩λοκτήτο[υ] τότε
[1.7 cm]..[..]νη [γ]υμνάcη Φρύγ[αc] Τ̣ύχ̣η̣.
[τότ' Ἕλ]ε̣νο[c] α̣ὐ̣τ̣ὴ̣ν̣ ἠθέληcε πρὸc γάμουc
ὁ̣ δὲ π̣ρ̣ό̣τλαc cτρατ̣ο̣[ῦ]
[λαβεῖν τε] κ̣α̣ὶ̣ Δ̣η̣ί̣φ̣ο̣β̣ο̣c̣. ἡ δὲ βάρβαροc
]..[.....]..η.c̣ε̣ γυμνάδοc πόcιν
[1.2 cm].ω̣ν̣ [δ' ἡ]θέληcε ⟦ἔνγαμον λαβεῖν⟧
[τὰ]νδρὸc πα[τὴρ] τ̣ότε τοῦτον ᾔνεcεν γάμον.
[ὀρ]γῇ δὲ λημ[φ]θ̣ε̣ὶc Ἕλενοc ὡc ὑβριcμένοc
[αὐ]τόμολοc εἰ̣[c Ἕ]λ̣ληναc ἐκπορεύετ⟨αι⟩
[cύ]μμαχοc ἐπ̣[ελθ]ώ̣ν· τίc Φρυγῶν ἐλπὶc ἦν [
[ἔπ]ειθε μήτε[.....] ε̣ὐ̣γενοῦ[c τ]α̣ύ̣της παρ[
[.....]μ..δ...[......]..μήτε.[.].ε̣ι̣ν Ἑλλά̣[δ]ο̣[c]
]..[
].[

Column 3
[Approximately 18 lines]

Supplements are gathered from Parca 1991 and the first edition.
Column 2
2. ⟦λ⟧π⟦ι⟧ο: c̣φαλω̣; the c written over β, Maresch. φαλ Parca (1991), Maresch. **17** Superscript: πλάνουc Fournet (personal communication). **21**. ἐμοὺ⟦c⟧ `θ'άμνῳ Van Minnen (personal communication). [πράγμ]ατα West. **23–25**. Parca reads:

..δ̣ε̣ι̣.ακαι.[..] προcπλάκεν̣τ̣α̣[
..[.].δι' ἦ̣ν̣...[............]ατα
[..] ἔσχατ̣ο̣ν̣ [....].φιλονείκοιc π̣άλιν
[...]α̣δο̣c̣ α̣[......]ι̣πον ἤλυθεν τέλοc

etc. The new restoration, however, has moved]ατα up a line, so that it completes verse 23; this moves the ends of the next four lines up a verse. 34. [τοῦτ]ο̣ν Parca (1991). 34. *Superscript*:].[.]...ἥ̣ρπ̣άc̣ε Parca (1991).

Translation

~~Or will the sons [of Priam allow to escape?] their hated enemy?~~ Or will I, a solitary man, ~~le~~ escape my foes, the many sons of Priam? [ ]

A certain fragrance [ ]
A friendly voice approaches

---

crown
Oh famous rampart, great among the gods, offspring of the head, Pallas, child of
arms-bearer
heaven, gorgon-killer and baleful eyed and gorgon slayer, with a brilliant shield, with serpents on your chest, with a spear in your right hand, well-armed and
motherless goddess
manly looking, with boots on your feet, unmarried (for indeed you live in a place unsuited to brides) maiden from a virgin birth; you bear, most powerful in all things, the beams of the Titan in the rays of your head and the disks of the moon in your divine cheeks and the world in your two hands. You rule over all. Because
, a wanderer of the depths
of you I behold the light, with you beside me I would escape even the wrath of the gods. Taking courage from your power even now I endure, ferrying letters to the Laconian woman among the Phrygians.

I will do everything: on the way I will alter my visage, I will hide my~~y~~ robes in
b for I have you also there again.
some ush. Taking heart, I go: for you, goddess, stand beside me.

---

because of

[ ] embroiling [ ] double [ ] again because of rivalry [ ]

final end has come [ ] noble city of the Trojans [ ] this is a welcomed thing [

Laconian

] hateful because of the athletic woman [ ] alone wants to kill Paris [

] the bow of Philoctetes [ ] a dreadful fate might harass the Phrygians [

] Then Helenus wanted to take her to wed, and also Deiphobus.
And the champion of the army [ ] the husband of the athletic woman.
And the barbarian woman [ ] wanted ~~to take to wed~~ [Deiphobus].
And the father of the man agreed to the wedding.
And Helenus, seized by anger as though he had been treated outrageously, set out to the Greeks as a deserter, and came to them as an ally. What hope was there for the Phrygians? [ ] of the noble [ ] of Greece.

Comments

This papyrus, from the third or fourth century CE, contains what appears to be a fragment of tragedy; its most recent editor identified the subject as Odysseus in disguise at Troy. Traces of three columns remain, but only the second, which contains forty-one iambic trimeters, is relatively intact.

Some of the poet's revisions correct what could be described as objective errors. So, for example, at line 21 he originally wrote "I shall conceal myy robes in a ush" (κρύψω πέπ̣λ̣[ο]υϲ δὲ τοὺϲ ἐμοὺυ̣ άμνω̣ τινι). "Myy" (ἐμουυ̣) should have been "my" (ἐμοῦ), and "ush" (άμνω̣) should have been "bush" (θάμνω̣), so the poet tried to fix them, crossing out the doubled υ and writing θ above the line. But corrections of errors like this are rare in the papyrus.

Equally rare are revisions accompanied by erasures. In fact, the poet does this only a few times in the surviving column. For example, the first two verses of the intact second column contain erasures and corrections that suggest that this was a first draft.

> ⟦ἢ πα̣ῖ̣δ̣ε̣ϲ̣ τ̣ὸ̣ν̣ [ἐ]χθρὸν π̣ο̣λ̣έ̣μ̣ιο̣ν̣ ω̣……ρ̣….[ ]⟧
> ἢ παῖδαϲ ἐχ̣θροὺϲ Πρ̣ι̣α̣μίδ̣αϲ ⟦λ̣⟧π̣⟦ι⟧ολ̣λ̣ο̣ὺ̣ϲ̣ μό̣ν̣[οϲ]
> ἐκφεύξομαι;

> ~~Or will the sons [of Priam allow to escape?] their hated enemy?~~ Or will I, a solitary man, ~~le~~ escape my foes, the many sons of Priam? (1–2)

The end of the first verse is not legible, but it seems to express the same idea as verse two, with the syntax altered. The poet initially meant to write something like "or will the sons of Priam capture their hated enemy?" – but then decided to make the speaker the subject of the sentence and the sons of Priam the object. He therefore crossed out the first verse, probably immediately after writing it, and put the second and third in its place. The next version was not finished. After writing "the sons of Priam" (Πρ̣ι̣α̣μίδ̣αϲ) he wrote λ̣ι̣-, possibly meaning to write λίπων ("leaving"), then changed his mind. He modified the λ to make an unusually written π, and wrote ο directly over the ι, before continuing with λ̣λ̣ο̣ὺ̣ϲ̣ μό̣ν̣[οϲ | ἐκφεύξομαι. These changes, which occur in the line, suggest an author composing – and altering his intentions – as he wrote.[8]

[8] The same impression is reinforced by the metrically incomplete line 5.

A third erasure occurs in line 8, where the poet initially wrote, redundantly, "Gorgon killer and baneful glancer and Gorgon slayer" (γοργοφόνε καὶ λοιγῶπε καὶ γοργοκτόνε). To fix this the poet crossed out γοργοφόνε and wrote ὁπλοφόρε ("arms bearer") above it. Redundancy is not an objective error like ἔμουυ, but the poet's momentary preference for the revision is indicated by his having struck out γοργοφόνε.

Revisions that are added after the original version is crossed out foster the impression that a text develops in a linear fashion, with later revisions replacing earlier ones. So too revisions that correct objective errors. But revisions in this papyrus do not usually correct errors or accompany erasures, and most of the time it is unsafe to presume that the new variant would eventually replace the original reading. So, for example, line 7 was initially written κρατὸϲ λόχευμα, Παλλάϲ, οὐρανοῦ θάλοϲ ("offspring of the head, Pallas, child of heaven"), but the author later wrote στέφοϲ ("crown") above θάλοϲ. The revision is not objectively preferable to the original, and the poet did not erase θάλοϲ – an indication that he had not decided between the two versions. Likewise in line 10, εὔοπλε κα[ὶ ἀρ]ϲ̣ε̣ν̣ωπέ, κρηπειδόϲφυρε ("well-armed and male-faced, boot-wearer") has ἀ̣μ̣[ή]τ̣ωρ θεῖε ("motherless goddess") written above καὶ ἀρ]ϲ̣ε̣ν̣ωπέ, which seems to imagine a new version εὔοπλε ἀμ̣[ή]τ̣ωρ θεῖε, κρηπειδόϲφορε ("well-armed motherless goddess, boot-wearer") but does not prefer it, as the absence of erasure indicates.

These are small, lexical alterations. Lines 33–37 saw a significant overhaul. This passage was initially composed as follows:

ἡ δὲ βάρβαροϲ
[1.2 cm].ω̣ν̣ [δ' ἠ]θέληϲε ἔνγαμον λαβεῖν.
[ὀρ]γῇ δὲ λημ[φ]θ̣ε̣ὶϲ Ἕλενοϲ ὡϲ ὑβριϲμένοϲ
[αὐ]τόμολοϲ εἰ̣[ϲ Ἕ]λ̣ληναϲ ἐκπορεύετ⟨αι⟩
[ϲύ]μμαχοϲ ἐπ̣[ελθ]ὼ̣ν·

> And the barbarian woman [ ] wanted to take to wed [Deiphobus]. And Helenus, seized by anger as though he had been treated outrageously, set out to the Greeks as a deserter, and came to them as an ally.

But the poet disliked ἔνγαμον λαβεῖν (34), and cancelled it. He replaced it with γυμνάδος πόϲιν ("the husband of the athletic woman"). The resulting sentence, which I cannot reconstruct but was certainly awkward, was then supplemented with a new (hard-to-decipher) version: ὁ̣ δὲ π̣ρ̣ό̣τλαϲ ϲτρατ̣ο̣[ῦ]/[ ]..[.....]..η.ϲ̣ε̣ ("and the champion of the army [. . .]"). Here the erasure indicates that the poet wanted to replace ἔνγαμον λαβεῖν; but the revision turned out to be much greater than he initially envisaged, and the new variants do not definitively replace the base text. This is indicated by the fact that he did not extend his erasure to include the whole sentence.

The revision here represents a significant change in sense. This section of text tells of the events leading to Helenus' defection to the Greek camp, a defection that leads to the theft of the Palladium and then the construction of the Trojan horse. Helenus defected because after the death of Paris he had sued for Helen's hand and lost it to Deiphobus. Lines 33–34 (in whichever version) tell of Helen's marriage to Deiphobus; lines 35–37 tell of Helenus' defection to the Achaeans in response. In the first version Helen is the subject of the verb and actively chooses Deiphobus. In the second revision Deiphobus is the most likely referent of ọ δὲ π̣ρ̣ọ́τλας ϲτρατ̣ọ[ῦ]. If the (lost) verb of the second version was ἥρπ̣άϲ̣ε (so Parca 1991 ad loc.), then Deiphobus did not *win* Helen in this revision; he seized her. This undoes the motivation for Helenus' defection: the Trojan prince had means of redress if Deiphobus had only stolen Helen away. Accordingly, the poet added a new line above 35, in which Priam consents to the marriage of Deiphobus and Helen ([τἀ]νδρὸϲ πα[τὴρ] τ̣ότ̣ε τοῦτον ἥ̣νεϲεν γάμον, "And the father of the man agreed to the wedding"). The affair concluded – and Helenus spurned – the defection is again motivated. Here we are dealing not only with different possible *texts*, but with different possible *stories*, a virtual field which extends to the level of narrative semantics.

III

PSI I 17 (MP³ 1608)

Fourth/fifth century CE (so Manfredi, PSI XV p. 42; Page 1941, Lodi 1912 proposed third century)

Seven Epigrams on the Death of Euprepios

First edition: Lodi 1912 (source of most of the supplements below). See also Calderini 1915 (copious commentary), Page 1941 no. 117.

This papyrus contains a series of epigrams on the theme of the death of a certain Euprepios. The poems are written without regard to metrical line end.[9] Each epigram is separated from the next by a strong and clearly visible paragraphos. Five of the epigrams are written on the back of the papyrus, and two are on the front. The front of the papyrus is a palimpsest.

9 For further examples of poems written without reference to meter, see Cribiore 1992: 259 n. 29; Fournet 2009.

There are traces of at least twelve lines, running in the opposite direction of the writing that now survives, of which I can only decipher A1 (Calderini 1915 saw much more).

Text

A. Front: Erased Underlayer

A1

ἡ θ⸤υγάτηρ δ’ἀνέθηκε τάδε⸥ | θ⸤ρ⸥ε̣π⸤τήρια δοῦσα⸥
    ⸤καὶ φθιμέ|νῳ̣·⸥ πᾶσαν τὴν χάριν ὧ⸤δ’⸥ ἀπέ|χω·
⸤Εὐπρέπιος μὲν ἐγών·⸥ | ὁ δὲ ⸤νήπιός ἐ⸥στι θυγατρός,
    Ἡράκλειος [ὁ παῖς]· Εὐπρ[επίου] | δὲ πα[

B. Back

B1

Ἀγ]γέ̣λλει τὸ σχῆμα καὶ ἴ[νδαλμ’] | [ο]ὐ̣ βραχὺν ἄνδρα·
    τοῦτο γ[....] |.ου τ̣ρ[.]ι ·δρυ[....] ἡ θυγά̣[τηρ] |
ἀ̣λλὰ̣ διαρρ̣ή̣δην [ἐ]πισημ[ότατον] |κ̣αὶ ἄριστον
    ὄλβῳ̣ καὶ πλ̣ο̣ύ[τῳ̣], | τοὔνομα δ̣’ Εὐπρέπιον. |

B2

ἐνθάδε μὲν κεῖ̣ται τῆς εἰκό̣[νος] | ἡ γ̣ραφ̣ὴ̣ αὕτη`:´

              ῳ̣
        Εὐπρεπίου· ψ̣υ̣χ̣ὴ̣ ἐν̣ | μ̣α̣κ̣ά̣ρων ἀγοραῖς·

οὐ̣ τι....
οὐ γάρ πω | τοιοῦτο̣ς̣ α̣ν̣ήλυθε̣ν̣ ε̣ἰς Ἀχ̣έ̣ροντα· |
        τῶν ὁς̣ίων ἀνδρῶν Ἠλύσιον τὸ τέλο̣ς̣, |
ἔνθα διατρίβειν ἔλαχεν πάλαι ἔκ̣ τι|νο̣ς̣ ἐ̣σθλῆς

                                        εται
    μοίρη̣ς· οὐ δὲ θανεῖν | τοὺς ἀγαθοὺς λέγομεν. |

B3

τόνδ’ ἐσορᾷς ὦ ξεῖνε τὸν ὄλβιον | ἀνέρα κεῖνον
    τὸν σ̣οφὸν Εὐπρ[έπ]ιον κ̣α̣ὶ̣ βασιλε̣ῦ̣σι φίλο̣ν̣
    πάντων ἁψάμενον | γεράων·
ἡ θυγάτηρ δ’ ἀνέθηκε τάδε̣ | θρεπτήρια δοῦσα
                    π⸤ᾶσ⸥α̣ν̣ τ̣ὴ̣ν χάριν ὧδ’ ἀ̣πέχω.
    καὶ φθιμένῳ̣· | χάριτος δ’ οὐδὲν ἔλειψεν ἐμοί̣. |

B4

εἰ κ̣αὶ τὴν φωνὴ[ν] ὁ ζωγράφος | ὧδ’ ἐνέ[θ]ηκεν,
    εἶπες ἂν ὡ̣ς̣ ἤδη | φθέγγεται Εὐπρέπιος·
εἰ γάρ τις παρι|ώ̣ν τῆς ε̣ἰ̣κ̣όνο̣ς̣ ἐγγύθεν ἔλθοι, |
                    περ
    οὔατα πα̣ρ̣θήσει ὡς καὶ ἀκουσόμενο̣ς̣. |

B5

δ’ ἔκ̣γο̣ν̣ο̣ϲ̣
⸤Εὐ⸥πρ̣έπιοϲ μὲν ἐγών· ὁ δ̣[ὲ] νήπιόϲ ἐϲτι̣ | θ̣υ̣γα⸤τρόϲ⸥
ο̣ι̣[..]ω̣ϲ̣
τ̣ρ̣ι̣ [approx. 8 letters] χ[.]ι̣ρω[
......[14 letters]ε̣υπρ̣ε[

C. Front: Overwriting
C1
Οὐ] γὰρ ἐν ἀνθρώ̣π̣ο̣ι̣ϲιν ἐὼν | [ἐ]βάδιζεν ἐκείνη̣ν̣
η
τὴ̣ν ὁ̣δὸ̣ν |ἣ̣ν ἀρετῆϲ οὐκ ἐκάθ[ε]ρε θ̣έ̣μ̣ι̣ϲ̣|
τοὔνεκα ῥηϊδίωϲ
ἔνθεν ἐ̣ϲ ἀθανά̣τουϲ καὶ̣ ἀε̣ί̣ζωον̣ | βίον ἦλθεν·
ἄϲμενοϲ οὐλομένη̣[ν] | ὠ̣ϲ̣ά̣μενοϲ γενεήν
τοῦτο τὸ μοχθηρὸν̣ | ϲῶμ̣’ ἀνα̣δ̣υ̣ϲάμενο̣ϲ̣.
φῦλον ἀπευ̣ξ̣άμενοϲ

Translation

A. Front: Underwriting
A1
My daughter set this up, giving payment for her upbringing even to the dead: thus I have received all gratitude in full. I am Euprepios and this is the child of my daughter, the boy Herakleios, of Euprepios . . .

B. Back
B1
The shape and form announce no small man; [. . .] daughter, but explicitly the brightest and the best in happiness and wealth. And his name is Euprepios.

B2

his likeness; Euprepios’
In this place is set a portrait of Euprepios; his soul is in the agoras of
Not [. . .]
the blessed. For never did such a man go to Acheron: Elysium is the destination of holy. He got it as his lot to live there long ago from some happy destiny, nor
is it said
do we say that good men die.

B3

Euprepios, wise and a friend to kings.
Here, Stranger, you look upon that blessed man who achieved every honor. My daughter set this up, repaying her debt for her upbringing, even though he is dead:
Thus I have received all gratitude in full.
No gratitude was lacking to me.

B4
If the painter had also put a voice in here, you would say that Euprepios was speaking now. For if a passer-by should come near the image, he will give ear
as though
in order to hear.

B5

and this is the offspring
I am Euprepios and the child of my daughter is . . .

C. Front: Overwriting
C1
When he was among mortals he did not walk the road which Themis had not
Wherefore he easily went to eternal life, glad to be pushing away this destructive race.
purified of virtue. From there he went to the immortals and eternal life, shrinking from this toilsome body
glad to be pushing away this toilsome race.

Comments

This text contains only one clear correction of an error (metrical): the other revisions explore new possibilities of expression. Throughout this papyrus, the poet adds alternatives without canceling the words he had originally written, further support to my contention that revisions create but do not prefer new versions. In a subsequent copy, the poet would have had to choose which variants to retain: we will see that he did not always prefer new versions to old.

Five epigrams (B1–B5) are written on the back of the papyrus; a sixth (C1) is written on the front. Still just discernible on the front of the papyrus are the traces of at least fourteen erased lines running in the opposite direction to epigram C1, of which one epigram (A1) is still legible. It appears that the process of writing was as follows: after writing out several epigrams on the front (A), the poet turned his papyrus over and wrote epigrams B1–B5 on the back. When he had filled the space on the back, he turned the papyrus over again and wrote C1 on the front, from which he had previously erased his earlier text.

In epigram C1, written on the newly erased front of the papyrus, the second couplet was originally as follows:

ἔνθεν ἐ̣c ἀθανά̣τουc καὶ̣ ἀε̣ί̣ζωον̣ | βίον ἦλθεν·
τοῦτο τὸ μοχθηρὸν̣ | cῶ̣μ̣ʼ ἀνα̣δ̣υ̣cάμενο̣c̣.

From there he went to the immortals and eternal life,
shrinking from this toilsome body. (3–4)

Then, two revisions. Beneath and slightly to the left of cῶμ' ἀναδυcάμενοc ("toilsome body") we find φῦλον ἀπευξάμενοc ("despising this race"), while above the last two lines the author wrote ἄcμενοc οὐλομένη[ν] | ὠcάμενοc γενεήν ("glad to be pushing away this destructive race"). This epigram is alone on the back of the papyrus, and the poet had ample space to make changes below the line. It would be natural to use this ample space first, so φῦλον ἀπευξάμενοc was probably the first revision. It envisaged a new final couplet:

ἔνθεν ἐc ἀθανάτουc καὶ ἀείζωον βίον ἦλθεν·
τοῦτο τὸ μοχθηρὸν φῦλον ἀπευξάμενοc.

From there he went to the immortals and eternal life, despising this toilsome race.

Unhappy with this, our poet came up with a whole new alternative which was placed directly above the original because the first revision had already taken up the space below. This envisaged yet a third version of the couplet:

ἔνθεν ἐc ἀθανάτουc καὶ ἀείζωον βίον ἦλθεν·
ἄcμενοc οὐλομένη[ν] | ὠcάμενοc γενεήν.

From there he went to the immortals and eternal life, glad to be pushing away this destructive race.

At some point, the first colon of line 3, ἔνθεν ἐc ἀθανάτουc, was given a superlinear variant τοὔνεκα ῥηϊδίωc. This could have been done at any point, and in my uncertainty I have left it out of the versions above. Editorial hygiene, however, would dictate that for each of the versions given above there should be a second possible version including the change to line 3. This is an excellent example of the role of probability not only in the revision process but also in editorial procedure: even my reconstructions of the genetic process are only probable.[10]

No erasures are present in the revisions to C1, and I do not think that we can safely guess which of these three versions the poet would ultimately have preferred. We will see shortly that in at least one other case he reverted to an older variant after trying something new.

B3.1–2 are almost certainly an example of initial composition. For reasons that will become clear shortly I give the lineation of the papyrus, which does not reflect metrical divisions, on the left:

[10] See Gurd 2005.

τόνδ᾽ ἐcορᾷc ὦ ξεῖνε τὸν ὄλβιον ἀνέρα κεῖνον πάντων ἁψάμενον γεράων·	τόνδ᾽ ἐcορᾷc ὦ ξεῖνε τὸν ὄλβιον | ἀνέρα κεῖνον πάντων ἁψάμενον | γεράων·

Here, stranger, you look upon that blessed man who achieved every honor.

There is a metrical error in the second verse: πάντων ἁψάμενον | γεράων ("who achieved all gifts of honor") is not a complete pentameter, and while πάντων ἁψάμενον will fit the meter, γεράων does not. Γεράων would fit the meter if this were a dactylic hexameter, however. I suspect that the poet's error here was simply to begin a hexameter when he needed a pentameter.

The moment he realized his mistake, he composed a new verse to replace it: τὸν̣ cοφὸν Εὐπρ[έπ]ιον κ̣α̣ὶ̣ βαcιλ̣ε̣ῦcι φίλ̣ο̣ν̣ ("Euprepios, wise and a friend to kings"). The papyrus now looked as follows (again, the diplomatic transcription is on the left):

τόνδ᾽ ἐcορᾷc ὦ ξεῖνε τὸν ὄλβιον τὸν c̣οφὸν Εὐπρ[έπ]ιον κ̣α̣ὶ̣ βαcιλ̣ε̣ῦcι φίλ̣ο̣ν̣ ἀνέρα κεῖνον πάντων ἁψάμενον γεράων·	τόνδ᾽ ἐcορᾷc ὦ ξεῖνε τὸν ὄλβιον | ἀνέρα κεῖνον τὸν c̣οφὸν Εὐπρ[έπ]ιον κ̣α̣ὶ̣ βαcιλ̣ε̣ῦcι φίλ̣ο̣ν̣ πάντων ἁψάμενον | γεράων·

We might expect the poet to have continued to the end of the epigram before returning to correct the metrical deficiency. But elsewhere corrections are placed directly over what they are meant to replace; here, on the other hand, there is no attempt to line up original with revision: the complete new pentameter is written above the line which contains the last two words of the first hexameter. The revision's location suggests that it was made immediately after the poet reached γεράων; the beginning of verse 3 had not been written yet, and rather than trying to write a superscript correction over empty space – here, unlike in the case of C1, he anticipated writing more – the poet wrote the new metrically correct verse over the previous line and then continued with the second couplet. But the revision represented a wholesale change in sense: the original thought was abandoned and a new idea – Euprepios' wisdom and high social position – was imported, in part from epigram B1. The location of the correction, its substantiality, and the metrical faultiness in the base text suggest that the author was not copying and correcting from an earlier text but composing a first draft.

The second couplet in the same epigram (B3.3–4), on the other hand, *is* the result of copying. Of the fourteen lines erased on the front of the papyrus, the substance of one epigram can still be discerned (A1). A1.1–2 were recycled as the second half of B3 (2–3). (A1.2–3 were recycled as the beginning of epigram B5. Between the end of B3 and B5 the poet inserted a new poem, epigram B4.) But B3.3–4 was not copied *exactly* from A1.1–2. On the contrary, he tried something new. A1.1–2 were as follows:

> ἡ θ⸤υγάτηρ δ᾽ἀνέθηκε τάδε⸥ | θ⸤ρ⸥ẹπ⸤τήρια δοῦςα⸥
> ⸤καὶ φθιμέ|νῳ̣·⸥ πᾶςαν τὴν χάριν ὧ⸤δ᾽⸥ ἀπέ|χω·
>
> My daughter set this up, giving payment for her upbringing
> even to the dead: Thus I have received all gratitude in full.

B3.3–4, as they were originally written down, were:

> ἡ θυγάτηρ δ᾽ ἀνέθηκε τάδẹ | θρεπτήρια δοῦςα
> καὶ φθιμένῳ̣· | χάριτος δ᾽ οὐδὲν ἔλειψεν ἐμοị. |
>
> My daughter set this up, repaying her debt for her upbringing, even though he is dead: No gratitude was lacking to me.

The end of the pentameter has been changed from the "earlier draft" represented by A1.1–2. But this revision was in its turn revised – or rather, it was rejected. Above χάριτος δ᾽ οὐδὲν ἔλειψεν ἐμọί (B3.4), the poet wrote π⸤ᾶς⸥α̣ν̣ τ̣ὴ̣ν χάριν ὧδ᾽ ἀ̣πέχω – the "original" words from A1.2. This is a particularly clear example of the fact that revisions are not permanent, and that the elaboration of a text is not a linear or progressive, still less teleological, process: authors can go backwards as well as forwards. Authors' latest revisions do not necessarily end up as their favored ones.

I have tried to show that in the traces of revision surviving from the literary papyri of Egypt it is usually possible to reconstruct the chronology of revisions in a document, and often possible to surmise the reasons: versions follow on each other in an order that could be called rational. Because of this, we can identify what I called "local" teleologies in which revisions recategorise earlier versions as momentarily counterfactual. But because most revisions do not correct errors, because many revisions are not accompanied by unambiguous cancellations of their earlier versions, and because – as we have seen once in the case of PSI I 17 – new versions can re-instantiate previously rejected versions, it is not safe to extend local teleologies to a global level: when an entire writing process is being described we should speak only of processes in which possibilities are proliferated or versions are hypothesized; this process ends only when a final version emerges, for reasons that are hard to know until after it has happened. It is this global virtuality that prompted me to characterize revision as a region where the themes of this volume take place as non- or pre-conceptual practices: here the likely, the possible, the "maybe" or "as if" that characterise all modal thought come *before* what there will be to be thought, talked about, or read.

CHAPTER 10

Likeness and likelihood in classical Greek art

Verity Platt

The majority of chapters in this volume trace how the concept and terminology of "likelihood" emerged in rhetorical, philosophical, and literary discourse at a critical moment in Greek culture, when the language of *eikos* gave voice to a particular set of challenges in the relationship between truth, knowledge, and critical reasoning. Is it possible to draw conclusions about the unknown – whether past or future events, the structure of the universe, or parallel fictional worlds – based on one's familiarity with what is plausible, appropriate, or truth-like? The concept of *eikos* holds out the possibility, for good or ill, of extending the benefits of knowledge into the less well-charted territories of belief and imagination. In this sense *eikos*, although a predominantly *verbal* phenomenon, also has the potential to apply to the realm of the visual. Indeed, the semantic and conceptual relationship that late classical texts observe between likelihood (*eikos*) and likeness (*eikōn*) suggests that *eikos* is relevant to the vocabulary of mimesis, and thus the concept of representation.[1] Accordingly, this chapter sets out to explore whether the notion of likelihood had any conceptual utility for Greek art of the fifth and fourth centuries BCE and, if so, what form it might have taken.

In grounding an approach to classical art in terminology of the period, I take my lead from recent scholarship that has advocated the interpretation of visual culture through terms and concepts that had specific meanings

I am grateful to Victoria Wohl, whose invitation to contribute to this volume and careful guidance as editor gave me the opportunity to explore familiar material through such a fresh and exciting lens. I would also like to thank Charles Brittain, Tad Brennan, Iakovos Vasiliou, and Nancy Worman for helping me tackle the daunting text of the *Timaeus*, Richard Neer for his help in accessing images, and Milette Gaifman, Jas' Elsner, and Jeff Rusten for their valuable feedback on the chapter.

[1] E.g. Pl. *Ti.* 29b–d and Xen. *Mem.* 3.10.8, discussed below. As Burnyeat observes (2005: 146), Plato's play on the relationship between the two in the *Timaeus* is possible because *eikōs* is a participial form of the verb *eoika*, "to be like."

and theoretical discussions within antiquity.[2] The family of concepts that clusters round the language of *eikos* offers an "emic" means of analyzing practices of representation that emerges naturally from the language employed by contemporary artists, patrons, and viewers.[3] It is thus part of a complex network of ideas and assumptions that constitute classical Greek visuality.[4] This should not give us too much cause for confidence, however: as Michael Squire has recently argued, the integration of, or comparison between, verbal and visual modes of expression is never straightforward, least of all in antiquity.[5] Imposing a concept grounded in linguistic usage directly onto visual phenomena risks eliding or oversimplifying what is proper to images *qua* images. A beholder's relationship with a work of art depends in part upon visual, spatial, and material properties that have only a tangential relationship to language and, as the ancient ekphrastic tradition repeatedly reminds us, words inevitably constitute an inadequate or partial response to physical objects.[6] Nevertheless, the application and relevance of *eikos* to a wide range of discourses and activities in the classical period suggests that it offered a powerful tool not only for theorizing the relationship between knowledge and reality, but also for mapping this relationship in a logical and systematic way onto that which is unknown and unseen. Images too have a part to play in both uncovering such possibilities and reflecting upon the process of doing so.

This chapter thus explores a range of visual applications of the concept of likelihood in accordance with two of the predominant verbal arenas in which *eikos* seems to have been at play. First, I address the use of *eikos* in forensic oratory, from the mid fifth century BCE, to mean what is "plausible" or "truth-like" in the absence of direct legal evidence. In such contexts *eikos* operates as a technical tool for the purposes of persuasion and finds its most direct visual analogy in the practical means by which images are

[2] See especially Perry 2005, Neer 2010 and, on late antiquity, Madero 2010. An exemplary model of this kind of approach is Donohue 1988, who conducts a systematic analysis of usage of the term *xoanon* across different periods and genres.

[3] Compare Elsner and Meyer forthcoming on the relationship between art and rhetoric in Roman culture. As Elsner comments in his introduction, "In general Classical archaeology and art history have not been very successful at finding 'emic' theoretical models or formulations – that is, conceptualizations of the material-cultural world which come from within antiquity itself and might have been recognizable by the ancients." On the emic-etic distinction in cultural studies, see Headland *et al.* 1990.

[4] On visuality as a set of "historical techniques" (as distinct from the mechanism of sight/vision), see Foster 1988: ix and Nelson 2000. Note that *eikos* does not receive an entry in Pollitt 1974, which concentrates on terms that are applied directly to ancient art rather than the broader semantic field related to practices of image-making.

[5] Squire 2009. [6] E.g. Elsner 2004, Platt 2011: 180–211.

presented as being "likely" or "plausible" as representations of their prototypes. In this sense, *eikos* corresponds to the stylistic strategies of what is known as classical naturalism and offers a productive, even refreshing, means of approaching this notoriously complex and well-trodden artistic phenomenon. Second, I draw upon the concept of the *eikōs muthos* (likely story), as presented in Plato's *Timaeus*, and the role that concepts of likelihood play in philosophical exegesis.[7] As a tool for analyzing the truth-value of cosmologies, the *eikōs muthos* or *logos* (Plato uses both terms) is conceptually bound to the ontology of images in the Platonic hierarchy of being. Most importantly, as Myles Burnyeat and Jenny Bryan have recently argued, *eikos* provides Plato with a means of reflecting upon the nature of a discourse that is itself applied to "likenesses" (*eikones*) but does not necessarily fall prey to the negative critique of mimesis that we find elsewhere in the Platonic corpus.[8] The relationship Timaeus develops between *eikos* and *eikōn*, I suggest, forms a fitting parallel to the proliferation of images in the fifth and early fourth centuries BCE that offer a visual commentary upon their own representational status, most often through the depiction of images within images (such as vase-paintings that depict statues), sometimes even juxtaposing images with their prototypes in a self-reflexive gesture to their status as both representation and visual exegesis. In this way, images demonstrate the suitability of visual media to address issues of ontology, testifying to the broader cultural concerns with which the Platonic project is in dialogue.

Naturalism and the *tekhnē* of likelihood

As Michael Gagarin discusses in his contribution to this volume, *eikos* arguments first come to the fore in forensic oratory of the mid fifth century BCE and are associated by Plato and Aristotle with the orators Gorgias, Corax, and Tisias.[9] As a rhetorical means of persuading one's audience in the absence of factual evidence, such strategies depend upon what is likely, usual, or plausible.[10] For Aristotle, rhetorical proofs fall into two categories: "artistic" (*entekhnoi*) arguments constructed by the speaker, including those

7 See especially *Ti.* 29b, 48d, and 55d (discussed below).

8 See Burnyeat 2005 and Bryan 2012: 114–91. On *eikos* in Plato's *Phaedrus*, see also Bryan in this volume.

9 Pl. *Phdr.* 273a–274a; Arist. *Rh.* 1402a: see Gagarin (this volume, pp. 15–16), with Radermacher 1951: 28–35 on evidence for Corax and Tisias.

10 On the "truthiness" of *eikos* arguments, see also Farenga in this volume. Note that Cicero translates the associated term *pithanos* ("persuasive") as both *probabilis* and *veri simile* (*Academica Priora* 31): see Glucker 1995.

that employ *eikos*, and "nonartistic" (*atekhnoi*) proofs, in the form of laws, witnesses, the testimony of slaves under torture, contracts, and oaths, which require no skill on the part of the speaker.[11] This emphasis on the language of skill (*tekhnē*) is telling: in the context of rhetoric, *eikos* offers a practical tool for crafting what appears to be "truth-like." It thus stresses the active agency of the orator in formulating an argument that will correspond, in its degree of plausibility, to the subjective knowledge of his audience, rather than to the objective facts of the case; as Aristotle comments, an argument that is persuasive is so in reference "to someone" (*pithanon tini*).[12] Importantly, the persuasive force of *eikos* arguments does not rest upon their being *identical to* the truth, but on being sufficiently *like* the truth in order to fulfill a specific legal and rhetorical function. To that extent, *eikos* can have a positive or negative value, and (despite Socrates' protestations to the contrary) does not automatically entail moral judgment.[13] Indeed, Aristotle distinguishes in the *Rhetoric* between an "apparent" likelihood (*phainomenon eikos*), which is knowingly false, and a "true" or "genuine" likelihood (*alēthes*), which is made in good faith.[14]

How does all this relate to classical naturalism? In expressing the way in which an "artistic" (or "skillful") argument is composed in line with the audience's knowledge of what is usual or plausible, *eikos* offers a rhetorical analogy to the visual strategies by which images persuade their viewers of their status as plausible representations of their prototypes: both are techniques for the generation of likenesses. The history of Western art has traditionally rested upon the assumption that, from the early fifth century BCE, the concept of likeness increasingly shaped artistic practice so that, in representations of the human body in particular, the correspondence between images and their referents became ever more persuasive.[15] In this sense, the development of so-called "naturalistic" modes of representation – the

[11] *Rh.* 1355b35, 1375a24. Compare Socrates' (explicitly negative) claim at *Phdr.* 272e that, for sophistic rhetoricians such as Tisias, "He who is to be an artist in speech (*tekhnēi erein*) must fix his attention upon probability (*to eikos*)." On the significance of *tekhnē* in this period, see Löbl 1997: vol. II, and most recently Angier 2010.

[12] *Rh.* 1356b26, discussed in this volume by Allen, pp. 55–56.

[13] At *Phdr.* 267a, Socrates claims that Tisias, like Gorgias, "saw that probabilities (*ta eikota*) are more to be esteemed than truths" and made "small things seem great and great things small by the power of their speech"; cf. 273a–273c. As Bryan discusses in this volume, however, even the *Phaedrus* holds out the possibility of a more positive model of *eikos*, in which knowledge of the truth informs the dialectician's persuasive presentation of what is similar to the truth.

[14] See *Rh.* 1402a17–23, in relation to Protagoras' claim to "make the weaker argument the stronger," with discussion by Allen in this volume, pp. 60–63.

[15] See esp. Gombrich 1960: 99–124, on the "Greek Revolution." For more recent discussions of the rise of naturalism (and a critique of Gombrich), see Neer 2002: 27–86; Elsner 2006; and (for a more sociologically grounded approach) Tanner 2006: 31–96.

"Greek miracle" – runs in parallel to the increasing prominence of *eikos* in Greek rhetoric and literature. Yet, crucially, the concept of likelihood or plausibility does not entail an *absolute* correspondence between image and reality. Rather, it suggests that an image should conform to its viewers' subjective experience and expectation of what is usual or plausible: representations, like rhetorical proofs, are only ever convincing "to someone."

Rather than subscribing to a model of representational objectivism (as has too often been assumed in classical archaeology), the notion of visual likelihood as used in antiquity offers a more expansive model of likeness that both sidesteps Platonic concerns over hierarchies of being (on which more below) and allows for a more flexible relationship between images and reality.[16] In this sense, *eikos* offers a more historically embedded and workable model of representation than that of "truth to nature," a concept which is largely the product of art history's Renaissance (rather than antique) inheritance.[17] Second, it allows for acknowledgement that the rendering and perception of likeness is itself culturally contingent, despite the visual rhetoric by which so-called "naturalistic" images claim an exclusive relationship to their prototypes.[18] Third, a model of representation based on *eikos* readily incorporates the idea of stylistic change, whereby the form and appearance of images need not conform to a progressivist model of increasing naturalism, but to shifting notions among artists, patrons, and viewers of what is a likely or plausible rendition of a given prototype. It is worth observing, in this regard, that *eikōs* is sometimes translated as "appropriate" or "fitting," and is thus connected to the concept of "appropriateness," which, through the Greek *to prepon* and the Latin *decor* had a far-reaching influence on the relationship between style, subject matter, and context in Greco-Roman art.[19]

The concept of a "likely" or "plausible" representation thus shifts the visual rhetoric of the image away from a notion of precise replication, suggesting instead that an image need only fulfill a minimum set of criteria for likeness in order to be convincing *as a likeness* to a given viewer. As James Allen points out in relation to Aristotle's discussion of

[16] For a critique of empirical or "quasi-scientific" approaches to naturalism, see Neer 2002: 28–29.

[17] See Neer 2002: 28–29, on the influence of Vasari, and Elsner 2006: 69–70.

[18] On naturalism and cultural relativity, see Layton 1977 (foundational), with Bryson 1983: 1–35 and Mitchell 1986: 77–94.

[19] Burnyeat 2005: 146–47, citing Plato's use of the phrase *eikos kai prepon onoma* at *Soph.* 225a. On appropriateness in Greco-Roman art, see esp. Pollitt 1974: 217–18, on *prepon*; and Pollitt 1974: 341–46, on *decor*; Perry 2005: 28–77 and Marvin 2008: 234–38, with e.g. Pl. *Resp.* 420c–d, on the importance of painting a sculpture with pigments "proper to each" specific body part (*ta prosēkonta hekastois*).

"the truth-like" (*to homoion tōi alēthei*) in the *Rhetoric*, such verisimilitude should be understood as the likelihood not that there is a truth to which x is similar, but that x "[exhibits] the features, characteristics, or marks that belong to the truth."[20] By analogy, the naturalistic image might purport to be compatible with or conform to patterns *characteristic* of reality but need not match a specific prototype in every detail.[21]

If a "likely" representation need only meet a minimum set of criteria for likeness, rather than the rigorous demands of a more exacting verisimilitude, the gap that opens up between image and prototype potentially allows for other elements to influence the object's appearance. This relational model corresponds rather well to what we find in classical Greek art. Unlike later trends in Hellenistic and Roman visual culture, classical (and later, classicizing) modes of representation do not necessarily aim at a form of exacting "realism."[22] The so-called "verism" of Republican Roman portraiture does not find analogies in Greek sculpture prior to the fourth century BCE, for example, while the concept of *rhuparographia* (the "painting of sordid subjects") is associated by Pliny the Elder with the Hellenistic painter Piraeicus.[23] In classical art the looser concept of likeness suggested by *eikos* shapes the relationship between image and prototype in multiple ways, affecting both the degree to which images reflect the general rules of observable reality (as in the depiction of anatomical detail or drapery) and the degree to which they depict specific models (as in the relationship between a portrait and its sitter). For example, while we must be wary of projecting an anachronistic model of idealism onto classical sculpture (whether Hegelian *Geist* or Gombrichian conceptualism), it is fair to say that the stylistic features of fifth-century male nudes such as Polyclitus' *Doryphoros* or the Riace Bronzes, for example, are not driven purely by mimetic goals (Figures 1–2). Rather, a careful process of selection, abbreviation, exaggeration, and refinement contributes to the creation of a "plausible" figure with a complex relationship to observed reality. As Guy

20 *Rh.* 1355a11: see Allen in this volume, p. 61.

21 As Neer 2002: 30 comments, "'Mimesis' as it is sometimes used in modern discourse – in the sense of perfect and flawless reproduction – was simply not an operative concept in early Greece."

22 *Pace* Childs 1988. This is not to deny that realism, too, is historically contingent and ideologically driven; rather, the "warts and all" visual rhetoric of Hellenistic depictions of age, infirmity, and deformity is rarely to be found in Greek art prior to the third century BCE. On these, see Pollitt 1986: 141–47; Garland 1995: 105–22; Sande 1995; and Stewart 1990: 224–28; with Zanker 1987 on Hellenistic "realism."

23 On Piraeicus, see Pliny, *HN* 35.112, with Squire 2009: 363–66. On verism in Roman portraiture (and its relationship to Hellenistic art), see Richter 1955; R. R. R. Smith 1981: 36–38; Giuliani 1986: 190–220; Tanner 2000: 35–40; and Fejfer 2008: 262–70. On the "'tyranny' of likeness" as an interpretive model in the study of Greek portraiture, see Dillon 2006: 9.

Figure 1. Riace Warrior A, Museo Archeologico Nazionale, Reggio Calabria.

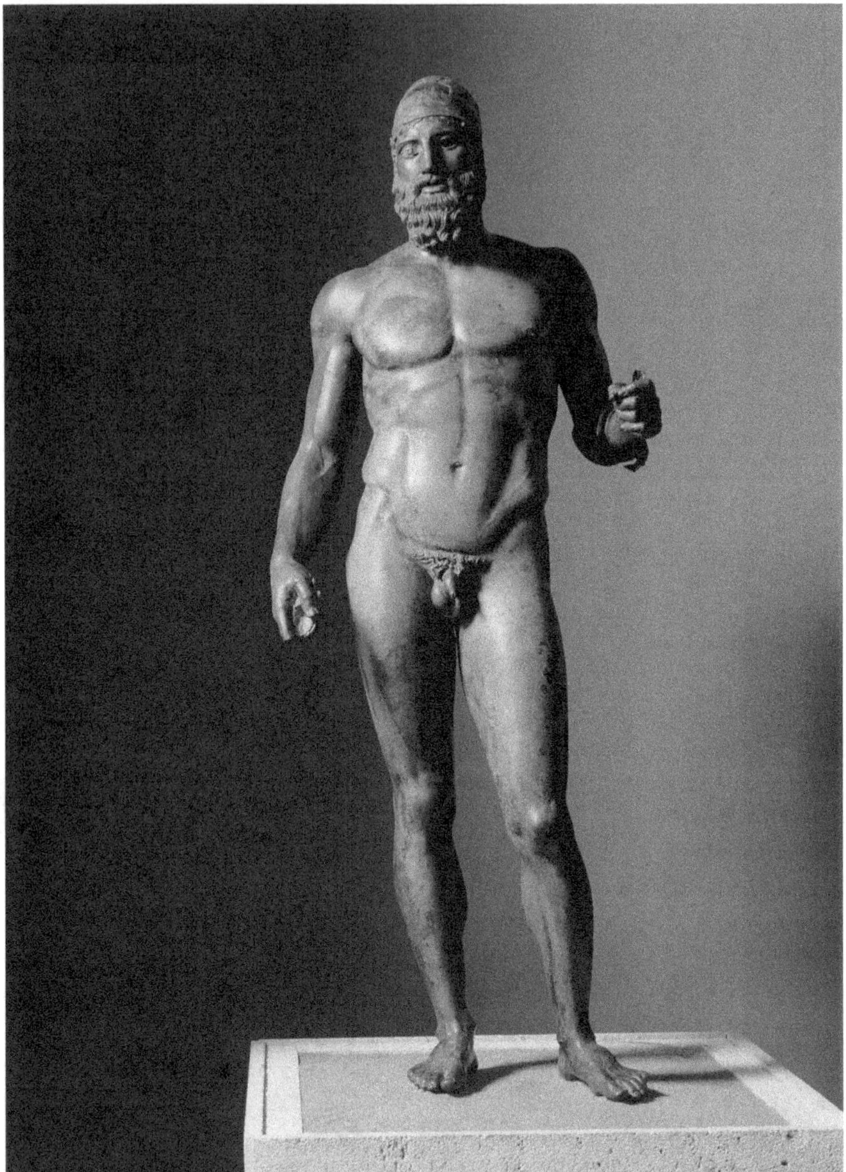

Figure 2. Riace Warrior B, Museo Archeologico Nazionale, Reggio Calabria.

Métraux (followed by Richard Neer) has observed, classical anatomy incorporates many surprisingly non-naturalistic aspects, including distortions of proportion (such as the elongated arms of figures); visual contradictions (such as immature genitals on mature masculine bodies); imaginary hypodermal structures (such as eight instead of six abdominal muscles, or the so-called "athlete's girdle," a line between the iliac crest and the inguinal ligament); and impossible contortions of the body (such as an excessive pronation of the forearms).[24] This is not a radical observation, of course; classical architecture is known for its use of optical refinements, while Plato rails in the *Sophist* against methods of visual distortion in painting and sculpture, such as the proportional adjustment of colossal works to allow for the position of the viewer.[25] In some cases these distortions work to draw attention to "naturalistic" aspects of the sculpture, such as the use of pronation, which in the case of the Riace Bronzes serves to highlight the bulging veins of the forearms; in others, they are driven by clear ideological factors, such as the relationship between the virtue of *sōphrosunē* (moderation) and small genitals.[26] Together, however, such features operate as part of the *tekhnē* by which the sculptor generates a persuasive and "appropriate" image. In this sense, they constitute a *visual rhetoric of likeness* rather than maintaining an exact relationship between image and prototype.

The concept of *eikos*, then, allows for the incorporation of less naturalistic features into the depiction of the human body. These may serve to enhance the naturalistic plausibility of the image, but in some cases they also highlight features that might be considered "implausible" or "unlikely," in order to emphasize the unexpected or even superhuman qualities of the individual portrayed. Indeed, it is arguably the subtle incorporation of implausible elements into a plausible frame that shapes the iconography of heroism and divinity in classical sculpture. If *eikos* offers a means of extending the benefits of knowledge into the less well-charted territories of belief and imagination, as I suggested above, then it also offers a means of applying the concept of plausibility to the depiction of beings who are thought to exceed the boundaries of human form in terms of size, beauty,

24 See Métraux 1995: 32–68 and Neer 2010: 147–55, who cites such features in order to refute the "empiricist" idea that the Riace Bronzes were cast from live models (on which see Konstam and Hoffman 2004).

25 Pl. *Soph.* 235e ff. On visual refinements and distortion in the classical period, see Haselberger 1999: esp. 60–67. Note also the claim, passed down by Pliny (*HN* 34.65), that whereas "the ancients made men as they are" (*quales essent*), the fourth-century BCE sculptor Lysippos represented them "as they appear to be" (*quales viderentur esse*). On Lysippos, see Moreno 1974: 127–43, testim. 18.

26 On genital size, see Stewart 1997: 92 and 95, who argues that the highly defined muscles of male nudes such as the Doryphoros "serve as a fetishistic substitute for the phallus."

Figure 3. Aphrodite of the Agora, Agora Museum, Athens (Agora S 1882).

strength, and power. This applies to the over-lifesize scale of statues of the gods, the excessive musculature of heroizing nudes (as in the case of the Riace bronzes), or the hyper-naturalistic depiction of drapery on the bodies of female deities, which often seems to confound the laws of physics in its busy play of concealment and revelation (as in the case of the so-called "Aphrodite of the Agora," Figure 3).[27] Such details cannot simply be

[27] Agora Museum, Athens, dated *c.* 420 BCE based on its similarity to the parapet of the Temple of Athena Nike and identified as an Aphrodite Ourania seen by Pausanias in the Athenian Agora (1.14.7): see Ridgway 1981: 217, who argues for a later date; Lapatin 2001: 91–94; and Osborne 2011: 208–209.

explained by means of an anachronistic model of idealization; rather, they contribute to the careful balance of plausible and implausible features that characterizes the concept of anthropomorphism within the Greek religious and cultural imagination.[28] Crucially, however, the stylistic framework of "likeness" within which such unlikely features are incorporated projects a visual rhetoric of *presence*, suggesting that such superhuman beings exist in a realm contiguous with – and even the same as – that of the spectator.

While the concept of likelihood or plausibility might help us understand the play of naturalistic and hyper-naturalistic features in classical depictions of the gods, it also has an important social function. In forensic oratory *eikos* is employed as a technical strategy by which specific character traits might be mobilized in support of an argument: in Plato's infamous example of Corax and Tisias' rhetorical method in the *Phaedrus*, a weak man who attacked a strong one is encouraged to defend himself with the false (but more plausible) claim that he was too feeble to have done so.[29] This principle of highlighting the most plausible or appropriate characteristics of an individual in order to convince an audience in the law courts of their behavior finds a visual parallel in the portrayal of individuals on Attic grave reliefs of the late fifth and fourth centuries, which have a commemorative function that is deeply embedded in polis ideology.[30] Although naturalistic in style (often employing relief so high that it verges on three-dimensional sculpture), funerary images of this kind do not operate as precise *mimēmata* of the people they commemorate but are more commonly described as "social role portraits," which employ a high degree of pictorial convention (Figure 4).[31] They are better understood as "plausible" or "appropriate" representations of individuals according to age, gender, and social status – that is, as images that "[exhibit] the features, characteristics, or marks" that are proper to or expected of certain social groups.[32] They therefore draw upon the stylistic rhetoric of naturalism in order to assert the continued presence of the individual portrayed, while inscribing him or her into a

28 Compare Pascal Boyer's theory of the play of intuitive and counterintuitive ontologies in the imaginative construction of the divine (Boyer 1994: 91–125 and 2001: 52–105). On the relationship between human and superhuman aspects in Greek conceptions of the divine body, see Vernant 1991.

29 Pl. *Phdr.* 267a, 273b–c; cf. Arist. *Rh.* 1402a17–23, with discussion by Allen in this volume.

30 See, e.g., Stears 2000 and Neer 2010: 182–214.

31 See, e.g., Meyer 1989; Stears 1995, 2001; Leader 1997; Osborne 1996, 2011: 65–84; Younger 2002; and Turner 2009 on the construction of gender roles in Attic grave reliefs, with Sourvinou-Inwood 1995: 328–37 on the notion of "social persona" in classical funerary art. Cf. Stewart 1990: 210, who comments that portraits of this period "generally favored convention over physical likeness in order to insert their sitters into this or that social niche."

32 *Rh.* 1355a11: see Allen in this volume, p. 61.

Figure 4. Grave stele of Pausimache, National Museum, Athens (3964).

suitable commemorative category through the use of iconographic formulae. In this way they visually yoke the concept of the "likeness" (*eikōn*) to that of the "likely" (*eikōs*) or "the appropriate" (*to prepon*). As the epigraph on the stele of a young woman called Pausimache, dedicated by her parents, states, "Here [stands] a memorial of your virtue and self-restraint for passersby to see" (σῆ[ς] δ'ἀρετῆ[ς μ]νημ[ε]ῖον ὁρᾶν τό[δ]ε τοῖς παριõσιν σωφροσύνη[ς] τ[ε]).[33] The relief is explicitly described as a *mnēmeion*, a "reminder" of the young woman's abstract qualities, rather than a precise depiction of her visual appearance and, sure enough, she is depicted in a highly conventional manner (Figure 4): her even features and voluptuous breasts and thighs (which borrow from the iconography of Aphrodite) identify her as a young woman on the brink of marriage, her desirability and femininity highlighted by the mirror in which she contemplates herself – a motif which underlines her transformation, in death, into an image open to the public gaze.[34] Viewed in tandem with Pausimache's epitaph, the image projects a compelling tension between virtue and sensuality that perfectly captures the delicate status of women of her age and social group, employing a visual rhetoric of plausibility which, allied to the naturalistic style of the relief (note her right foot breaking out of the frame in a gesture of presence), underlines the tragedy of her loss to both her family and the polis community as a whole.

The more flexible concept of "likelihood" or "plausibility" encapsulated by the term *eikos* thus applies to the operations of classical naturalism in several ways: it allows for a looser bond between an image and the individual it depicts, reflecting the use of iconographic formulae dependent upon social convention rather than mimetic verisimilitude (as we saw in the case of Attic grave reliefs); it also applies to the relationship between image and body more generally, whereby a spectrum of plausibility in the image's correspondence to observable reality allows for the incorporation of unnaturalistic or hyper-real elements into an overall stylistic framework of "likeness" (as we saw in relation to heroizing male nudes and representations of the gods). Both of these impulses have often been collapsed under the notion of classical "idealism," whereby social and physical forms of idealization come to stand for a unified and abstracted Greek aesthetic

33 *CEG* 518, Clairmont 1970: no. 13 and Kaltsas 2002: 160 no. 317, dated 400–375 BCE, from Charvati in Attica, now in the National Museum, Athens (NM 3964). See Leader 1997: 693, with Burton 2003, Tsagalis 2008: 155–57, and Bruss 2010: 400–1 on the inscription.

34 On the role of mirrors in Attic grave reliefs, see Leader 1997: 693; Younger 2002: 183–85; Bectarte 2006; and Neer 2010: 298–300; with Balensiefen 1990 and Frontisi-Ducroux and Vernant 1997 on mirrors in Greek art more generally, and Taylor 2008 on Roman examples. On the influence of heroic and divine iconography on late classical grave reliefs, see Himmelmann 2000.

that ignores the historical complexities at work within specific contexts of representation and beholding. Recast according to the concept of *eikos*, however, the degree to which naturalistic art mirrors observable reality – its visual "truthiness" – proves to be dependent upon cultural ideas and expectations that are continually re-forged in dialogue with various forms of collective knowledge. The image stimulates a process of evaluation in the viewer, prompting an active engagement with the *tekhnē* of representation (rather than mere absorption in naturalistic illusionism). As we shall see in the following section, this self-conscious examination of the plausibility and persuasiveness of depiction is one of the most striking features of classical Greek art.

Ontologizing *eikos*

In offering a model of representation that frees classical naturalism from a rigid "truth to nature," the concept of *eikos* suggests alternative ways of thinking about the status of the image *qua* image – that is, its *ontology*. As Jenny Bryan demonstrates in her contribution to this volume, the philosophical implications of *eikos* are repeatedly addressed by Plato, who contrasts arguments from likelihood with those based on truth and proof in the *Theaetetus* and the *Phaedrus* and mounts an ethical critique of their exploitation in the context of law.[35] As a sophistic strategy that can potentially mislead an audience through its persuasive divergence from the truth, *eikos* stands as an important Platonic parallel to the concept of mimesis, and the manufacturing of mere appearances which lack "reality or truth" in the representational arts.[36] In generating effects that have a plausible "similarity to the truth" (*homoiotēta tou alēthous*) rather than constituting the truth itself, both rhetoric and the arts have the capacity to deceive.[37]

However, just as Plato's attitude to mimesis is famously more nuanced and difficult to pin down than any simple summary of his position might suggest, so is his treatment of *eikos*.[38] Both concepts find their most sustained and complex interrogation in his late work the *Timaeus*, in which the titular character employs the device of a "likely tale" (or *eikōs muthos*) to examine the mimetic status of the entire cosmos, which he describes as a

[35] Pl. *Tht.* 162e2–163a1; *Phdr.* 267a6–7, 272d2–274a5.

[36] Pl. *Resp.* 596e4 (*phainomena, ou mentoi onta ge pou tēi alētheiai*); cf. *Soph.* 265b.

[37] Pl. *Phdr.* 273d4–5.

[38] See Bryan 2012: 114–91, and in this volume. On the complexities of Plato's treatment of mimesis, see in particular Halliwell 2002: 37–149.

perceptible "likeness" (*eikōn*) of the transcendent Forms, created by a divine Demiurge. Here, just as the creation of likenesses (*eikones*) is transformed from a misleading and inferior category within the Platonic hierarchy of being to the process by which the material cosmos itself has been generated, so the concept of likelihood (*eikos*) is elevated from a cynically employed tool of sophistic rhetoric to a critical component of cosmological exegesis. Indeed, Plato explicitly connects the two terms in Timaeus' statement at 29b3–c3:

> ὧδε οὖν περί τε εἰκόνος καὶ περὶ τοῦ παραδείγματος αὐτῆς διοριστέον, ὡς ἄρα τοὺς λόγους, ὧνπέρ εἰσιν ἐξηγηταί, τούτων αὐτῶν καὶ συγγενεῖς ὄντας. τοῦ μὲν οὖν μονίμου καὶ βεβαίου καὶ μετὰ νοῦ καταφανοῦς μονίμους καὶ ἀμεταπτώτους, καθ' ὅσον οἷόν τε καὶ ἀνελέγκτοις προσήκει λόγοις εἶναι καὶ ἀνικήτοις, τούτου δε μηδὲν ἐλλείπειν. τοὺς δὲ τοῦ πρὸς μὲν ἐκεῖνο ἀπεικασθεντος, ὄντος δὲ εἰκόνος εἰκότας ἀνὰ λόγον τε ἐκείνων ὄντας. ὅτιπερ πρὸς γένεσιν οὐσία, τοῦτο πρὸς πίστιν ἀλήθεια.
>
> With regard to a likeness (*eikonos*) and its model (*paradeigmatos*), this distinction should be drawn: that accounts (*logous*) are kin (*sungeneis*) of the subject matter of which they are interpreters (*exēgētai*). So: accounts of what is permanent and stable and evident with the help of reason are stable and permanent, insofar as it is appropriate for accounts to achieve irrefutability and incontrovertibility and not to fall short of this standard, but accounts of that which is created to be like (*apeikasthentos*) that [which is stable and unchanging], being of a likeness (*eikonos*), are likely (*eikotas*) and stand in proportion to those others. As Being is to Becoming, so truth is to conviction.[39]

Following his declaration that *logoi* must be "kin" (*sungeneis*) of the entities that they describe, Timaeus advances the notion that a *logos* about a likeness must therefore have the quality of likelihood, juxtaposing *eikonos* and *eikotas* in a playful pun on the terms' shared derivation from the verb *eoika*, "to be like."[40] He then goes on to state at 29d1–3 that, given the impossibility of applying *logoi* that are consistent and accurate (and thus more applicable to abstract reasoning) to perceptible and unstable entities such as the created universe, it is fitting "to accept the likely story (*eikota muthon*) about these matters and search for nothing more."

Debate over the truth-value that Plato assigns to Timaeus' cosmology is notoriously complex, as is its problematic status as both *logos* and *muthos*; for the purposes of this chapter, I follow recent scholars of the *Timaeus* such as Jenny Bryan and Sarah Broadie in accepting Myles Burnyeat's argument

39 Text and translation (with minor modifications) from Bryan 2012: 115–16.
40 Burnyeat 2005: 146. On the pun's problematic significance, see Rowe 2003b.

that, in the case of *logoi* about the sensible world (the realm of Becoming), "being *eikōs*" is a standard which an account is expected to *live up to*, rather than a symptom of deficiency, whether of certainty or truth.[41] Although it may not have the rigor of a mathematical proof (unlike *logoi* of "stable and permanent" entities within the realm of Being), an *eikōs logos* is a plausible and appropriate account of an *eikōn* which, in its order and rationality, is itself a likeness of the order and rationality that characterize the Forms.[42] As Bryan has suggested, Timaeus' attempt to yoke the concepts of *eikōn* and *eikos* together facilitates "a positive representational relation" between these two realms, emphasizing that "the perceptible world is of ontological and epistemological value, albeit in some qualified way."[43]

Here, it is important to point out that the cosmos that forms the subject of Timaeus' *eikōs muthos* is described not only as an *eikōn* but also as an *agalma*, a term derived from *agallein*, "to adorn" or "to honor," which is more commonly applied to statues of the gods.[44] As an object that gains its power and authority through the instrumental role it plays in establishing reciprocal relations with the divine, the *agalma* indicates a bond between image and model that is grounded in practices of dedication and ritual viewing. Rather than simply occupying an inferior position as a representation of a god in an ontological hierarchy based on the concept of mimesis, the *agalma* has a specific *function*, pointing us toward something separate from itself but to which it has a certain relation (thereby sharing its sanctity). Timaeus' use of the term *agalma* thus shifts attention away from a critique of the necessary insufficiency of representation to a theologically grounded theory of the image that is arguably more positive in its formulation of the relationship between the material and transcendent (and thus more appropriate for a cosmology).[45] To that extent it is in

41 Burnyeat 2005: 150, followed by Bryan 2012: 114–91 and Broadie 2012: 31–38; on Burnyeat's interpretation of *eikos*, see also Betegh 2010. On the use of *eikos* to express the uncertainty and subjectivity of cosmological discourse by Presocratic philosophers such as Xenophanes and Parmenides, see Bryan 2012: 6–113.

42 See, e.g., Bryan 2012: 119: "For an account to be 'likely' in this sense is for it adequately to explain the nature of its subject-matter as having been likened to a model."

43 Bryan 2012: 5.

44 On the use of *agalma* in Plato's *Phaedrus* and *Timaeus* and its relationship to the concept of *theōria*, which also has important religious connotations, see Nightingale 2004: 163–72. On the ritual significance of the term *agalma*, see Platt 2011: 90–91; with Gernet 1981: 73–111; Scheer 2000: 8–18; Bettinetti 2001: 27–37; and Nick 2002: 12–15. On *eikones* as visual representations, see Vernant 1990, 1991: 164–85; Halliwell 2002: 182–84; and Tanner 2006: 106–108.

45 Compare Pl. *Symp.* 215a–b, where Alcibiades compares Socrates to the crude *eikōn* of a Silenus that opens up to reveal golden *agalmata* within, with Steiner 2001: 132–33. Likewise, as Burnyeat 2005: 149–50 points out, Timaeus' discussion of the truth-value of *logoi* describes them as "interpreters" (*exēgētai*), suggesting an exegetic mode of discourse that is explicitly sacred in tone.

keeping with the more positive model of mimesis proffered by the *Timaeus* as a whole, offering a means of bypassing the ontological impasse found in earlier Platonic discussions of representation.

What, then, might Timaeus' use of *eikos* mean for our understanding of contemporary image-making? Crucially, his proemium develops the concept of the "likely" or "appropriate" account as a tool for analyzing the status of a likeness *as a likeness*.[46] This usage has a meta-representational, self-reflexive quality that is concerned less with the insufficiency of *eikones* when *compared* to their models (for, as we are told in the *Cratylus*, likenesses must be different from their originals to some degree, otherwise they would simply be replicas), than with the *internal* characteristics that allow an image to be successfully and plausibly representative of its model while being necessarily different from it.[47] *Eikos* as presented in the *Timaeus*, then, offers a serious means of thinking about representation *as representation*, and does not directly apply to the ontology of *eikones* themselves, but to the nature of *discourses that address the nature of eikones*. In allowing us to take a step back from the relationship between image and prototype in order to examine the means by which we address images *qua* images, Timaeus' use of *eikos* encourages an awareness of the potential for representations to operate discursively, stimulating analysis of their own nature and function.

In reflecting on the meta-representational potential of eikastic discourse, the *Timaeus* echoes a striking feature of fifth and early fourth-century Greek art: that is, the depiction of images within images in such a way that their identity as likenesses is highlighted for the viewer, thereby generating a visual commentary on the act of depiction that draws attention to the representational operations of the entire work. This is particularly true of vase-painting, a medium well known for stylistic experimentation and self-reflexive artistry (as demonstrated by early classical examples such as the Foundry Cup), but it is also found in sculpture, including votive reliefs and friezes.[48]

In the case of vase-painting we find a fascinating shift with the move to red figure in the first half of the fifth century. Depictions of the rape of Cassandra as she claims asylum at the statue of Trojan Athena, for example,

46 As suggested by Betegh 2010: 218 and Bryan 2012: 144–45.

47 Pl. *Crat.* 432b2–c6; cf. *Soph.* 270a7–b2, discussed by Bryan 2012: 139–44.

48 On the self-reflexive impulse of vase-painters from the late archaic period onwards, see Neer 1995, 2002 and Lissarrague 2001, with Mattusch 1980 and Neer 2002: 77–84 on the Foundry Cup (Berlin Staatliche Museen F 2295, from Vulci; Beazley, *ARV*² 400.1, dated 490–480 BCE). On images within images in Greek vase-painting more generally, see De Cesare 1997, Oenbrink 1997, Neils 2004, and Marconi 2011. On meta-representational features in fourth-century votive reliefs and architectural sculptures such as the Bassae frieze, see Platt 2011: 31–50, 114–19.

Figure 5. Attic red-figure calyx krater attributed to the Altamura Painter.

no longer feature an animated depiction of "the goddess herself" (as we find in earlier black figure) but draw attention to Athena's presence in the form of a *statue*, identified by its base and a style that is more stiff and self-consciously archaizing than the fluid forms of the scene's mortal protagonists.[49] On a calyx krater attributed to the Altamura Painter, dated *c.* 465 BCE, this stylistic contrast serves to highlight the mimetic potential of naturalism and the innovative *tekhnē* of the painter (Figure 5).[50] When compared with the archaic frontality and rigid drapery of Athena's image (here the armed *Promachos* type), the soft, fleshy form of Cassandra, with its three-quarter torso and dramatically foreshortened calves, makes a claim for the status of the depiction as a plausible likeness, or *eikōn*, of the female body, in all its mortal vulnerability. Yet, while it draws attention to the naturalistic potential of painting, the krater also mounts a visual interrogation of the form and function of sacred images, or *agalmata*. The question of how material representations of the gods provide access to their

49 For the depiction of sacred images on Greek vases, see Schefold 1937, Alroth 1992, Hölscher 2010, Marconi 2011, and Gaifman forthcoming a and forthcoming b. On the significance of the Athena statue in vase-paintings of the rape of Cassandra, see Moret 1975; Matheson 1986: 103–7; H. A. Shapiro 1989: 36–37; Connelly 1993: 90–109; Mazzoldi 1997; Oenbrink 1997: 34–72; G. Ferrari 2000; and Gaifman forthcoming a, with *LIMC* s.v. Aias II nos. 16–108. On a parallel shift in the depiction of Athena alongside "Ajax and Achilles Playing Dice," see Gaifman 2006: 264–67.

50 Boston, MFA 59.176, *ARV*² 590, 11, dated second quarter of the fifth century; see also discussion in Platt 2011: 93–96.

divine prototypes is a particularly urgent one in the case of Cassandra; indeed, not only her own fate but also that of Ajax and the ensuing mythic cycle of the Greeks' return from Troy depends upon it. Yet facing out of the scene, Athena's statue does not respond directly to Cassandra's supplication (indeed, the position of the goddess's spear almost suggests complicity with Ajax's attack). Instead, the device of frontality is employed to engage the external beholder, encouraging him or her to consider the relationship between deities, their statues, and their mortal worshipers.[51] The painting may purport to offer us a plausible likeness – or *eikōs eikōn* – of Athena's statue but leaves open the question of how far the statue itself might be understood as a plausible likeness of the goddess. Its marked stylistic contrast to the scene's "living" figures means that, according to the visual logic of the krater, the statue does not make the same *claim* to likeness, suggesting that in the case of talismanic objects such as the Trojan Athena the relationship between image and model should be construed rather differently: instead of focusing on the statue's mimetic relationship to its prototype (i.e. its status as an *eikōn* of Athena), we are encouraged to focus on its ritual efficacy (i.e. its status as a sacred *agalma*).[52] Through its juxtaposition of different media and styles, then, the krater not only prompts questions about the mythical narrative it depicts but also makes a claim for its own status as an appropriate vehicle through which to examine practices of representation.

Objects such as the Altamura Painter's *Iliupersis* krater demonstrate how a growing interest in the visual dynamics of mimesis in the fifth century went hand in hand with a self-conscious experimentation in the power of art to explore its own analytical potential. In particular, the ways in which vase-paintings draw attention to shifting visual ontologies through the device of *mise en abyme* make clear how – in the language of the *Timaeus* – the development of naturalism gave artists the visual and conceptual tools with which to realize the image's power to be its own *exēgētēs*. Visual exegesis has the power to captivate its audience precisely because it shares the same representational challenges as its subject: like Timaeus' *eikōs mythos* and the *eikōn* it describes, image and prototype are "kin" (*sungeneis*), for the painted scene and objects depicted within it both exist within the realm of representation, wherein they are subject to parallel affordances and constraints.[53]

[51] On the relationship between archaic frontality and classical absorption, see Elsner 2006.

[52] On the theological implications of such scenes, see Gaifman 2006, forthcoming a, forthcoming b; Hölscher 2010.

[53] Arguably, Classical vase-painting asserts its suitability to comment on a whole range of mimetic activities, including that of theater, as demonstrated by the self-reflexivity and meta-theatricality of objects such as the Pronomos Vase: see Taplin and Wyles 2010.

Figure 6. Fragment of an Apulian red-figure calyx krater attributed to the Painter of the Birth of Dionysus, Allard Pierson Museum, Amsterdam (2579).

Given the ontological issues at stake in these self-reflexive practices, it is not surprising that they are applied with particular enthusiasm to statues of the gods.[54] Indeed, by the early fourth century BCE we find that the narrative framework of scenes such as the Rape of Cassandra is sometimes abandoned altogether; instead, statues of deities are accompanied by divine and mythical beings who offer a more concentrated visual commentary upon the relationship between sacred images and their divine prototypes. Take, for example, the series of well-known Apulian vases in which cult statues rendered in a naturalistic style are pointedly juxtaposed with the deities they represent: in Figure 6 (dated 400–385), an oft-cited example of late classical self-reflexivity, a bronze statue of Apollo holding a phiale and bow within a temple is accompanied by a depiction of the "real" Apollo playing his lyre, as if the god were present at his own shrine, his identity confirmed by the label APOLLON; in Figure 7 (dated *c.* 375–50), a painter applies finishing touches in encaustic to a marble statue of Heracles, while the hero himself, together with Zeus and Nike, looks on with a gesture of wonder, as if bearing witness to his own apotheosis in image form.[55] In both cases, the vase invites a comparison between statue and deity, emphasizing the materiality and facture of the sculpted image in contrast to its "living"

[54] See above, n. 49.

[55] Figure 6: Amsterdam, Allard Pierson Museum 2579, from Taranto, attributed to the Painter of the Birth of Dionysus; see Trendall and Cambitoglou 1978: vol. 1, 34–36, no. 10; Schneider-Hermann 1972: 31–34, who identifies the Apollo within the temple as the classical "Kassel" type; Oenbrink 1997: 126–27; Platt 2011: 120–22; and Gaifman forthcoming a. Figure 7: New York, Metropolitan Museum of Art 50.11.4; see Trendall and Cambitoglou 1978: vol. 1, 266–68, no. 47; Trendall 1989: 78; Todisco 1990; De Cesare 1994; Oenbrink 1997: 389–90; Marconi 2011, with further bibliography, and Gaifman forthcoming a. On Heracles' gesture as an expression of *thauma*, see Schneider-Hermann 1972: 40; Moreno 1984: 94; Oenbrink 1997: 150; and Marconi 2011: 155. On the sophisticated treatment of representation in Apulian vase-painting, see Lissarrague 2008.

Figure 7. Apulian red-figure column-krater attributed to the Group of Boston, Metropolitan Museum of Art, Rogers Fund (50.11.4).

prototype.[56] This attention to the processes of production and mimesis thus draws attention to the status of the vase itself as a painted object, bracketing the entire exercise within the category of "representation." On one hand, both vases make a claim for the plausibility (or "likelihood") of the anthropomorphic system: each statue shares key iconographic details with its deity, while the presence of both gods within their sanctuaries confers authority upon their images, confirming their status as authentic likenesses (or *eikones*). On the other hand, the self-reflexivity of each scene alerts us to the fact that although it may confirm the plausibility of the statue as a likeness of the *painted* god, it makes no such claim about the plausibility of the painted god as a likeness of the *god himself*, who remains unseen and unknowable, except through his *eikōn*. Indeed, in Figure 7 the painter has his back to Heracles and is unable to take part in the process of verification that the vase makes accessible to its external viewers, as if to alert us to the epistemological gap that necessarily lies between sacred images and their divine prototypes.

[56] In Figure 6 the bronze of Apollo's statue is indicated through the use of a paler slip and white highlights, while in Figure 7 the marble of the statue is indicated through the use of a white slip, and the application of wax pigments to Heracles' lion skin is indicated by the painter with his spatula and bowl and his assistant (to the left of the scene), who prepares rods in a charcoal brazier in order to apply heat for an even finish: see Marconi 2011: 150–51.

Figure 8. Reverse side of Figure 7.

The question of divine visibility is explored still further on the opposite side of the Heracles vase, where we encounter Athena in conversation with one of the Dioscuri, who is depicted in his familiar guise as a young man with spear and *petasos* hat (Figure 8). The Dioscuri usually appear as a duo, so where is the god's twin? Significantly, he is accompanied not by his double, but by a star, an allusion to the brothers' association with the Gemini.[57] In this way, the conventions of anthropomorphism are juxtaposed with the manifestation of deity as cosmic phenomenon, prompting us to ask which is the more appropriate *eikōn* of the god: his visible heavenly body, or his man-made image?[58] In encouraging the viewer to compare and evaluate different strategies for depicting divine bodies, the vase thus asks us to assess their relative plausibility *qua* images, for the circular relationship that the Heracles scene asserts between god and statue reminds us that, for all its suggestions of divine presence, we remain within the realm of representation. Reading through the lens of the *Timaeus* (which, incidentally, is usually dated *c.* 360 BCE, thus exactly

[57] On the thematic use of Dioscurean doubles in explorations of the relationship between epiphany and representation in antiquity, see Platt forthcoming.

[58] On the notion of the stars as "visible gods" in the *Timaeus* (40a, 41a), see Nightingale 2004: 172; with Gaifman 2012: 103–108 on a parallel distinction between visible astral gods and invisible deities worshiped in the form of *agalmata* at Pl. *Laws* 931a.

contemporary with this vase), we might claim that while the deities beyond the vase-painting are assigned a parallel status to the elusive Platonic Forms, the vase itself, like Timaeus' account of the cosmos, claims only to offer a "likely account" of the likenesses we encounter in the realm of Becoming. The painting thus draws attention to the ontological dilemmas intrinsic to the problem of representing invisible or transcendent prototypes, while at the same time asserting its own suitability as a medium through which to explore the problem.

Conclusion

My discussion here has merely scratched the surface of the complex theological and representational issues to which such scenes allude. What I hope to have demonstrated, however, is the sheer confidence and vitality with which classical images interrogate the epistemological foundations of their own mimetic enterprise. They do not function as a direct "illustration" of the philosophical problems explored in the Platonic corpus, so much as point to a broader contemporary interest in the ontological status of cultural representations and the discursive modes that can be deployed to examine them. The intense attention to this theme in vase-painting of the first half of the fourth century BCE is a natural outcome of the trends that had already been set in motion by the development of naturalism in the fifth century, which provided artists with a stylistic toolbox that equipped them to probe the relationship between mimesis and ontology. Objects such as the Apollo and Heracles vases discussed above seem to acknowledge the recursiveness and contingency of human discourse and the elusiveness of certain types of knowledge (such as that of the gods' true nature). At the same time, they assert the importance of *hypothetical* knowledge (such as the iconographic conventions of anthropomorphism), which forms the epistemological armature upon which much of human activity depends, including ritual communication with the divine. In this, classical artists demonstrate a deep traditionalism allied to a spirit of experimentation and inquiry that is surely a product of the same cultural forces that made the concept of *eikos* so key to the philosophical, rhetorical, and literary practices of their contemporaries.

CHAPTER 11

"Why doesn't my baby look like me?" Likeness and likelihood in ancient theories of reproduction

Daryn Lehoux

English, like Greek, has an interesting root to its word for probability. Likeliness, just like *eikos*, is derived from a word for similarity. In the Greek case, *eikos* is the neuter participle of the perfect-tense verb *eoika*, "to resemble." In his 2004 book, *Plato's Natural Philosophy*, T. K. Johansen played on just this overlap of meanings between likelihood and likeness to draw out an interesting epistemological position in which Plato's *Timaeus* could find grounding.[1] Traditionally, the truth-value of the account Timaeus offers us in the dialogue was seen as probabilistic, where it was supposed to have been offered as being a more or less *likely* truth. Johansen turned this account beautifully around, however, arguing that Timaeus was not thinking of his story as probable, but instead as *similar to* what truly happened at the creation of the cosmos, about which we can have no direct experience.

A near-identical set of slippages happens when we come to think about the paradigmatic instance *within* the cosmos of the copying of models, that of biological reproduction, where organisms produce offspring that are like themselves: simians producing simians, but lichens producing lichens. Why should that be the case? More to the point, what mechanism can we find to account for the passing-on of traits or essences? Asking this question from a history-of-science point of view brings to light a secondary issue that gestures tantalizingly at the other meaning of *eikos* in Greek. In thinking about inheritance, we in the early twenty-first century automatically look back to two distinct historical theories as the twin foundations of our own account. The later of the two is the discovery of the chromosomal mechanism for inheritance by Sutton and (independently) Boveri in the early twentieth century. In its turn, this discovery rested on the other of our two foundations, all but forgotten until Sutton and Boveri's day: Mendel's

[1] Johansen 2004.

studies on pea plants. Mendel's work would eventually come to form the basis of our understanding of inheritance outcomes, an understanding that is, at its core, probabilistic.

Thus we take it for granted now that doctors will recommend (or counter-recommend) tests to pregnant women based on the statistical chances that their babies will have certain birth defects. People not uncommonly make informal Mendelian calculations of the chances the baby will have blonde hair or blue eyes. "She'll probably be tall," they suppose. What is striking is that we do not find the discussion framed in quite this way in ancient sources. Perhaps we should not be surprised; after all, mathematical methods for measuring and expressing probability were only invented in the early modern period.[2] Nevertheless not all of the ways we commonly have of expressing our expectations about offspring find themselves mathematized. "She'll probably be tall," is one such case (and perhaps even the most common type). Nothing in theory prevents the ancients from talking in this way, even if we find that for some reason they very rarely do so. What we find instead, though – and it is indicative of the fact that they do still think about reproduction in terms of likely versus unlikely outcomes – is a good deal of talk about why things sometimes turn out differently than expected. When the baby does not look like the parents (or more pointedly perhaps, like the father), then we see expressions of surprise, and the search for retroactive explanations begins to take center stage. But what we tend to find in the answers provided is not for the most part the provision of mechanisms of *unlikeness per se*. Instead, we see (sometimes quite counterintuitively) explanations that appeal to mechanisms for *likeness*. Expectations foiled by improbable outcomes, then, tend to root themselves in explanations for how likeness, *eikos*, is passed on, by pointing to the multiplicity of potential sources for likeness in the (surprisingly complex) process of generation.

Thus the driving question of this chapter might be most simply framed by asking how the ancients thought that babies came to look like who they came to look like. To give a sense of just how wide open the field of explanation might be, let me quote from a passage in the Pseudo-Aristotelian *Problems*, where the author poses the following (partly counterfactual) conundrum:

> διὰ τί, ἐὰν μὲν ἐκ τοῦ σπέρματος τοῦ ἡμετέρου γένηται τὸ ζῷον, τοῦτο ἡμέτερον ἔκγονόν ἐστιν, ἐὰν δὲ ἐξ ἄλλου τινὸς ἢ μέρους ἢ ἀποκρίσεως, οὐχ ἡμέτερον;

[2] See Hacking 1975, 1990; Daston 1988.

> Why is it that if a living thing is born from our seed it is our child, but if from another part or secretion it is not ours?[3]

This is nothing short of asking why any maggots that may grow on our excrement are not considered to be "our babies." It is, on the face of it, rather a shocking question, but not in fact a bad one. Why should we favor one bodily secretion over another in just this way?

The answer that the author of the *Problems* comes up with roots itself in a pair of definitions concerning those bodily secretions that are properly "ours," *hēmeteros*, versus those that are not ours, *allotrios* (literally, "otherly"). But there is another aspect to the problem that this author flags for us, which is the non-trivial fact that the animals that come to be from our seed are *like* us (*hoion hēmeis*), whereas the animals that come to be from excrement are different (*allotrios* again, *Pr.* 878a3–5).

Two issues raise their heads immediately. One is that many young animals are not superficially "like" their parents at all (bees are a good example). Another, perhaps just a modification of the first, is that likeness is always relative. Mammal babies generally look like their parents, to be sure, but then they never look exactly like either parent precisely, nor a partwise-identical assemblage with features borrowed from one parent here, the other there. So too, there are sometimes even greater differences, whether by deformity, questionable paternity, or some other cause. The apparently simple and everyday fact of "family resemblance" turns out to be significantly vaguer than we might like, and questions of individual difference pose more than a few puzzles for the ancients. We tend to be dazzled by the elegance and simplicity of our modern zipping and unzipping chromosomes, of Mendelian Punnett squares and what we are told about the nigh-schematic colors of peas, and so we are fooled into thinking that the problem is simpler than it is. If we look at the question from the Greek point of view there are some serious hurdles to overcome. Think of individual body parts, for example. How is it that someone could "have" their father's nose? Like the ancients, we know that the father's seed comes from the testicles. The testicles are distantly connected by means of an intricate network of blood and other vessels to the nose, but there seems no simple way to explain how the shape of the father's nose got down into the seed and then collected there to be passed on along with some – but never all – of his other traits. So, too, with the mother's hair color, or even more distantly plausible characteristics such as her manner of speech. Worse still,

[3] *Pr.* 878a1–3. All translations are mine unless otherwise indicated.

how is it that I could have my grandfather's nose if my father did not have such a nose himself?

Generation in Aristotle

The most detailed account we have of generation before Aristotle – and one that Aristotle seems to have been engaging with closely – is that found in the Hippocratic treatises *On Generation* and *On the Nature of the Child*.[4] There the author argues that the resemblances of a child to its parents can be accounted for by the fact that the seed of each parent (and it is significant that both parents contribute seed) "comes from" all of the parts of each parent's body. The father's nose can be inherited because his seed comes, in part, from that very nose. The mechanism by which the child acquires one parent's nose and not the other's has to do with which parent contributes the most seed from that body part. The smaller volume is said to be "conquered" or "dominated" by the larger (*krateisthai*, a political or military analogy). Conquering is also an important mechanism for sex determination, which is, notably, not a resemblance relation. Instead, sex determination depends on the "strength" or "weakness" of the seed contributed by each parent (again with any larger volume conquering the smaller), and weak or strong seed can be contributed in larger or smaller quantity by either parent. Finally, any respects in which the child resembles neither parent are chalked up to mechanical problems in the womb. So in the case of sickly children born to two healthy parents, the author tells us that the problem may have been that the womb was too "open," allowing crucial nutrition to escape during gestation. Similarly, birth defects are said to be the product either of an overly constricted womb (the author uses the analogy of placing a container around a growing cucumber to restrict its shape – let us call this the "fruit-mould" theory), or else of some kind of impact on the belly of the mother during fetal development.

What the author of the Hippocratic treatise does not offer, though, is any clear mechanism by which traits could skip a generation. Since likeness happens because the seed comes from the actual parts of each parent's body to which the child has a resemblance, it does not seem possible for a grandmother's hair color, for example, to be passed on to a child unless

[4] These two short texts are now widely thought to have originally comprised all or part of one single work, and, for simplicity's sake, I will hereafter refer to the composite text as simply *On Generation*. See Lonie 1981, G. E. R. Lloyd 1978 (note that the title of *Peri Gonēs* is there translated as "The Seed"). For the question of inheritance in Aristotle see Cooper 1988, Coles 1995, Henry 2006.

the child's parent also had that same color. Grandparents, accordingly, are not discussed in *On Generation.*

Aristotle, by contrast, goes to some length to allow inheritance from grandparents and earlier ancestors, but in order to do so he must abandon (in fact invert) the Hippocratic account of the seed. The seed cannot, he says, come from the parts. In order to make that point, he first canvasses the arguments that would lead one to believe that it did in fact come from all the parts of the body. In the first instance, he says, sex is very pleasurable, and so it must involve the whole body in some way. Secondly – and here he breaks rather abruptly with a claim in the Hippocratic text – he says that "maimed" parents (using his word: *koloboi*) produce maimed children.[5] For the third argument Aristotle invokes a central example from *On Generation*, that children resemble their parents not just in general but in specific parts. The fourth argument that might lead one to believe that seed comes from the whole body works on the grounds of consistency: the parts should be formed as the whole is formed, and if we believe that the whole is formed from seed, then we should believe that there is a seed for each of the parts as well.

In opposition to these arguments, and as the building blocks on which his own theory will stand, Aristotle offers the following: first, children do resemble their parents in the parts, but they also resemble them in manners and habits, and there is no way for the seed to "come from" the manners and habits. Secondly – and I find this an interesting and difficult argument for him to put to his opponents – children can inherit characteristics that show up in the parents only after the child is grown, such as a mother's prematurely grey hair or a father's receding hairline, which a child can acquire even though the parents did not yet show (or perhaps did not even yet know about) those traits when the child was conceived. Thirdly, Aristotle thinks that to say someone resembles their father "in their face" or "in their hands" introduces a whole host of difficulties, since the face and hands are themselves made up of a number of more basic parts, such as skin, bones, hair, and muscle, which are themselves compounds of the four elements. There is also the question of how all the parts stay in their proper relation to each other in the seed – if we push it too far we may be forced to suppose that the seed is actually a "tiny animal" (*zōion mikron*), as the Dutch microscopist Antoni van Leewenhoek would much later believe.

[5] It is unclear whether by "maimed" he means those with birth defects, or those who have met with accidents or amputations, or perhaps more likely both (he tells us in a different context about a man from Calchedon who had been branded on the arm, and a version of the brand appeared on his child); Arist. *Gen. An.* 721b33.

A final reason to suppose that the seed does not come from the parts is the problem already flagged with the Hippocratic account: children do sometimes resemble grandparents or other ancestors rather than the parents themselves. The example he offers us is the case of the "woman from Elis" who conceived a child with a black man.[6] The woman gave birth to a daughter who was not black. The daughter, though, later had a son who was black like his grandfather. This, Aristotle says, simply cannot have happened if the traits in the seed come partwise from the traits of the parent. As with the father's hairline, the inheritance of traits not actually manifest in the parent would seem impossible.

Nevertheless, it is clear that seed must have something to do with the makeup of the parts in the parent's body and with the individual characteristics of those parts. If the seed can not come from the parts, then what else could it be doing? At this point Aristotle thinks of the way in which the body is fed and maintained in a continuous process of nutrition coming from food. Food goes into the body, becomes concocted into blood, and the blood serves as nutrition for the parts of the body. Since the blood nourishes the parts such that they grow and are maintained slightly differently in each person (everybody has their own individual characteristics), Aristotle concludes that the seed must be some kind of residue from this process of nutrition. Individual likeness is carried by the seed because it is a residue of the nutrition going out to the parts of the parent's body, nutrition that is so constituted as to maintain or build the individual characteristics of that individual parent – Socrates' food turns into Socrates' particular nose and so the residue from that food, Socrates' seed, also carries in it "Socrates' nose-ness." The color of a parent's eyes or the shape of their hands, it would seem, is then inherent in how that parent turns his or her food into his or her body parts on a day-by-day basis, and the leftovers from that process of blood formation, the seed, then also carries the same color of eye for the child to inherit.[7] To put this in Aristotelian terms: the nutrient is *potentially* what the body parts of the individual are *actually*, and that potentiality applies whether the nutrient goes to make up the body

[6] Arist. *Gen. An.* 722a10. See Henry 2006.

[7] Aristotle worries at some length over whether the seed is left over in the same way as excrement is left over from nutrition (*suntēgma*), or whether it is an "excess" (*perittōma*) above and beyond the nutritional process. He settles on excess on the grounds that "a waste product is one of the things that are contrary to nature" (*ta de suntēgmata tōn para phusin ti, Gen. An.* 725a1). That the seed is in fact the most refined type of nutrient is proved in part, he thinks, by men's exhaustion after orgasm – loss of a very small amount of fluid should not be so tiring unless that fluid was particularly rich in some way (*Gen. An.* 726a25). For an alternative reading of Aristotle's account of seed, see Coles 1995, critiqued (I think rightly) by Henry 2006.

of the parent, or whether (as an excess residue) it goes into the making of offspring. Aristotle draws a helpful analogy: "It is like with painters of portraits, generally something similar to what has been used up remains behind" (οἷον τοῖς γραφεῦσι τοῦ ἀνδρεικέλου πολλάκις περιγίνεται ὅμοιον τῷ ἀναλωθέντι, Arist. *Gen. An.* 725a27). (And notice the triple likenesses invoked here: the seed is like [*hoion*] the paint left on the palette, the paint left on the palette is like [*homoion*] the painting, and the painter is himself a painter of "man-likeness" [*andreikelou*, compounded on an archaic form of *eikos*].) The idea here is that the colors that went into creating the painting also remain on the palette at the end and could, one supposes, be used to make another painting, more or less similar to the first. But the "could" in that sentence brings to the fore an important aspect of Aristotle's analysis: the paint on the palette is only potentially a painting, not yet actually a painting, and the transformation from potential to actual requires the action of a painter. That he conceives of this process as involving an actor working on a material to produce something new is significant.

The Hippocratic author of *On Generation* thought that both men and women contributed seed to the formation of the baby, and although Aristotle has redefined what seed is in reaction to something very like the account we find in *On Generation*, he agrees with that author that females as well as males do contribute to the formation of the child. But what the female contributes is in several important respects different from what the male contributes, and Aristotle is often hesitant to call her contribution "seed," even if sometimes he still does so.[8] In the first instance, her contribution is incompletely concocted. Blood, as we have seen, is concocted food for Aristotle, and semen is concocted blood. But, he thinks, it is clear from observation that what corresponds to seed in the woman is just the menstrual blood (it appears at about the same age in females as semen does in males, for example). Insofar as her blood never reaches the second stage of concoction that the man's does, which is to say it never turns into semen *per se*, then the process of seed formation has in the woman fallen short of its total completion. There are all kinds of theoretical reasons that support this idea for Aristotle. In the first instance, women were near-universally thought by ancient medical writers to be "colder" and "wetter" than men. Since concoction as a process requires heat (depending on the author it is thought to be akin to either cooking or fermentation), it only makes sense

[8] Among the reasons Aristotle gives for women not contributing seed *per se* is that they can conceive without having reached orgasm (*Gen. An.* 727b9). But see 716a3–16 and 766b12–15, where female *sperma* seems to be grammatically implied: *to tou arrenos sperma . . . to de tou thēleos*. Compare also Arist. *Metaph.* 1049a3 and 1049a15, on which see M. L. Gill 1989, Connell 2001.

that women would be less able to concoct their food properly. In fact, for Aristotle, the inability of the female to concoct her food completely into semen becomes in a certain sense the defining characteristic of what it means to be female.[9] Core Aristotelian ideas about the workings of "the active" and "the passive" in causation also play important roles. Aristotle argues repeatedly in his scientific texts that all change involves two conceptually distinct entities, one acting, the other being acted upon. When a plant grows, its internal nature acts to change its material structure from one thing (or size or configuration) into another thing (or size or configuration). When an animal moves, its soul acts to change the position of its body. When a rock falls from a cliff, natural motion brings its heavy, earth-like material closer to the center of the Cosmos. In each of these cases, we see a mover of some sort moving a moved (acted-upon) body. Thus, in generation, Aristotle sees the semen as mover, causing change in the (passive) blood of the female to produce offspring.

That the semen of the male is the active mover and the menstrual blood of the female is the moved material is central to how Aristotle understands generation. In fact, he thinks that the male seed contributes no material at all to the child – all of the material in fact comes from the mother in the form of her menstrual blood. The male seed, semen, simply causes motion in the blood that forces it to change into offspring and the semen apparently evaporates materially after this process is complete. Aristotle likens what happens to the early stages of cheese-making.[10] In order to make cheese from milk, it is necessary first to clump the fats and protein in the milk together into a more solid mass. To accomplish this, cheese-makers introduce small quantities of rennet or something similar which produces clumping and separates the curds out from a watery liquid called whey. This process of coagulation of a liquid is, Aristotle thinks, very similar to the early stages of the formation of a fetus from menstrual blood. The solidification of a body from the introduction of a small quantity of foreign liquid, though, is not merely analogous; the process is in fact related, since menstrual blood and milk are in some ways the same kind of thing for Aristotle: they are both, after all, residues left over from the production of food in a mother. The semen, then, imparts change in the material strata, the menstrual blood, and is thus seen by Aristotle as imparting "form" in the offspring.

There is one other function to male seed, closely related to its role as form to the female's matter, which is that male seed is responsible (and solely

[9] Arist. *Gen. An.* 728a18. [10] Aristotle is not alone in this analogy. See Ott 1979.

responsible) for the imparting of the *psukhē*, the soul, to the offspring. It is not that the mother does not have a soul herself (Aristotle explicitly says that she does and that it is the same as the soul in males), but that because of the incomplete concoction of her seed she is unable to impart an actuality, to use the Aristotelian terminology, to the matter she supplies. On an Aristotelian understanding, matter considered just as (abstract) matter – what philosophers call "matter *qua* matter" – has no particular shape or qualities. Babies, by contrast, each have a particular shape and quite a number of perceptible qualities. They come to this condition for Aristotle by having a form, an actuality, imposed on the wider range of potentials represented by the raw matter itself. This actuality – what makes the individual living thing what it is *as* a living individual – is its soul. The details of what this means are difficult and often obscure to the modern reader (Aristotle dedicates an entire treatise to the subject) but we can take for our present purposes just the idea that the soul individuates the fetus and is responsible for the motions that are its living functions (eating, growing, breathing, moving, thinking). This soul is imparted to the offspring by the male.[11] Part of the reason why Aristotle has the woman providing only matter and potential, and the male providing no matter but actuality, is in order to explain why it is that we do not see parthenogenesis in nature.[12] If the woman could provide actuality (soul) to the fetus, then there would be no reason why she could not give birth in the complete absence of male seed. The fact that babies always require two parents, and that those parents must always be a male/female pairing, means that each must be contributing something distinct to the offspring to make it viable, and Aristotle finds the answers in his already developed, exhaustive, and sophisticated pairwise theories about change, causation, and the soul (active/passive, form/matter, actual/potential).

Aristotle on likeness

Having laid out the general Aristotelian account of generation, we can now look in more detail at how likeness obtains between children and parents. As with the rest of Aristotle's theory of generation, this aspect too relies on a considerable theoretical background developed across the Aristotelian

[11] Although in the case of spontaneously generated animals it is imparted by the heat of the sun. Aristotle tells us that the *pneuma* that is the soul-stuff in semen is "analogous" to the *aether* that makes up the sun and the stars (*Gen. An.* 736b35). See also Lennox 1982.

[12] Spontaneous generation is not parthenogenesis, and Aristotle says that parthenogenesis has never been observed, although he does speculate about the possibility in "some fish" (*Gen. An.* 741a33).

corpus. In the first instance Aristotle conceives of each parent both as a specific individual and as a human being or animal more generally. The words "specific" and "general" in this instance need to be understood as Aristotle understood them, as being derived from "species" and "genus." For Aristotle, genus and species have nothing to do with a formal system of biological taxonomy in our sense but instead were used much more broadly, and as applicable to any situation where we want to talk about the relationships between an individual thing or concept as distinct from the more general kind of thing it is.[13] An individual person is, on his understanding, a *species* of a larger genus. Socrates is thus a species of the genus "human" (or "male," or "Athenian," depending on how we want to talk about him), "human" is a species of the genus "animal," "animal" a species of the genus "living thing," and so on. We may also, if we like, talk about couches as a species of the genus furniture, which shows that genus and species are logical categories rather than biological.[14]

To account for the actual transmission of particular traits from the parents (as well as for the failure of this transmission), Aristotle borrows an important idea we already saw in the Hippocratic *On Generation*, that of mastery or conquering. So, for Aristotle, the male sperm can be mastered or conquered by the mother's blood (and he uses the same word, *krateisthai*), and so traits from the father can fail to be passed on. The mechanism is subtly complicated in Aristotle, however, by the addition of a second kind of possible failure in transmission, something he calls "being loosened," *luesthai*. The sperm of the father or the blood of the mother can be conquered, loosened, or both, so traits can be passed on – or fail to be passed on – in more than one way, each of which has its own complications.

To see this in action, we first need to determine what exactly happens when the semen meets the blood in the womb and coagulation of the fetus takes place. At first blush, since the semen is providing all the motive force for the coagulation or actualization, one might expect that the child should then look exactly like its father in all respects (the mother, after all, is providing only the raw material which is – in theory at least – being wholly actualized by the semen). This may mean that in the ideal case, the child should not only resemble the father in all its partwise traits, but also with respect to its sex (contrast this with the earlier Hippocratic

13 See, e.g., Balme 1987.

14 Victoria Wohl has pointed out to me that it is tempting to think of likeness as the paradigmatic operator linking genus and species in all instances (joining members of – and as – a genus), even if Aristotle himself never talks about it in this way.

account where sex was never a resemblance relation).[15] We thus should get a boy identical to his father. Of course, this never happens, so how to explain what really occurs? Aristotle sees the reality of trait inheritance, where some traits come from the mother, some from the father, some from grandparents, and some from no one in particular, as a series of departures from the ideal boy-just-like-his-father case. "Departure" is, in fact, not quite a strong enough word. Aristotle sees it as a series of small *terata, monstrosities*. The reason for his use of such a loaded word is that he seems to be looking farther down the explanatory chain when he looks at departures from fatherlikeness. He wants a theory that can explain not just femaleness or mother's-nose-ness but that can explain – using the same set of simple mechanisms – any departure at all from resemblance to the parents, including what the ancients would call monstrosities. He then weakens the definition of a monster to such a degree that it includes *any* departure from fatherlikeness. The birth of a female child, different from her father with respect to sex, is just this kind of departure:

> καὶ γὰρ ὁ μὴ ἐοικὼς τοῖς γονεῦσιν ἤδη τρόπον τινὰ τέρας ἐστίν· παρεκβέβηκε γὰρ ἡ φύσις ἐν τούτοις ἐκ τοῦ γένους τρόπον τινά. ἀρχὴ δὲ πρώτη τὸ θῆλυ γίγνεσθαι καὶ μὴ ἄρρεν.
>
> Indeed, what is not like the parents is in a certain sense a monster, for its nature has deviated in some way from the type (*genus*). The first beginning [of this] is the generation of a female instead of a male. (*Gen. An.* 767b6–8)

One may legitimately recoil at the sentiment expressed here,[16] but it is worth attending to the mechanism that lies behind it, since at its heart stand the central ideas of Aristotle's theory of reproduction. It appears that because of the Aristotelian mechanism of inheritance and the respective roles of the two parents, likeness to the father should require no explanation at all. Nevertheless, we never see perfect likeness to the father, so the

[15] For the ideal case in Aristotle, see *Gen. An.* 766a19; Balme 1987.

[16] I have often touted the methodological importance of the historian's suspending his or her disbelief and giving the ancient arguments as much play as they deserve (see Lehoux 2012), but this somehow seems easier with astrology and giant grasshoppers than it does with the astoundingly sexist attitudes of many ancient biological authors (all male, not coincidentally). Perhaps part of the problem is that these latter attitudes did more than just deceive people; they in fact caused real harm to real people in the world and have exerted a negative influence even up to the present day. Nevertheless, I will throughout this chapter treat the ancient attitudes as matter-of-factly as I can. They are what they are, and I trust that modern readers will see them for what they are without repeated comment or criticism on my part. The literature on sex and gender in antiquity (and in Aristotle in particular) is massive, but for illustrative purposes I might here cite N. D. Smith 1983, Laqueur 1990, Dean-Jones 1994, Blundell 1995, Freeland 1998, King 1998, Connell 2000, Flemming 2000, Mayhew 2004, Kosman 2010.

ideal situation seems never quite to apply – an important caveat. What is interesting is just that Aristotle still parses the problem set, this unlikeness to the father, as *likeness* to others rather than as unlikeness in itself, no matter how far the divergence in question. Thus it is on these grounds that he looks for some additional mechanism to account for likeness to the mother, likeness to grandparents, or to others, and this mechanism seems at least partly to accede some active or quasi-active role to the female's matter, clouding the purest version of his initial matter/form and active/passive distinctions.

In other works, most notably for our purposes the *Metaphysics* and *On the Soul*, Aristotle troubles himself at length with the puzzles and problems thrown up by his form/matter distinction. Modern commentators similarly have long argued over the precise meanings of many of Aristotle's terms. In general, though, the arguments about whether or in what senses we can speak about form and matter as being one or as being inseparable, have confined themselves to Aristotle's discussions about physical objects such as bronze mirrors or about already-formed living individuals such as Socrates. In the case of the bronze mirror, the matter is bronze and the form is flat, circular, and polished. Its form, however, can change, through reworking, melting, or corrosion. Nevertheless, the form at any given moment determines the actuality of the various potentials inherent in the matter at that time (mirror-ness rather than statue-ness or sword-ness). Importantly, though, we never have formless bronze – such a thing exists only conceptually. We may find it convenient sometimes to talk of bronze just as bronze in the abstract, but in fact bronze only exists in the world as bronze *things*.[17] Form and matter, then, are always one, even if they are conceptually or logically distinguishable for some purposes. In the case of Socrates, the situation is a little more complex insofar as many of the changes that happen to Socrates come not at the hands of other people (as in the case of crushed or melted bronze), nor just at the hands of external physical causes (as in the case of corrosion). Instead, many of the changes that happen to Socrates are internally caused: he grows up from a child, becomes hungry, moves around in the world, learns things, grows old. In the case of living things, we need to account for what acts to cause these changes and the answer that Aristotle settles on is that the *psukhē* is the agent causing motion and change in the body, which is to say it is the *form* of the body's matter. Like with the form and matter of the bronze mirror,

[17] "Things" here include ingots and lumps. On issues of unity in form and matter, see Shields 1988. See also Lennox 1997, Bolton 1997.

the psyche and the body are always coexisting, at least insofar as the body is a living and self-moving organism.

At the point of generation, though, which is to say at the moment of conception, the motion of the child's soul has an absolute beginning, where what were previously two distinct and non-living fluids, the menses and the semen, come together and become something very different from what they were, which is to say they become a single self-guided, self-developing fetus. We can talk about the forms and the matters of each of the fluids that come to make the fetus, but neither of them is living, neither of them is ensouled in the same way as the baby will be, so something happens when they come together that changes both fluids into something new and remarkable, containing its own internal principal of change and motion, its own soul. Aristotle says:

> τοῦ δὲ σπέρματος ὄντος περιττώματος καὶ κινουμένου κίνησιν τὴν αὐτὴν καθ' ἥνπερ τὸ σῶμα αὐξάνεται μεριζομένης τῆς ἐσχάτης τροφῆς, ὅταν ἔλθῃ εἰς τὴν ὑστέραν, συνίστησι καὶ κινεῖ τὸ περίττωμα τὸ τοῦ θήλεος τὴν αὐτὴν κίνησιν ἥνπερ αὐτὸ τυγχάνει κινούμενον κἀκεῖνο.
>
> The semen, being a residue and being moved with the same motion by which the body grows (being partitioned from final nourishment), when it has arrived in the womb it sets the residue of the female and causes in it the same motion by which it happens to be moved itself. (*Gen. An.* 737a18–22)

The semen, then, "is moved" and causes motion in the menstrual blood. In both cases the motion is characterized with an identical formula that serves to set up a hierarchical sequence running "motion-causing" from the soul of the father to the semen and finally to the residue in the womb (the stepwise formulaic explanation is reminiscent of the kinds of "*a* is to *b* as *b* is to *c*" claims that we find so commonly in ancient science, and which ultimately gesture back to mathematical proportion theory). At the end of this process, we have a self-moving and self-growing fetus with its own independent nature, form, or soul.

It is in this process that we finally see the mechanics of partwise trait determination. The parts as they end up coming to be in the baby will all have been caused by the motions instantiated at this moment in time. As I said above, there are two possible ways in which the fetus can depart from the type of the father: the semen can be mastered (*krateitai*) or it can be loosened (*luetai*). Of these two, loosening is, at least in some ways, the lesser failure (for lack of a better word) of the semen to impart the traits of the father to the baby, and mastering is the stronger. If the motion imparted by the semen is mastered, says Aristotle, the characteristic that should have otherwise been passed on instead *inverts* to its exact opposite characteristic. Thus the birth of a female child is a case of the motion in the semen having been mastered and so the form of the father *qua* male inverts to its opposite in the baby, with the result that she will be a female (and here the opposites must have always been present in potential for this to occur). When the semen is thus mastered, says Aristotle, it is usually (but not always) the case that she will also generally take after her mother, since the father *qua* male and the father *qua* individual-with-certain-features tend to be mastered together.

Where things begin to get more interesting is with respect to the phenomenon Aristotle calls "loosening." Being a smaller departure than mastery – and it is interesting that he thinks of it as a smaller departure – loosening does not cause inversion to the opposite, as mastery does, but instead causes a reversion to something "closer" to the individual parent. Aristotle now clearly shows that he is thinking of the parents in terms of the full flowering of his ideas about genus and species. The parents are individuals (species), who are themselves the children of other parents, which makes them members of families (genus), which are in turn *individual* families (species) of the genus human, and humanity is in its turn a species of the genus animal. Loosening allows traits from a larger genus to manifest in the baby in place of a trait from the individual – specific – parent. A loosening, then, can cause a baby to look like its father's father (the individual closest to the father in the father's own genealogy). A greater loosening can allow for more generic or more distant resemblances: the baby can look like the father's grandfather, for example. So, too, the semen can be mastered with respect to sex *and* loosened with respect to some traits, thus allowing resemblance to the mother's mother, or her mother's mother's mother. The baby can even look like no one in particular in any or many of its features, which Aristotle sees as reversion away from the specific parents and their forebears, to the even broader genus "human." Serious birth defects he parses in these same terms, where the loosening is so great that the resemblance of the part or parts reverts so far as the genus "animal."

For all the examples he gives of loosening in these important passages of the *Generation of Animals* (IV 3), the lines along which the resemblance runs seem at first sight to be curiously gendered. Loosening results in a male child looking like (*eoikos*) his father's father "or an earlier ancestor" or else results in a female looking like her mother's mother "or an earlier [ancestor] (*Gen. An.* 768a31–37)." He offers no examples in these passages of a female child looking like her father's father or her mother's father, or of a male child looking like his mother's mother or his father's mother. Commentators have sometimes wondered about how deeply Aristotle meant to entrench the absence of cross-generational cross-gendered likenesses here – was it really an impossibility on his theory?[18] In fact, it seems not to be, at least not if we take his earlier example of the woman from Elis seriously. Recall that there we saw a white daughter being born to a black father and a white mother, and this white daughter later gave birth to a black son. Although Aristotle does not elaborate on the mechanism for inheritance in his discussion of this case (it was invoked in a very different context), it does at least serve as a clear instance in the *Generation of Animals* of a child, the black son, taking after his mother's father, thus loosening to a male ancestor on the mother's side. It seems that in the technical discussion of loosening in Book IV 3, although Aristotle uses only the phrases "father's father," "mother's mother," and "more distant ancestors" (gender neutral), he must have meant to include "any ancestor" under the various grades of likenesses accounted for by loosening.

The mechanics of loosening are also an interesting case where Aristotle seems to allow at least some degree of agency to the female menses. He likens the situation to the case where a knife becomes dulled by what it is cutting or a hot object cooled by heating up something else. The material acted upon in these cases thus also acts back on the agent of change. The point is particularly clear in the Greek where he uses active and passive participles and the phrasing becomes something like "the cutting-thing is dulled by the being-cut-thing, the heating-thing cooled by the being-heated-thing," which we are to understand as the active being acted upon by the passive.[19] It is nevertheless significant that Aristotle always parses the biological cases as instances of *failure* (*elleipsis*) in the active principal, the semen. He tells us that there are two ways in which the semen can fail: either because it has a deficient *dunamis* in and of itself, or else because it is simply overwhelmed by the volume and coldness of the female menses (heat being one of the main motive vectors of Aristotelian semen).

With this powerful and flexible explanatory framework, Aristotle is able to explain, using only mechanisms for *likeness*, how unlikeness (even extreme unlikeness) to either or both parents can happen. The baby looks unlike the father? This is explainable by likening it to the mother, or a grandparent, or to a more remote ancestor, perhaps even one we do not

[18] See, e.g., Cooper 1988: 28; Henry 2006.

[19] *Gen. An.* 768b16–18: τὸ τέμνον ἀμβλύνεται ὑπὸ τοῦ τεμνομένου καὶ τὸ θερμαῖνον ψύχεται ὑπὸ τοῦ θερμαινομένου.

recall. No family resemblance whatsoever? Then the cause becomes likeness to a larger genus, "human" generally or even "animal" more remotely.[20]

Unlikely likeness

Another interesting aspect of the case of the "woman from Elis" is how remote it is from Aristotle's own experience and just how much of an overkill it is for the point he wants to make in the passage in which he discusses it, which is no more than the idea that any mechanism of inheritance needs to be able to skip a generation. Could Aristotle not have simply pointed out that we often see children resembling grandparents and leave it at that? Why use hearsay evidence from the farthest reaches of the Peloponnesus? Perhaps Aristotle felt that the general proposition of likeness to grandparents might be controversial (the Hippocratic author did not even discuss it, after all). Perhaps Aristotle was taking advantage of differing skin color as a way of leveraging an objective-looking measure of quantities as nebulous as likeness and difference. Furthermore, such distantly mixed-race children were probably much less common in fourth-century Athens than they would be today, and the difference instanced in this case perhaps made it both memorable and dramatic. Indeed, the drama that results from unexpected color became something of a trope in later Greek and Latin literature and would have a long literary history into the early modern period.

Witness the case discussed in the *Declamations* of Calpurnius Flaccus where "a married woman bore a black son" (*matrona Aethiopem peperit*: we are left to assume from the context and perhaps from the class overtones of *matrona* that the woman must not have been black herself).[21] The drama of the passage arises from the immediate suspicion that falls on the woman (and in which it seems Calpurnius wants us to share, at least provisionally) when her baby takes after the wrong race. The prosecution, accusing her of adultery, makes the point that it is a law of nature, *legem esse naturae*, for children to take after their parents and so the blackness of the baby is taken to be likeness to an unnamed black father.[22] Here the binary skin colors, as in Aristotle, serve to underscore the drama and establish the measure of

[20] In the wake of Darwin, though, we need to be careful not to see these latter genus-reversions as being *genealogical* reversions in the same way as reversion to grandfathers or great-grandfathers was. Reversion to "animal" works because humans are members of the genus animal, not because humans *came from* animals in the same way as babies "came from" their grandparents.

[21] *Declamation 2*, dating from the late first or early second century CE. I thank Raffaella Cribiore for bringing this passage to my attention.

[22] On laws of nature in antiquity, see Lehoux 2012.

extreme difference. It leaves no appeal to the subjectivity of likeness that may be invoked had the baby been simply said "not to look like his father," or to have "resembled one of the slaves." The dichotomy of black skin versus white skin seems to offer not just memorability, but also a measure of objectivity. So, too, the prosecution's case and the reader's presumed expectations both make it clear that the woman's defense is meant to be, on the face of it at least, beyond the pale. And an audacious defense it turns out to be, one that happens to hark back to the Hippocratic "problems-in-the-womb" account of dissimilarity, arguing that the baby came out black because it had been burnt by excessive heat in the womb (the defense counsel even hints that we should not be surprised if the problem recurs with her future pregnancies). Indeed, he even tries a clever inversion by suggesting that the very fact that the woman brought the baby to term proves she must have had a clear conscience about its conception.

Much more commonly, though, this literary trope of the baby's skin color comes bundled with a different explanation, one rooted in likeness after all.[23] Witness the character Persinna, queen of Ethiopia in Heliodorus' third-century CE *Aethiopica*, who offered the following explanation to her daughter concerning the girl's unexpected (in this case, white) skin:

> οἱ δὴ τὰς βασιλείους αὐλὰς κατὰ καιροὺς ἱδρυσάμενοι . . . τοὺς δὲ θαλάμους τοῖς Ἀνδρομέδας τε καὶ Περσέως ἔρωσιν ἐποίκιλλον. ἐνταῦθά ποτε ἡμᾶς, δεκάτου παρήκοντος ἔτους ἐξ οὗ με γαμετὴν Ὑδάσπης ἐγνώρισεν, οὔπω τε παίδων ἡμῖν γεγονότων, ἠρεμεῖν τὸ μεσημβρινὸν συνέβαινεν ὕπνου θερινοῦ κατακλίναντος, καί μοι προσωμίλει τότε ὁ πατὴρ ὁ σός, ὄναρ αὐτῷ τοῦτο κελεύειν ἐπομνύμενος, ᾐσθόμην τε παραχρῆμα κυοφορήσασα τὴν καταβολήν . . . ἐπειδὴ δέ σε λευκὴν ἀπέτεκον, ἀπρόσφυλον Αἰθιόπων χροιὰν ἀπαυγάζουσαν, ἐγὼ μὲν τὴν αἰτίαν ἐγνώριζον ὅτι μοι παρὰ τὴν ὁμιλίαν τὴν πρὸς τὸν ἄνδρα προσβλέψαι τὴν Ἀνδρομέδαν ἡ γραφὴ παρασχοῦσα καὶ πανταχόθεν ἐπιδείξασα γυμνήν, ἄρτι γὰρ αὐτὴν ἀπὸ τῶν πετρῶν ὁ Περσεὺς κατῆγεν, ὁμοιοειδὲς ἐκείνῃ τὸ σπαρὲν οὐκ εὐτυχῶς ἐμόρφωσεν.

> Those who in the course of time came to build the royal palace . . . made use of the romance of Perseus and Andromeda to adorn the bedchambers. It was there one day that your father and I happened to be taking a siesta in the drowsy heat of summer. Ten years had passed since Hydaspes [the king] made me his wife, and still no child had been born to us. But that day your father made love to me, swearing that he was commanded to do so by a dream, and I knew instantly that the act of love had left me pregnant . . . But you, the child I bore, had a skin of gleaming white, something quite foreign

[23] For a discussion of other instances see Sussman 1994.

> to Ethiopians. I knew the reason: during your father's intimacy with me the painting had presented me with the image of Andromeda, who was depicted stark naked, for Perseus was in the very act of releasing her from the rocks and had unfortunately shaped the embryo to her exact likeness.[24]

This is a great departure from both Aristotle and the Hippocratic *On Generation*, insofar as likeness, *homoioeides*, comes in from the outside, through the queen's eyes. Persinna, however, worries that no one will believe her version of events, and we too may think that such a far-fetched causal account must have been conceived *post hoc*. As it turns out, however, Heliodorus is not alone in invoking the likeness-inducing power of the eyes. We find a similar causal account in the second-century CE medical writer Soranus of Ephesus:

> τί δεῖ λέγειν, ὅτι καὶ τὸ ποιὸν τῆς ψυχῆς κατάστημα φέρει τινὰς περὶ τοὺς τύπους τῶν συλλαμβανομένων μεταβολάς; οὕτως ἐν τῷ συνουσιάζειν πιθήκους ἰδοῦσαί τινες πιθηκομόρφους ἐκύησαν· ὁ δὲ τῶν Κυπρίων τύραννος κακόμορφος ὢν εἰς ἀγάλματα περικαλλῆ κατὰ τοὺς πλησιασμοὺς τὴν γυναῖκα βλέπειν ἀναγκάζων ὁ πατὴρ εὐμόρφων ἐγένετο παίδων· οἱ δ' ἱπποτρόφοι κατὰ τὰς ὀχείας ἔμπροσθεν τῶν θηλειῶν εὐγενεῖς ἱστᾶσιν ἵππους. ἵνα οὖν μήτε ἄμορφον ἀποτελεσθῇ τὸ γεννώμενον ἀλλοκότους φαντασίας ἐν τῷ μεθύειν τῆς ψυχῆς ὑπομενούσης, νηφέτωσαν αἱ γυναῖκες ἐν τοῖς πλησιασμοῖς, εἶθ' ὅτι καὶ πρὸς τὰς μητέρας ὁμοιότης τις οὐ κατὰ σῶμα μόνον, ἀλλὰ καὶ κατὰ ψυχὴν ἀναφέρεται τῶν γεννωμένων. καλὸν οὖν εὐσταθούσῃ τῇ ψυχῇ καὶ μὴ παρακόπῳ διὰ μέθην ὅμοιον ἀποτελεσθῆναι τὸ γεννώμενον.
>
> What is one to say concerning the fact that various states of the soul also produce certain changes in the shape of the fetus? For instance, some women, seeing monkeys during intercourse, have borne monkey-shaped children. The tyrant of the Cyprians, who was misshapen, compelled his wife to look at beautiful statues during intercourse and became the father of well-shaped children. Also horse breeders, during breeding, place noble horses in front of the females. Thus, in order that the offspring may not be rendered misshapen, women must be sober during coitus because in drunkenness the soul becomes the victim of strange images; this is because the offspring bears some resemblance to the mother as well, not only in body but in soul. Therefore it is good that the offspring be made to resemble the soul when it is stable and not deranged by drunkenness.[25]

Soranus here goes beyond Heliodorus by ascribing causation not to the image coming into the woman's eyes directly, but to the state of her soul

[24] Heliodorus, *Aethiopica* 4.8.3–5, Morgan 1989, trans., slightly modified. My thanks to Victoria Wohl for the example.
[25] Soranus, *Gynaecology* 1.39.1–2, Temkin 1991 trans., slightly modified.

as affected by what she was seeing (and although questions of *seemliness*, another shade of meaning behind *eikos*, may well have underlain the tension in the other, skin-color examples, here we see them brought straight to the fore).

It is worth noting that the causal mechanism in these instances seems to be limited to women's roles in conception. Just as with Persinna, all of Soranus' examples focus on the images that females encounter during intercourse. The tyrant of the Cyprians had his wife, not himself, look at beautiful statues. So, too, noble horses are said to be placed in front of the mares, not the mating pair as a whole. Similarly, Soranus has no parallel injunction against the man's being drunk during conception. At least a partial answer may be hidden in his slightly unusual word choice in the first sentence of this passage: he does not quite say that the child is affected by the state of the woman's soul, but the *tupos*, the "mould" or "impression," reminiscent perhaps of the fruit-mould theory of the womb that we saw in the Hippocratic author.

The second interesting aspect of Soranus' use of mares is the reminder that such breeding could be deliberately manipulative. In the examples of skin color above, it is the counter-expectancy that gives them their pointedness. In the case of horse breeders, expectations and desires for deliberate outcomes instead give rise to purposeful constraints on reproductive practices (Soranus' example of the tyrant of Cyprus is similar in this regard). Chasing down his hint, we do find in ancient veterinary manuals techniques for influencing the characteristics of foals (in fact their color) that use the same visual mechanism to influence the mother during conception.

> ὁπόταν βουληθῇ πῶλον ἐπιμελούμενος τῆς ἀγέλης ἀνὴρ μέλανα ἢ ξανθὸν καὶ ποικίλον ἢ λευκὸν ἵππον ἀπαστράπτοντα ταῖς ξανθιζούσαις θριξὶν ἐπὶ ταῖς λευκαῖς <γενέσθαι>, ἐν τοῖς χωρίοις τοῖς χλοεροῖς τῆς συνήθους πλησίον πηγῆς εἰκόνα ἐκ βαφῶν ποικίλων συγκερασάμενος στησάτω, ἣν βλέπουσα ἡ θήλεια ἐν τῷ πίνειν ὄμμασι διψῶσιν ὥσπερ ποτοῦ καὶ ἐρώτων ἀπεικαζομένοις γάμοις ὅμοιον ἂν ἀποτέκοι τῷ ὁρωμένῳ ἵππῳ. πάντα γάρ, ὅσα ποικίλματα ὁ νοῦς αὐτῆς διὰ τῆς ὁράσεως κατέπιε, τούτοις ἀπεικάσματα καὶ ἡ γαστὴρ ἀπερεύξεται καὶ γεννήσει.
>
> When the man caring for the herd wishes a black, or yellow and dappled foal to be birthed, or a white horse shining with yellow hair upon the white, let him set up an image (*eikona*), having painted it colorfully, in the grassy fields near the usual watering holes. Seeing the image while drinking with thirsty eyes – like a drunk's and in those marriages that produce copies (*apeikazomenois*) of what [someone] lusts after – the female may give birth

> to a likeness of the horse she beheld. Indeed, such patterns as her mind drank through the eyes, the belly will also engender and send forth copies (*apeikasmata*) of them.[26]

Immediately, we notice the prominent use of no fewer than three words built on the root *eikos*: in the image the farmer sets up, in the copies borne by the mare, and in the copy-producing marriages of the drunken humans. So, too, the cross-fertilization of the metaphors with those in Soranus is prominent. Where Soranus likened the human reproduction of likenesses to what happens in horse breeding, we see the author of the *Hippiatrica* likening horse-breeding to human marriages. Also as in Soranus, drinking seems to be an important proximate cause for some reason, and the author here stretches it from the tippling of human couples to the thirst for water in the mare. And, again, the focus is entirely on what the female sees, not on the male.

What really leaps out as new in this passage, though, is the expressly qualified certainty offered by the author. Our previous authors have talked about reproduction and the causes of the different kinds of similarity in the complete absence of any explicit talk of probability (even if probability seems often to be implied in expressions of expectation and surprise). The basic model has been to explain either what we know in general to happen (babies resemble or fail to resemble parts of each parent), or to explain remarkable occurrences (historicized and involving named individuals), such as when specific babies turned out the wrong color. In the *Hippiatrica* we see expectation and desire moved to the forefront. The author of this passage is not writing a treatise on causation, but instead instructing the reader in a *tekhnē*. He is offering us a method for achieving a precisely defined and desired result: a horse of a certain and (one presumes, expensive) color.[27] Given the kinds of concerns and explanations at play in our previous authors it seems clear that they never thought any particular instance of generation was guaranteed to produce any particular result, so ideas of probability always underlay their expectations just as improbability underlay the surprises. But here we see the author explicitly invoking the likelihood of the outcome with his use of the grammatical construction of

[26] *Hippiatrica Cantabrigiensia* 44.4. Compare also *Hippiatrica Berolinensia* 14.9 and *Geoponica* 16.21, where we are told to clothe the male in a blanket of the color we wish the foal to be (again, the female's eyes are the key).

[27] Even Soranus' account of the case of the Cyprian tyrant does not say that he compelled his wife to look at beautiful statues *in order to* have well-shaped children; he compelled her to do so *and then had* well-shaped children. We may well see his action as purposive, but it is not quite worded that way in Soranus and the inference is left to the reader.

an plus the optative, *apotekoi*. She *may* produce a resemblance of the *eikona* she was looking at. Exactly how likely he thinks the outcome to be is not clear, though it would seem from other references to similar practices that expectations of success were at least seen as commensurate with the efforts and rewards.

Conclusion

I have already remarked that this last case, unlike most of the other instances of image copying, is marked by its goal-directedness. The other difference, closely related to this, is that here the change in color is presented as inherently desirable, whereas many of our other instances of dissimilarity hinged on the unseemliness (*aeikēs*, though they do not use that word) of the baby's difference.[28] For Persinna and Calpurnius' *matrona*, the skin color of their babies puts their lives in direct danger from charges of adultery. So, too, in Aristotle and the Hippocratic authors, words denoting undesirability or unseemliness begin to cluster around their causal descriptions in exactly the same way as the baby's likeness becomes more and more distant from the parents themselves. "Monsters" and "maimings" come to be contrasted with "healthiness." Words like "sickly" and "crooked" (which in Greek carry the same moral connotations as the English) are used freely. It is interesting that this clear discomfort with increasing perceptions of unlikeness comes to be physically handled by the same mechanisms that also seek to assuage male anxiety about paternity, no matter how implausible those mechanisms may seem to be (not only to us, but clearly also to the ancients themselves if the dangers to Persinna or the *matrona* mean anything).

Likeness for offspring, then, comes to seem rather extraordinarily likely, insofar as even what are taken to be the extremes of unlikeness come to be handled under the rubric of likeness to something beyond the parents themselves (although the case of Calpurnius Flaccus shows that such a move was not necessarily universal, for he reverts to a fruit-mould theory rather than distant similarity to account for extreme difference). On the other hand it also seems clear that – thinking about it another way – the paradigmatic instance of (apparently undesirable) unlikeness in color was seen as unlikely enough for individual cases to be surprising and suspicious. But this point seems only to hold for human reproduction: in veterinary and agricultural texts, where color changes are represented in many cases as inherently desirable, the probability of achieving the end seems to be

[28] Again, with thanks to Victoria Wohl.

thought of as being within reasonable reach, even if it was not quite taken for granted, and indeed it is exactly in this context that we see the one explicit expression we find describing the probability of the outcome.

One now begins to speculate on why it should be the case that likeness to things seen should be thought of as more likely in animals than in humans. Is the end easier to achieve in beasts because they are more easily fooled by what they see? Here one may want to note that the influence of vision in humans is, in all of our cases, limited to human females, so perhaps ancient attitudes about gender and reason should come into play. On the other hand, perhaps the hint in Soranus about the danger being to the *tupos*, the mould of the baby, is significant – is this somehow just a (likeness-seeded) fruit-mould theory? In any case, the fruit-mould theory shares with the image-likenesses the idea that the cause of extreme difference is located in the female. Only Aristotle, with his mechanism of loosening-to-genus, allows the chain of causation for difference to run on the male side as well as the female.

CHAPTER 12

Galen on the chances of life

Brooke Holmes

On the subject of nature, Galen was an indomitable optimist. He embraced the claim made in Plato's cosmological magnum opus, the *Timaeus*, that the world and everything in it were created by an intelligent Demiurge, and he never grew tired of admiring the creator's handiwork. He accepts, with equal enthusiasm, Aristotelian teleology, according to which each part of an organism is oriented toward an end (*telos*) – namely, the life proper to that organism. For Galen, every part of the human body has been created "so that it would not have been better had it come into being differently" (ὡς οὐκ ἄν, εἴπερ ἑτέρως ἐγεγόνει, διέκειτ' ἂν ἄμεινον).[1] His monumental hymn to Nature, *On the Usefulness of Parts*, is nothing less than an attempt to prove this claim. Throughout his vast corpus of writings Galen is passionately committed to the intelligence of Nature and the boundless foresight with which "she" enables us to live and thrive.

Yet, at the same time, Galen did not believe that the Demiurge had created a perfect world. In a well-known critique of the God of Moses, one that echoes the *Timaeus*, he emphasizes that the Demiurge achieves the best results he can, given the material conditions under which he has to work.[2] These conditions mean, among other things, that the organism's continued survival is never guaranteed in advance. It is always, rather, only a likely outcome, and therefore firmly grounded in the domain of probabilities. As it turns out, there are many ways in which things can go wrong. To the

I am grateful to audiences at the *Eikos* workshop in Toronto, the "Approaches to Ancient Medicine" conference in Exeter in August 2011, and the "Medical Powers" conference held at Oxford in September 2012 for their feedback on earlier versions of this study, and especially to Anna Marmodoro, Jim Hankinson, Philip van der Eijk, and Chris Gill. I would also like to thank David Leith, David Kaufman, and Victoria Wohl for their helpful responses to the written chapter.

1 Gal. *UP* 1.5 (III 9 K = 6,21–22 Helmreich). Abbreviations for Galen's texts follow Hankinson 2008: 391–97, which also includes information on modern editions.

2 Gal. *UP* 11.14 (III 906 K = 158,23–26 Helmreich). On Galen's teleology, see Hankinson 1989, Schiefsky 2007. For the application of the "best possible world" principle to the human body in the *Timaeus*, see Burgess 2000.

extent that Galen was, first and foremost, a physician, he was exceptionally well acquainted with the malfunctioning of the human organism.

Galen navigates, then, between an unwavering belief in Nature's intelligence and self-sufficiency and a professional interest in the rich spectrum of organic failure. These apparently conflicting commitments can be better understood as mapping the two poles of the natural world as Galen sees it: on the one hand, a deck stacked in favor of life by virtue of the Demiurge's intentions; on the other, the labile materiality of mixtures. Between them lies the terrain where the flourishing of life is a probable outcome. It is a space of dynamic open-endedness, the space of the live body, not the dead one, of physiology, not anatomy.[3] If we are interested in the probabilities that govern human life, we should begin here, with the conditions that make life itself probable in the first place.

All organic life is open-ended. Yet open-endedness is *especially* true of human life. The reason, for Galen, as for a number of ancient medical and moralizing authors, is that the space of probable flourishing is open to our interventions in it and, more specifically, our impact on the material conditions of life. What the embodied agent does or does not do – for example what he eats, how he exercises, how he sleeps, whether he worries or grieves excessively – can tilt the odds toward flourishing or disease, life or death. The odds are affected, too, by the actions of the physician. The success of these interventions depends, in part, on the self-control of the embodied agent (whether he indulges his appetites, whether he keeps to a regimen). Here, open-endedness correlates with ethical possibility. But success also depends on how much we as humans can ever know about how to maintain our vital being. Here, the open-endedness of organic life remains stubbornly open because of the constraints on our capacity for mastery and, more specifically, our capacity to understand the workings of life.

Galen was, in fact, highly self-conscious about what he perceived as the limits on our understanding of certain things, such as the nature of the soul and the translation of Demiurgic intentions into the immanent teleological intelligence of nonconscious life. In these areas he is aware that our accounts are, at best, plausible, but still only probable, falling short of certainty.[4] More than once he qualifies his own account by invoking

[3] See Debru 2008: 280–81. As Debru observes, the more Galen moves from anatomy to physiology, the more his accounts "become nuanced, complex and plausible only."

[4] The account that is plausible, probable, persuasive, reasonable – for which Galen prefers to use the language of *to pithanon*, rather than *to eikos* – is sometimes opposed to what is mistaken (*mokhthēron*), as at *SMT* 2.5 (XI 474 K). But the language of plausibility can also be negatively inflected. It may designate something that is plausible but untrue, as at *Prop. Plac.* 11 (182,23 Boudon-Millot and

the *locus classicus* of the "likely" story (*eikōs muthos*, *eikōs logos*), that is, the moment in Plato's *Timaeus* when the eponymous narrator offers up a likely story of the creation of the cosmos.[5] On other occasions he despairs of finding even a probable story to tell about how the plan of the Demiurge is actualized in living beings. The question of probability thus concerns not just the success or failure of organic life, but also the stories we tell about it.

In what follows I first demonstrate Galen's commitment to probable flourishing against models of nature and the body that privilege the forces of chance and necessity, considering in particular his struggle to define the nature of the purposefulness of nonconscious or what I will sometimes call vegetal life. Such an inquiry is important to understanding how Galen uses teleology to account for the likelihood of flourishing. I go on to consider the reasons why such immanent purposefulness is not sufficient to guarantee life. The most important consequence of such insufficiency is the emergence of a space of ethical agency, where human beings take responsibility for the care of their own lives by making use of the resources of medicine to mimic and complete the technical labor of Nature.[6]

It is precisely the space of ethics that defines human life against vegetal life, where life does not *choose* to live, even as it unfolds with a sense of purpose and "technical" skill.[7] Yet, at the same time, human life comprises vegetal life – that is, the nonconscious body – and depends on it. In the final

Pietrobelli); *Sem.* 2.5.15 (IV 629 K = 182,16 De Lacy). At *Sem.* 1.1.2 (IV 512 K = 64,9–11 De Lacy), Galen seeks arguments that rely not on plausibility but on clear demonstration (*ex enargōn te kai di' enargōn*). Galen draws a similar distinction between logical demonstration and arguments from *eikos* at *Pecc. Dig.* 7.10 (V 101–102 K = 67,21–23 De Boer).

5 Gal. *Prop. Plac.* 13 (IV 759 K = 186,37 Boudon-Millot and Pietrobelli): *akhri tou pithanou kai eikotos autēn proerkhesthai phēsi.* He uses the same expression (*[akhri] tou pithanou kai eikotos*) three times at *PHP* 9.9.3–7 (V 792–93 K = 598,9, 20, 26 De Lacy), where he quotes from the *Timaeus*. On the *eikōs muthos*, see esp. Pl. *Ti.* 29c4–d3. Several recent interpretations (Burnyeat 2005, Bryan 2012: 114–60) have sought to show that the "likely story" should be understood as a positive achievement, rather than something "merely" probable; but it seems that for Galen, the probable account represents an (undesirable) epistemological limit. For the expression *akhri tou pithanou*, see also *Nat. Fac.* 1.14 (II 55 K = 141,17 Helmreich); *PHP* 9.7.12 (V 794 K = 600,22 De Lacy); *Prop. Plac.* 13 (IV 760 K = 186,37 Boudon-Millot and Pietrobelli), 15 (IV 764 K = 190,6 Boudon-Millot and Pietrobelli). Note that while I translate *to pithanon* as "probable," in some contexts it may mean something closer to "persuasive."

6 I have argued elsewhere that such an ethical space first appears in Hippocratic medicine: Holmes 2010: 192–227.

7 Galen tends to work with the Stoic-inspired division of living into two categories: the works of nature (*phusis*) and the works of the soul (*psukhē*): see von Staden 2000: 102, 107–11. Indeed, *On the Natural Faculties* begins with this division: *Nat. Fac.* 1.1 (II 1 K = 101,4–5 Helmreich). Galen was aware that what he was calling nature could also be called the "appetitive" soul (by Platonists) and the "nutritive" soul (by Aristotelians): see, e.g., *Foet. Form.* 3.13 (V 665 K = 68,12–17 Nickel); *PHP* 6.3.7 (V 521 K = 374,13–19 De Lacy); *Prop. Plac.* 3 (174,16–19 Boudon-Millot and Pietrobelli), 13 (185,38–186,1 Boudon-Millot and Pietrobelli).

section I take a look at some of Galen's reflections on the discontinuity of vegetal life and ethical life within human nature. I conclude by arguing that such discontinuity leaves its trace in the probable nature of Galen's stories about the nature of immanent intelligence, as well as in the uncertainty that characterizes the relationship of every embodied subject to his or her vital being.

Immanent teleology

We saw above that Galen professed throughout his career a strong commitment to the idea that Nature and individual natures are organized by a guiding intelligence. What Mark Schiefsky has labeled Galen's "biological functionalism" is especially visible in the massive tome *On the Usefulness of Parts*, in which Galen finds purpose in every last structure of the human body.[8] It is equally visible in *On the Natural Faculties*, where Galen goes beyond anatomical structure to inquire into the faculties (*dunameis*) that sustain the organism. If I focus here on *On the Natural Faculties*, it is because it offers one of the clearest pictures of the larger stakes underlying Galen's philosophical position, as well as some of the difficulties that position poses.

The subject matter of the treatise appears, at first glance, dry: the threefold activities of nature (*phusis*) in a living body, generation, nutrition, and growth. Yet it is not long before we realize that these subjects touch a raw nerve for Galen. Having embarked on the topic of nutrition, Galen veers into a heated discussion of the two opposed sects of those with something to say about nature:[9]

> τίνες οὖν αἱ δύο αἱρέσεις αὗται καὶ τίς ἡ τῶν ἐν αὐταῖς ὑποθέσεων ἀκολουθία; τὴν ὑποβεβλημένην οὐσίαν γενέσει καὶ φθορᾷ πᾶσαν ἡνωμένην θ' ἅμα καὶ ἀλλοιοῦσθαι δυναμένην ὑπέθετο θάτερον γένος τῆς αἱρέσεως, ἀμετάβλητον δὲ καὶ ἀναλλοίωτον καὶ κατατετμημένην εἰς λεπτὰ καὶ κεναῖς ταῖς μεταξὺ χώραις διειλημμένην ἡ λοιπή.
>
> What, then, are these two sects and what is the logical outcome of their hypotheses? The first type of sect posits that all substance subject to genesis and corruption is at once continuous and capable of alteration. The other believes that substance is unchangeable, unalterable, and subdivided into fine particles and separated by empty spaces in-between. (Gal. *Nat. Fac.* 1.12 [II 27 K = 120,14–21 Helmreich])

8 Schiefsky 2007: 388–91.

9 Gal. *Nat. Fac.* 1.12 (II 27 K = 120,8–9 Helmreich). On the "historiographical strategy" in play here, see Vegetti 1999: 389–95. See also, e.g., *Caus. Morb.* 1 (VII 1–2 K).

The "logical outcome" of these respective ontological positions is far-reaching. The corpuscular theorists, Galen argues, deny that we can assign a role to Nature or Soul. Everything depends, rather, on small bodies interacting below the threshold of perception. These theorists hold that we are nothing more than cattle, enslaved to the impressions of our senses. They deny dreams, birds, omens, and the whole of astrology: the gods, after all, could not care less about us. They think wisdom and temperance and filial love are all nonsense (*lēros*). By contrast, the other sect puts Nature first and foremost ontologically. It is Nature who creates plants and animals, Nature who endows creatures with the faculties that they need to survive and flourish according to their respective natures, including the faculties of caring for offspring, forming friendships, and developing moral judgments. Nature, according to this philosophy, does everything with skill and order (*tekhnikōs kai dikaiōs*), always keeping in mind the best outcome for the organism.[10]

Galen, unsurprisingly, is not shy about pledging his allegiance to the second sect. In other treatises he argues that the very structure of the human body gives a clear indication (*endeiknusthai*) of the wisdom and power of the one having created it, making the existence of the Demiurge a necessary conclusion.[11] Needless to say, Galen finds fatally improbable the arguments of those who claim that the body was created by "unskilled and irrational chance" (*kata tina tukhēn atekhnon kai alogon*).[12] It is not "by chance or without reason" (*ou . . . hōs etukhen oud' alogōs*), he argues in *On the Natural Faculties*, that Nature has created the uterus in such a way that it contracts around and retains the embryo for a certain amount of time, nor is it "at random and by chance" (*ou . . . eikēi ge kai hōs etukhen*) that there are fossae of a certain narrowness between the ventricles of the heart.[13] The foresight and the purpose of the Demiurge are evident even in the slime of the body.

The strongest evidence by far, however, for the intelligence and purposefulness of Nature comes from the activities (*energeiai*) responsible for creating and sustaining life – genesis, growth, and nutrition – and the "natural faculties" (*phusikai dunameis*) that lie behind them. It is just these faculties that are denied by corpuscular theorists, who argue that nature

[10] Gal. *Nat. Fac.* 1.12 (II 29 K = 122,9 Helmreich).

[11] Gal. *Foet. Form.* 6.2 (V 687 K = 92,3–5 Nickel). See also *Foet. Form.* 6.19 (V 695 K = 98,24–26 Nickel); *PHP* 9.8.3–27 (IV 783–91 K = 590,20–596,29 De Lacy); *UP* 17.1 (IV 360–61 K = 447,16–448,3 Helmreich), with *endeixis*.

[12] Gal. *PHP* 9.8.4 (IV 783 K = 590,22–23 De Lacy).

[13] Gal. *Nat. Fac.* 3.3 (II 148 K = 207,25–26 Helmreich), 3.15 (II 208 K = 252,11–12 Helmreich).

does not pre-exist and direct the formation of a complex body but rather emerges from a conglomeration of primary elements that lack any creative powers, that is, the powers to shape, increase, and nourish the organism. Moreover, these elemental micro-bodies are impassive and unchanging, making genuine genesis, growth, and nourishment impossible.[14] Those who would defend a corpuscular position are forced to accept that living beings come together randomly and the processes that sustain life are without *tekhnē*. Everything here, Galen alleges, is chance and necessity.

How do the corpuscular arguments work? Galen's attacks here are directed mostly toward two physicians who explained activities such as nutrition and urinary secretion in terms of overarching mechanistic principles, rather than innate faculties: Asclepiades of Bithynia (late second or early first century BCE) and Erasistratus of Ceos, the third-century BCE Alexandrian anatomist. Asclepiades reduced everything that happens in the body to the movement of invisible corpuscles (*onkoi*) through invisible passages or, more likely, voids (*poroi*).[15] The dominant principle governing these small bodies is what he called "movement towards what is fine" (*pros to leptomeres phora*). In the case of urinary secretion, for example, Asclepiades treats the bladder as nothing but a sponge that absorbs the fluid we drink in the form of vapors.[16] In other words, he finds the material composition of the bladder (its being porous) sufficient to explain the fact that urine collects in it: no immanent skill or intelligence necessary. Matter is, at its heart, undifferentiated, unshaped by purpose, and responsive only to the dictates of mechanistic necessity.[17] Erasistratus' position is more complicated, insofar as he seems to have adopted some form of Aristotelian teleology.[18] Still he too, according to Galen, eschewed the natural faculties, relying instead on the principle of "movement toward the void" (*pros to kenoumenon akolouthia*) to account for everything that happens in the body.

Galen does not deny the Erasistratean principle of *horror vacui* altogether. Later in *On the Natural Faculties* he accepts that there are two kinds of attraction: one by which a vacuum is filled, the other by which particular types of matter are attracted by particular bodies or parts of bodies. The

[14] Gal. *Nat. Fac.* 2.6 (II 100–101 K = 174,13–23 Helmreich). On the impassivity of the corpuscles, see also Gal. *Hipp. Elem.* 2.17 (I 418–19 K = 62,4–13 De Lacy), with Leith 2013.

[15] Leith 2012 convincingly argues that Asclepiades is referring to interstitial void with the term *poros*. See further, on the *onkoi*, Leith 2009.

[16] On the permeability of the parts of the body in Asclepiades, see Leith 2012: 183–84.

[17] See Vallance 1993: 699–701.

[18] Gal. *Nat. Fac.* 2.6 (II 102 K = 175,5–6 Helmreich), 2.7 (II 106 K = 178,13 Helmreich). On the difficult question of Erasistratus' relationship to Aristotle and the Peripatetics, see esp. von Staden 1997, Cambiano 2000. See further Lonie 1964: 441–43, Scarborough 1985.

first type is illustrated by air drawn into a bellows; the second, by the lodestone's attraction of iron.[19] Both kinds of attraction are used by Nature in arranging the body of the animal for the best. But it is the second type of attraction, that based on "appropriateness of quality" (*oikeiotēs poiotētos*), that most spectacularly exhibits Nature's providential reasoning. And to explain this type of attraction you need the natural faculties.

So what are the natural faculties? For one thing, unlike anatomical structures, they cannot be seen directly. Galen classifies them formally as a type of cause (*aitia*), the presence of which has to be inferred through its visible products (e.g. flesh or blood) and the activities (*energeiai*) it makes possible.[20] In this respect, the faculties resemble the Demiurge himself, who is glimpsed in the goal-directed behavior of his creations. The resemblance goes deeper still. Like the master creator, the essence of the natural faculties is unknown.[21] In fact, the relationship between the Demiurge and the natural faculties is not just analogical but direct; for it is through the activities of the faculties – and especially the attractive faculty and the expulsive faculty – that the Demiurge's plan comes to be enacted within the life of an individual organism.

What allows the natural faculties to translate the Demiurge's creative intelligence into the immanent purposefulness of nonconscious life is their capacity to differentiate between kinds of matter and, more specifically, between what is appropriate (*oikeion*) and what is foreign (*allotrion*) to the organism; for these capacities are crucial to the organism's basic task of self-maintenance, given that it is in a constant state of flux. If it is to succeed in conserving its particular being, a body needs to know how to replace what is lost with matter it can assimilate to itself (*to oikeion*). At the same time, whatever it cannot make its own or whatever has become not-self (*to allotrion*) has to be eliminated. Without these powers of discrimination, life, Galen believes, is impossible to sustain.

Yet does the discriminating work of the faculties exhibit actual intelligence or intentionality? The difficulty one runs into with such language is made clear in an analogy from the second book of *On the Natural Faculties*, in which Galen is trying to explain the generative power of the male seed. He starts off by likening the seed to the famous sculptor Pheidias. Both possess the faculties of *tekhnē*, which are activated when they meet with the

19 Gal. *Nat. Fac.* 3.15 (II 206 K = 251,5–6 Helmreich).

20 Gal. *Nat. Fac.* 1.2 (II 6–7 K = 105,13–16 Helmreich).

21 Gal. *Nat. Fac.* 1.4 (II 9–10 K = 107,14–20 Helmreich). For Galen's uncertainty about the nature of the Demiurge, see *Foet. Form.* 6.2–3 (V 687–88 K = 92,5–9 Nickel), 6.20–23 (V 695 K = 98,26–100,10 Nickel).

proper material (wax and female generative matter, respectively). Pheidias is external to what he creates, and so is the seed, at least in the beginning. But here is where the analogy starts to run into trouble; for the seed, as soon as it begins to work on the matter, *becomes* the nature of the animal, no longer external to the material.[22] And this makes it very different from the craftsman:

> ἡ γὰρ διαπλάττουσα τὰ μόρια φύσις ἐκείνη καὶ κατὰ βραχὺ προσαύξουσα πάντως δήπου δι᾽ ὅλων αὐτῶν ἐκτέταται· καὶ γὰρ ὅλα δι᾽ ὅλων οὐκ ἔξωθεν μόνον αὐτὰ διαπλάττει τε καὶ τρέφει καὶ προσαύξει. Πραξιτέλης μὲν γὰρ ἢ Φειδίας ἤ τις ἄλλος ἀγαλματοποιὸς ἔξωθεν μόνον ἐκόσμουν τὰς ὕλας, καθὰ καὶ ψαύειν αὐτῶν ἠδύναντο, τὸ βάθος δ᾽ ἀκόσμητον καὶ ἀργὸν καὶ ἄτεχνον καὶ ἀπρονόητον ἀπέλιπον, ὡς ἂν μὴ δυνάμενοι κατελθεῖν εἰς αὐτὸ καὶ καταδῦναι καὶ θιγεῖν ἁπάντων τῆς ὕλης τῶν μερῶν. ἡ φύσις δ᾽ οὐχ οὕτως, ἀλλὰ τὸ μὲν ὀστοῦ μέρος ἅπαν ὀστοῦν ἀποτελεῖ, τὸ δὲ σαρκὸς σάρκα, τὸ δὲ πιμελῆς πιμελὴν καὶ τῶν ἄλλων ἕκαστον· οὐδὲν γάρ ἐστιν ἄψαυστον αὐτῇ μέρος οὐδ᾽ ἀνεξέργαστον οὐδ᾽ ἀκόσμητον.
>
> For this Nature which forms the parts and adds to them little by little is most assuredly extended throughout their whole substance. Yes indeed, she [i.e. Nature] forms and nourishes and increases the parts through and through, not only on the outside. For Praxiteles and Pheidias and all the other sculptors used to fashion their materials only on the outside, insofar as they were able to touch them, but they left the inner depth unfashioned and unwrought and unaffected by art or foresight, because they were not able to get down into it and reach and touch all the parts of the material. But this is not the case with Nature. Rather, she succeeds in making every part of bone bone, every part of flesh, flesh, every part of fat, fat, and so on and so forth for the rest. Indeed, there is no part that is untouched, unworked, or unfashioned by her. (Gal. *Nat. Fac.* 2.3 [II 82 K = 160,10–24 Helmreich])

In the shift from the craftsman-seed to Nature, then, we move from an externalized skill to immanent knowing.

Yet even before this shift, there is a problem in equating the seed with the craftsman. Galen compares the two as artificers, insofar as each determines how much of the material in question is appropriate, since neither the blood nor the wax can discover how much of itself to contribute. But the seed's capacity to attract an appropriate amount of material falls short of an actual judgment. We must be on guard, Galen cautions, "lest we attribute some kind of reasoning and mind to the seed" (μή πως λάθωμεν τῷ σπέρματι λογισμόν τινα καὶ νοῦν χαρισάμενοι).[23] That attribution, he

[22] Gal. *Nat. Fac.* 2.3 (II 83 K = 161,16–17 Helmreich).
[23] Gal. *Nat. Fac.* 2.3 (II 85 K = 162,15–16 Helmreich).

goes on, would be a category mistake. For we would be talking not about the seed or Nature but about the living animal itself (*zōion auto*).[24] Instead of mind, Galen turns to an attractive faculty just like that exhibited by the lodestone when it attracts iron to explain the apparently "knowing" behavior of the seed in relationship to the generative material. The seed continues to drive the creation of the embryo through the alterative and shaping faculties.

It should be said that Galen was not himself always entirely happy with such an explanation. In *Formation of the Fetus* he struggles to accept that the natural faculties can fully account for the sperm's capacity to shape – or rather, to become – a nature, precisely because they are not just unwise but entirely without reason (οὐ μόνον οὐκ οὖσαν σοφήν, ἀλλὰ καὶ παντάπασιν ἄλογον).[25] The gap between Demiurgic intelligence and the natural faculties here seems too great: some mediating skill or agent is needed, yet Galen struggles to supply it. Galen faces a similar problem in *On My Own Opinions*, where he wonders what allows plants to discriminate between what is appropriate and what is foreign in the process of maintaining their being (an activity explained by the natural faculties elsewhere).[26] He concludes that plants must possess a rudimentary form of sensation (*aisthēsis*), recalling Plato's position in the *Timaeus* (77a), against centuries of Aristotelian and Stoic thinking on the subject, as well as his own thinking in *On the Natural Faculties*.[27] Here again, the knowingness exhibited by the plant eventually prompts Galen to find ways to bring the vegetal stratum closer to something like mind.

In *On the Natural Faculties*, however, Galen relies on the natural faculties to do the work of maintaining life at the vegetal level. He accounts for their

24 Elsewhere, he notes that there are physicians who do believe that parts of the body (e.g. muscles) function like animals: Gal. *Foet. Form.* 5.8–10 (V 690–91 K = 94,21–96,4 Nickel). Though Galen does not accept this position, he expresses some sympathy with it at *Foet. Form.* 6.23 (V 696 K = 100,17–20 Nickel).

25 Gal. *Foet. Form.* 6.31 (V 700 K = 104,19–20 Nickel). The chapter offers the richest discussion of how we get from the Demiurge to immanent purposefulness. Galen there also rejects the model whereby the Demiurge is like a craftsman who devises theatrical effects, transmitting motions to a machine that then moves by his design for a limited amount of time (*Foet. Form.* 6.5 [V 689 K = 94,1–2 Nickel], 6.12 [V 692 K = 96,14–16 Nickel]). Galen finds the theory improbable for the reason that he thinks it unlikely irrational substance could conserve by itself the transmission of skilled movements so perfectly. On "wind-up machines," see also *Sem.* 1.5.24 (IV 533 K = 84,1–4 De Lacy).

26 Gal. *Prop. Plac.* 15 (IV 764–66 K = 189,17–190,16 Boudon-Millot and Pietrobelli). For these faculties in plants, see Gal. *Prop. Plac.* 3 (174,5–9 Boudon-Millot and Pietrobelli), 9 (181,21–25 Boudon-Millot and Pietrobelli).

27 Gal. *Nat. Fac.* 1.1 (II 1–2 K = 101,1–5 Helmreich). On Aristotle, see Arist. *De An.* 414a32–b2, *Part. An.* 666a34. On the Stoics: Clem. Al. *Strom.* VIII 10.4, Diog. Laert. 7.86. See further Nutton 1999: 216–17.

capacity to discriminate between different types of matter by invoking, albeit implicitly, a pair of concepts that had found widespread applicability by the second century CE – namely, sympathy and antipathy. Galen's choice of such a model has important consequences for how the purposefulness of the natural faculties is conceptualized.

Sympathy plays a starring role in the confrontation that Galen stages between the continuum theorists and the atomists. There, Galen pits the body that "breathes together and flows together and suffers together," an image he gets from the pseudo-Hippocratic treatise *On Nutriment* (*c.* third century BCE or later) and proudly labels as Hippocratic,[28] against the fragmented worlds, both macrocosmic and microcosmic, of the atomists.[29] Galen makes the affirmation of a "kind of unity of substance" (*henōsis tis tēs ousias*) one of the prerequisites for recognizing Nature's "technical" power; sympathy is part of what guarantees this unity.[30] As Armelle Debru writes, "Galen's thought is shot through with the notion that the general intercommunication within and synergy of actions in the organism creates from it a unity, which accounts for our being able to speak of it as a 'system'."[31] By contrast, the *denial* of sympathy sums up the problems with a corpuscular ontology.[32] In this respect, sympathy is a consummate expression of what Christopher Gill has called the "high naturalism" that Galen shares with the Stoics, that is, a philosophy of nature committed to an organizing intelligence in the cosmos and resistant to assigning much weight to chance and necessity.[33]

[28] See *Nutr.* 23 (IX 106 L = 143,1–3 Joly): "There is one confluence; there is one common breathing; all things are in sympathy. All the parts as forming a whole, and severally the parts in each part, with reference to the work" (ξύρροια μία, ξύμπνοια μία, ξυμπαθέα πάντα· κατὰ μὲν οὐλομελίην πάντα, κατὰ μέρος δὲ τὰ ἐν ἑκάστῳ μέρει μέρεα πρὸς τὸ ἔργον). Note that virtually all modern scholars concur that the text is Hellenistic or later, in part because of the Stoic influence: see Diller 1936, Deichgräber 1973: 69–75, Joly 1975, Jouanna 1999: 401. Galen, however, seems not to have doubted its Hippocratic provenance: see also the citation of *Nutr.* 34 (IX 110 L = 144,25–145,3 Joly) at *Protr.* (I 25 K) – a reference I owe to David Leith – and the citations in the next note.

[29] See Gal. *Nat. Fac.* 1.12 (II 29 K = 122,7–9 Helmreich), 1.13 (II 38 K = 129,7–9 Helmreich), cited again at 3.13 (II 196 K = 243,10–13 Helmreich). Galen returned to the maxim often: see *Caus. Puls.* 1.12 (IX 88 K); *MM* 1.2 (X 16 K); *Trem. Palp.* (VII 616 K); *UP* 1.8 (III 17 K = 1.12 Helmreich), 1.9 (III 24 K = 1.17 Helmreich). On the importance of continuity in Galen's philosophy of the body, see De Lacy 1979, Jouanna 2003, Debru 2008.

[30] Gal. *Nat. Fac.* 2.6 (II 100 K = 174,6–7 Helmreich).

[31] Debru 2008: 275.

[32] According to Asclepiades, "nothing is naturally in sympathy with anything else, all substance being divided and broken up into inharmonious elements and absurd 'molecules'" (κατὰ δὲ τὸν Ἀσκληπιάδην οὐδὲν οὐδενὶ συμπαθές ἐστι φύσει, διῃρημένης τε καὶ κατατεθραυσμένης εἰς ἄναρμα στοιχεῖα καὶ ληρώδεις ὄγκους ἁπάσης τῆς οὐσίας, Gal. *Nat. Fac.* 1.13 [II 39 K = 129,9–12 Helmreich]).

[33] Gill 2010: 77–79.

Moreover, sympathy is important for explaining how Nature expresses its unique "technical" intelligence in relationship to the living being:[34]

> καὶ ἡ φύσις ἅπαντα τεχνικῶς καὶ δικαίως πράττει δυνάμεις ἔχουσα, καθ' ἃς ἕκαστον τῶν μορίων ἕλκει μὲν ἐφ' ἑαυτὸ τὸν οἰκεῖον ἑαυτῷ χυμόν, ἕλξαν δὲ προσφύει τε παντὶ μέρει τῶν ἐν αὑτῷ καὶ τελέως ἐξομοιοῖ, τὸ δὲ μὴ κρατηθὲν ἐν τούτῳ μηδὲ τὴν παντελῆ δυνηθὲν ἀλλοίωσίν τε καὶ ὁμοιότητα τοῦ τρεφομένου καταδέξασθαι δι' ἑτέρας αὖ τινος ἐκκριτικῆς δυνάμεως ἀποτρίβεται.
>
> And Nature accomplishes everything in an artistic and just manner, possessing faculties according to which each of the parts attracts to itself the humor that is proper to it, and then in turn attaches it to every portion of itself and entirely assimilates it, while that which has not been mastered in this humor nor is capable of being totally altered and assimilated to what is being nourished is expelled by means of another faculty, an expulsive one. (Gal. *Nat. Fac.* 1.12 [II 29–30 K = 122,9–16 Helmreich])

Galen seems to suggest here that the attraction of what is proper and the expulsion of what is foreign are simply expressions of the more general forces of sympathy and antipathy in nature. The cosmic sympathy most associated with Stoicism is subtly recast, then, in terms of the binary forces of sympathy and antipathy, without losing the Stoic overtones of a webbed world shot through with a kind of vital intelligence.

One of the clues that Galen is conceptualizing the attractive and expulsive faculties in terms of sympathy and antipathy is his frequent recourse to the lodestone, one of the paradigmatic models of sympathy in antiquity, in order to talk about attraction based on "appropriateness of quality," rather than the principle of *horror vacui*.[35] Other examples that Galen gives to prove attraction based on quality also assimilate the natural faculties to the principles of sympathy and antipathy, such as the capacity of emetics to draw out certain kinds of humors and the efficacy of antidotes to snake venom.[36] The work of the natural faculties is thus implicated in a broader network of physical forces that put bodies and qualities in particular relationships with one another, rather than subjecting them, *qua* material bodies, to a single set of mechanical principles.

[34] See also Gal. *Nat. Fac.* 1.13 (II 38–39 K = 129,3–9 Helmreich).

[35] See esp. Gal. *Nat. Fac.* 1.14 (II 44–51 K = 133,11–138,14 Helmreich), a long excursus on the lodestone and the reason for its particular powers. See also 2.7 (II 106 K = 178,5–7 Helmreich) and Holmes 2012 on the use of the lodestone to illustrate sympathy in Galen's commentary on *Epidemics* II.

[36] On poison and antidotes to poison as powerful illustrations of sympathy and antipathy, see Gaillard-Seux 2010: 311–16. On the presence of sympathy as a principle in Galen's pharmacology, see Keyser 1997.

By assimilating the natural faculties in the body to larger forces in nature as a whole, Galen, as we saw earlier, circumvents the need to posit intelligence or mind in accounting for what they accomplish. The examples of the lodestone and antidotes make clear that sympathy and antipathy cut across nature as a whole, encompassing both animate and inanimate life. The difference with Galen's position in, say, *On My Own Opinions*, is clear. Whereas the attribution of sensation to plants destabilizes the boundary between plant life and animal life respected elsewhere in ancient philosophy by bringing plants up to the level of animals, the model of sympathy and antipathy blurs the line between living and non-living, rational and irrational things by extending a kind of knowingness throughout all of nature.

Such a strategy means that *On the Natural Faculties* yields a strong contrast between the nonconscious, unthinking work of vegetal life and conscious, reasoned decision-making based on consideration of what is the best option. In a passage describing the action of the stomach on ingested food, for example, Galen stresses that the stomach does not contract for the sake of (*dia touto*) preparing the nutriment for the rest of the body. If it did, it would become – much like the seed likened to Pheidias risks becoming in the example we saw earlier – an animal in possession of reason (*logismos*) and intelligence (*nous*) and so capable of *choosing* the better of two options (*hōs haireisthai to beltion*).[37] Elsewhere, Galen playfully rejects the idea that the urine sets off of its own accord for the kidneys, "considering this the better course" (*touto beltion einai nomizonta*), much as we do when we go to the market.[38] Galen succeeds in explaining these bodily activities as purposeful but not intentional because of the natural faculties and, behind them, the broader principles of sympathy and antipathy.

In sum, then, Galen is navigating a difficult space in *On the Natural Faculties*. On the one hand, he stands firmly against the idea that life is either wholly random or wholly deterministic. On the other hand, he distinguishes the purposiveness that makes it likely to succeed from the *deliberate* pursuit of life. But at this point, we may ask: why does life become the responsibility of deliberate agents at all? Why is it not enough for the organism to survive by means of the ongoing, nonconscious actions of sympathy and antipathy? The answer to these questions returns us to the chanciness of life under even the best of circumstances. I turn now to consider how the possibility of error within the natural functioning of

[37] Gal. *Nat. Fac.* 3.13 (II 199 K = 245,21–23 Helmreich).
[38] Gal. *Nat. Fac.* 1.15 (II 57 K = 142,20–23 Helmreich).

life creates the idea that flourishing is only a probable outcome, one that must be secured by intelligent, conscious, knowledgable agents acting on the body as an object of care.

Doubling nature: the technical agent

Galen's larger view of nature, as we saw at the outset, is that our world is the best of all possible worlds. But that view does not mean that everything always turns out for the best. In the case of living things, the translation of the Demiurge's intelligence into something immanent within a physical body opens up flourishing as a probable outcome, something that is likely to be realized but cannot be guaranteed. That is to say, life is not something that happens at random or by chance. Yet neither is it a necessary outcome, like the attraction of matter into a void. It is something made possible and, indeed, *encouraged*, by the arrangement of matter in a certain way. Yet the very involvement of matter in the success of the outcome also opens life up to failure. In Plato's *Timaeus* we already encounter the idea that matter often proves an obstacle to the realization of the Demiurge's plans. The materiality of the created organisms can also interfere with their ongoing success. For Galen, too, the success of the natural faculties is always dependent on their material conditions, and especially the proportions of the underlying mixture of hot, cold, wet, and dry (the mixture is specific to different parts, depending on their function).[39]

In *On the Natural Faculties* we see such a principle expressed most clearly in Galen's recognition that a part functions successfully only if the mixture underlying the part is well balanced. If the qualities contributing to this balance are not measured appropriately, the activity proper to the part is compromised or destroyed: a faculty lacking its proper material is literally inoperative (ἅπασα γὰρ δύναμις ἀργεῖ ἀποροῦσα τῆς οἰκείας ὕλης).[40] This means that no faculty is immune to the vagaries of the mixture underlying it; no faculty is "tough as steel and unaffected by circumstances" (ἀδαμαντίνη τις ἡμῖν αὕτη μόνη καὶ ἀπαθής ἐστιν).[41] Moreover, the constant change in the material substratum of parts makes the individual parts dependent on the overall health of the organism for the conditions of their success. It is to some extent because no part functions independently of

39 E.g. Gal. *Nat. Fac.* 2.9 (II 126 K = 192,18–25 Helmreich), *Opt. Corp. Const.* 2 (IV 741–42 K), *Temp.* 9 (I 563–64 K = 34,20–35,3 Helmreich), on the mixture of skin.

40 Gal. *Nat. Fac.* 2.3 (II 83 K = 161,19–20 Helmreich). See also Gal. *Nat. Fac.* 2.8 (II 118 K = 187,3–8 Helmreich), with Debru 2008: 273.

41 Gal. *Nat. Fac.* 2.8 (II 112 K = 182,22–23 Helmreich).

the other parts that the success of a given faculty on a given occasion is governed by conditions of probability.[42]

If the parts are dependent on the whole for success, the whole is itself dependent on another set of factors for its well-being; for the body that, guided by an innate teleology, strives to maintain life must contend with another force – namely, the person himself, whose choices and actions always have an impact on the material conditions of the faculties. The decisions that the embodied subject makes about what to eat and drink, for example, are an integral part not just of the process of nourishment but of the maintenance of overall health.

The word that matters here, of course, is "decision." After all, as we have just seen, what distinguishes the work of living at the level of the subject, as opposed to the level of immanent life, is the capacity to make choices. It is true that Galen sometimes seems to imagine a continuum between the sympathies and antipathies of the organs and more conscious actions. The stomach "longs after and tends toward what is advantageous and proper to it and loathes and rids itself of what is foreign" (τῆς γαστρὸς . . . ὀρεγομένης μὲν καὶ προσιεμένης τὰ χρήσιμά τε καὶ οἰκεῖα, δυσχεραινούσης δὲ καὶ ἀποτριβομένης τὰ ἀλλότρια); in so doing, it determines whether one swallows quickly or with difficulty, sometimes even snatching food it deems appropriate out of the mouth without the person even wanting to swallow.[43] Nevertheless, Galen is elsewhere clear that choice defines the activity of the conscious actor vis-à-vis the activities of the faculties.[44] What is more, the very prevalence of disease suggests that the choosing agent can – and often does – override the tendencies of the faculties, undercutting their odds of success.

Take, for example, a passage where Galen is talking about the value of knowing about the nature and effects of bile. Having noted the deleterious effects that an excess of bile has on the overall constitution of a human being, he asks: "Would it not be absurd, then, for someone to *choose voluntarily* those things that contain more bile rather than those containing less?" (πῶς γὰρ οὐκ ἂν εἴη γελοιότατος, ὃς ἂν ἑκὼν αἱρῆται τὰ πλείονα χολὴν ἐν ἑαυτοῖς περιέχοντα πρὸ τῶν ἐναντίων;).[45] But, of course, such

42 Gal. *Nat. Fac.* 3.13 (II 190 K = 238,23–239,8 Helmreich), with the language of *eikos*.

43 Gal. *Nat. Fac.* 3.8 (II 172–74 K = 225,23–226,10 Helmreich). Galen goes on to give examples of animals whose stomachs are like hands, actually reaching out of their mouths to grab food. On the continuum of nonconscious and conscious appetite in the Hippocratic authors, see Holmes 2010: 196–200.

44 The working distinction between automatic actions and actions undertaken by a desiring, choosing agent goes back to the Hippocratics: see Holmes 2010: 171–77 and Holmes 2013.

45 Gal. *Nat. Fac.* 2.8 (II 114 K = 184,6–8 Helmreich).

a choice is only absurd if the choosing subject (or his physician) *knows* the negative effects bile will have on the body. For neither we nor our physicians can improve our bodies without knowing what is appropriate and inappropriate for them. Such knowledge complements the stomach's innate tendencies to seek what is proper and what is foreign. At the same time, if it is absent or inoperative, the embodied subject may make choices that undermine the striving toward life at the nonconscious, vegetal stratum of the self.

Unfortunately, just as the material conditions of life make success only a probable outcome, so the acquisition and effective use of knowledge about the body and what is proper to it is not a situation guaranteed in advance. We may be creatures fashioned with the tools to be stewards of our own health. More often than we would like, however, we fail to carry out that role. But, whereas the chanciness of the first scenario represents the irrepressibility of matter within the plan of the Demiurge, the chanciness of the second points to a different kind of erring; for it arises from an ignorance and irresponsibility which it is *in our power* to combat. Galen was deeply committed to this axiom throughout his career. In *On the Natural Faculties* he writes "it is in our power to alter and transmute morbid states of the body – in fact, to give them a turn for the better" (ἀλλοιοῦν γὰρ δήπου καὶ μεταβάλλειν οἷοί τ' ἐσμὲν καὶ τρέπειν ἐπὶ τὸ βέλτιον ἀεὶ τὰς μοχθηρὰς καταστάσεις τοῦ σώματος).[46] Insofar as we are rational agents, we have the power to educate ourselves to make the choices required to ensure the optimal condition of flourishing for our own bodies within the best possible world created by the Demiurge. If we fail to make the right choices, either because we do not pursue the appropriate knowledge or because we do not act on it, that is *our* failure. The practice of life becomes deliberate, reasoned. The chanciness of flourishing at the level of the plantlike body creates a space where rational agents succeed or fail to live.

We can see how probability at the nonconscious level helps to secure this ethical space in another anti-Erasistratean polemic from the treatise *On Antecedent Causes*. Here, rather than attacking Erasistratus for failing to grant causal force to the purposefulness and skill of Nature, as in *On the Natural Faculties*, Galen takes him to task for, among other things, failing

[46] Gal. *Nat. Fac.* 2.8 (II 114 K = 184,12–14 Helmreich). For Galen's confidence about what can be known regarding the body and its behaviors, see Hankinson 2009: 210–13. Elsewhere, he shows himself more sensitive to the limitations and difficulties involved in perceiving pathological states and correcting them. The "stochastic" nature of medicine was recognized as early as the Hippocratics. For the perception of medicine as a probabilistic art in the Corpus, see von Staden 2002.

to secure health as an ethical domain. Erasistratus' overriding error is that he refuses to recognize the antecedent or "procatarctic" cause, the concept that Galen is defending here.

By antecedent cause, Galen means an external factor antecedent to the disease that is in some way responsible for the disease process unfolding within the body; its effects may be felt long after the cause has faded away.[47] The example to which Galen returns time and again in the treatise – and which seems to have been the favored example of his opponents – is that of a crowd exposed to heat in an open-air theater.[48] For many members of the audience, the heat will not be sufficiently strong to trigger disease on its own. But in the case of bodies already weakened, say, by bad regimen, it is very *likely* to trigger a causal series that leads to fever. For Galen, this is enough to make the heat a crucial part of the etiological narrative. Erasistratus and his followers disagreed, for at least a couple of reasons.[49]

First, Erasistratus, at least on Galen's almost certainly skewed account of his views, denied that external factors, especially heat and cold, could *ever* form part of the causal picture of disease, thereby drawing a firm boundary between the outside and the inside of the body.[50] But second and for our purposes more important, Erasistratus refused to see *any* factor that does not invariably lead to a disease – say, fever – as a cause, according to Galen. That is, he eliminates from the roster of causes anything that has the potential to have a harmful effect or is likely to have that effect but does not, in fact, *always* have that effect on *everyone*. Therefore, in addition to heat and cold, internal causes such as exhaustion and repletion are eliminated from the causal portrait because they do not invariably lead to disease. Erasistratus is left with one and only one true cause of fever: the leakage of blood from the veins to the arteries (*paremptōsis*). For *paremptōsis* always leads to inflammation, and inflammation always leads to fever. The "always" guarantees the status of *paremptōsis* as a cause.

Galen at times seems to suggest that Erasistratus would permit a more nuanced picture of causality. Erasistratus seems to have been willing to grant, for example, that certain conditions, such as *plethora* – the overloading of the veins with undigested food – tend to lead to *paremptōsis*, thereby allowing probability back in.[51] But he apparently did not want to call such

47 See Hankinson 1998: 24–26, 45–48.

48 Gal. *CP* 2.11 (72,17–26 Hankinson), 8.100 (104,7–17 Hankinson), 10.126 (114,12–24 Hankinson).

49 Erasistratus is the only named opponent in the treatise, but the Methodists also seem at times to be encompassed by the attack, insofar as they denied the relevance of external causes.

50 Hankinson 1998: 31–35, 46–47.

51 Gal. *CP* 14.174 (134,13–25 Hankinson), with Allen 2000: 87–88.

conditions causes. These passages imply a simple terminological dispute. Galen, however, insists that the rift with his opponent goes much deeper. I would suggest that Galen's unease with Erasistratus' position arises from his own need to demarcate and secure space for complexity and probability within the realm of causality; for the problem with limiting cause to the single mechanical event that necessarily results in fever is that it threatens to leave everything else shrouded in randomness and chance, much as we saw in *On the Natural Faculties*. It is as if, for Galen, Erasistratus' refusal to extend the title of "cause" to even the likely factors in fever abandons those factors to indiscriminate chaos.

The result of such a scenario, on Galen's analysis, is the complete elimination of the ethical space where the chances of flourishing are negotiated by rational agents; for if we deny that heat is a cause of fever because it does not necessarily cause fever, we fail to see the very thing on which the outcome depends – namely, whether a given body is in robust health or whether it is already in a state of imbalance, conditions determined largely (for Galen) by whether the embodied subject is taking care of himself.[52] The cost of occluding the embodied subject from the causal picture is steep. If we accept, with Erasistratus, that repletion is not a cause of illness, Galen argues, we will be led to abandon an ethics of care, thinking that there is no connection between our actions and our diseases:

> Quid igitur prohibet omnes repleri cibis simul et potibus? Cur autem removemur ab indigestione, cur vero <moderate> exercitamur, cur autem ordinate dietamur? Haec igitur deponamus eruditi ab Erasistrato et frigus et estum negligamus, similiter autem et vigiliam.
>
> What then is there to prevent us from filling ourselves up with both food and drink? Why do we guard against indigestion, why exercise moderately, why run our lives in a well-ordered way? So, let us learn from Erasistratus and put them all aside, taking no notice of cold and heat; and the same goes for sleeplessness. (Gal. *CP* 15.187 [142, 3–6 Hankinson]; trans. Hankinson 1998)

That is, if everything except the necessary local event loses meaning, then we are facing the demise not only of a theory of causality but also of an ethics of taking care of life.

The category of procatarctic causes has strong associations with the Stoics. Yet, for the Stoics, the chain of antecedent causes leading up to its effects produces a deterministic understanding of fate, a causal story

[52] See also, e.g., Gal. *Diff. Feb.* 1.6 (VII 289–91 K), with Nutton 1983: 3–9.

marked by necessity.[53] By contrast, in Galen, antecedent causes are the backbone of a story about possible outcomes and the ethical opportunities they open up. By that I do not mean that Galen gets rid of necessity. For the weakened person who comes down with fever in the theater, the causal chain accumulates outcomes that are necessary if nothing intervenes. But Galen sees this necessity in terms of the local physical conditions, rather than some predetermined fate.

More importantly, Galen emphasizes the need to recognize as causes those factors whose pathological potential is realized *only some of the time*. Whether the negative outcomes are realized seems to be something that is largely up to us. It depends on whether we maintain the mixture of the body under optimal conditions, especially through the regulation of our lives in ways that are both medically and morally appropriate. Galen upholds, then, a view of the world not just as causally complex but as susceptible to intentional manipulation.

The possibility of taking care is occasioned, it is worth remembering, because of an undecideability at the heart of the body organized by Nature, the vulnerability of immanent intelligence to the vagaries of matter and mixture. It is dependent on the knowledge of what is appropriate or foreign to the human body (or a particular human body). That is, it depends on rational agents making choices that mimic the work of the natural faculties. The acquisition of knowledge has, on such an occasion, an instrumental purpose, insofar as it facilitates the pursuit of life by supplementing the work of Nature's immanent intelligence.

And yet, in some sense, such "technical" intelligence is alien to immanent intelligence in ways similar to the differences between the *tekhnē* of a sculptor like Pheidias and the *tekhnē* of Nature. I want to close by reflecting briefly on the nature of knowledge about the natural faculties themselves and how Galen's puzzlement about the immanent purposefulness that brings the organism into being and ensures its survival exposes the defining rift within the human organism – namely, the rift between conscious (deliberate) and nonconscious (automatic) life.

Conclusion: the limits of likely stories

Galen, as we have just seen, is confident that human beings have the capacity to acquire enough secure knowledge about the body and the

[53] Hankinson 1998: 25; Frede 2006.

things that affect its health to safeguard the flourishing of the organism. But there are some things in nature that fall outside the parameters of certain knowledge. Galen, as we saw at the beginning of this chapter, at times refers to such knowledge as knowledge that goes only so far as what is credible or plausible (*akhri tou pithanou*), and he characterizes speculation of this kind on a couple of occasions by appealing to the concept of the "likely stories" from Plato's *Timaeus*.[54]

In an extended discussion at the end of *Formation of the Fetus*, we see another category open up: subjects on which Galen cannot even reach a plausible explanation.[55] The subject at hand is one that Galen puzzled over, it seems, for most of his life: the nature of the soul.[56] More specifically, he is expressing his failure to understand how animals are created and how they sustain life. The quandary points to the elusiveness of Nature's immanent knowingness.

As we saw earlier, in this treatise Galen resists the idea (which he entertains in *On the Natural Faculties*) that the natural faculties are responsible for the formation of the fetus because, as he says, they lack reason and intelligence. Such faculties seem incapable of carrying out the plan (*logos*) of the Demiurge.[57] Yet he finds it hard to accept that the rational soul we possess after birth is responsible for the construction of the fetus – that is, that a single soul governs our creation and, indeed, continues to ensure the functioning of the parts. The problem lies precisely in the discontinuity between the action of those parts and our rational understanding of how they work; for what speaks against a single unitary soul, for Galen, is the fact that "the soul that manages us has no knowledge of the parts that obey its urges" (ἡ ἄγνοια τῆς διοικούσης ἡμᾶς ψυχῆς τῶν ὑπηρετούντων ταῖς ὁρμαῖς αὐτῆς μορίων).[58] He marvels at the fact that children can speak without any understanding of how the muscles involved produce the sounds or, for that matter, the work done by the nerves. And, even as adults, we do not have any knowledge of the parts of our bodies or their activity before the study of anatomy, a shortcoming that does not prevent us from, say, moving those parts.[59] If a single soul pervaded the entire body,

54 See above, nn. 4–5.

55 Gal. *Foet. Form.* 6.30 (IV 700 K = 104,14 Nickel): *oud' akhri tou pithanou proelthein dunamenos.* See also *Foet. Form.* 6.1 (V 687 K = 92,1–3 Nickel), 6.16 (V 693 K = 98,3–4 Nickel).

56 See, e.g., Gal. *Prop. Plac.* 14 (188,7–13 Boudon-Millot and Pietrobelli), with Donini 2008.

57 On the plan of the Demiurge, see Gal. *Foet. Form.* 5.9 (V 682 K = 82,16 Nickel).

58 Gal. *Foet. Form.* 6.25 (V 697 K = 100,28–29 Nickel).

59 See also, e.g., Gal. *Loc. Aff.* 6.6 (VIII 445 K), where he invokes the "Hippocratic" theme of "untaught nature." On untaught nature, see also Gal. *Caus. Symp.* 2.5 (VII 178 K), *Loc. Aff.* 4.6 (VIII 443 K), *PHP* 5.5.3 (IV 459 K = 316,30–31 De Lacy), *Sem.* 2.6.6 (IV 643 K = 198,7–8 De Lacy). On the Hippocratic source, see [Hpc.] *Nutr.* 39 (IX 112 L = 145,12 Joly).

Galen believes, the workings of the body would be entirely transparent to us.

The fact that the governing soul is ignorant of the nature of the very parts it commands points, for Galen, to a gulf between the two kinds of intelligence at work in the organism, one immanent in the parts, the other concentrated in the rational mind. Galen is mostly interested here in explaining how it is that parts of the body, such as muscles, respond to our desires. But one can extend the argument to the "knowing" functioning of the natural faculties themselves. The point is that, when pressed to account for such knowingness, Galen chose to mark the limits of rational inquiry.

We might see Galen's appeal to *aporia* as in some sense performative; for it enacts quite literally the problem that he describes in the speaking child. That is because the breakdown of even probable stories about the technical intelligence behind the fashioning and the functioning of the parts bears witness to the discontinuity between that intelligence and our own rational endeavors to account for the world and our own bodies. That failure suggests a truth that Galen sidesteps: that we are creatures designed around a fundamental blindness to our own nature, our own vital being. The life of the vegetal stratum and ethical life are resolutely irreconcilable, even as each depends on the other for the odds of its success.

In the end, then, Galen sees the pursuit of life unfolding on two levels: that of the natural faculties and that of the rational, conscious actor. At each level, life is the likely outcome, given that we live under optimal conditions for flourishing. Yet it is not a necessary outcome. What endangers success at the level of the faculties is the labile nature of the mixture, which renders them vulnerable to disruption. The trouble at the level of the person is the fact that our understanding of our bodies is neither given nor guaranteed, meaning that it is up to us to acquire the knowledge that allows us to make the choices that enable life to sustain itself. The very possibility that we may not make the right choices is what makes the pursuit of life at this level an ethical endeavor.

But the ethical nature of taking care is secured at an even more basic level by the very non-transparency of nature to the rational soul. That non-transparency is, in the end, absolute. By this I do not mean that we cannot gain a plausible or, in some cases, even a secure knowledge of how the parts work (how we digest food, how we excrete urine, how we speak). We can acquire such knowledge and, in so doing, we gain the skill to act, like a sculptor, on the body, and therefore the ability to shape our lives in such a way as to maximize our potential to flourish. But we can never internalize this knowledge so that it forms a continuum with the work of the natural faculties. However much, then, we seek to secure conditions under which nature has a good chance of succeeding in its aims, we remain estranged from – and, perhaps, surprised by – each moment at which merely probable life is actualized.

CHAPTER 13

Afterword

Catherine Gallagher

This rich and varied collection provides a wide panorama of ancient Greek intellectual life through the lens of its interest in *eikos*, or probabilistic thought. For those of us outside the field of Classics, it resonates with other works that have looked far beyond the major philosophers when assessing the intellectual legacy of classical Greek thought. Plato and Aristotle may have codified the demarcation separating philosophy, logic, and mathematics, on the one hand, and rhetoric, politics, myth, and art, on the other, as they launched a discipline that elevated reason over emotional persuasion, argument over narrative, certain knowledge of unchanging truths over judgments about what is most likely, and syllogistic demonstrations over exchanged opinions. However, the vitality of the concept of *eikos*, which is associated with all of those disparaged terms, indicates not only that Greek thought was hardly monolithic but also that even the philosophers did not always limit themselves to a single-minded "quest for certainty" (to use John Dewey's phrase).

Among the previous reminders of the diversity of our classical Greek inheritance belongs a counter-tradition inside Western philosophy that includes thinkers as dissimilar as Giambattista Vico, Friedrich Nietzsche, John Dewey, Hannah Arendt, and Jacques Derrida, who have called attention to the importance and intellectual prestige of the very discourses denigrated by the philosophers. For example, in the early sixties Arendt praised the ability to form representative opinions in Greek democratic deliberation, and she contrasted it to the classical philosophical operations that stressed logical demonstration. Philosophical demonstrations must be accepted by every rational person, she claimed, so they are *ipso facto* coercive and opposed to the diversity of views entertained in truly political debate. The expression of common-sense judgments, on the other hand, encouraged probabilistic thinking and allowed for the inter-subjective understandings to arise that are necessary for collective action.[1] The chapters in this volume that deal with persuasion in the law courts and the political arena – especially those of Vincent Farenga, Michael Gagarin, and Robert Tordoff – seem to agree with Arendt's general view that the trading of opinions was central to classical Greek institutions, and the authors present detailed evidence that arguments and narratives relying on probability were essential elements in those exchanges.

A different kind of immanent critique, focused on the self-contradictions and ambivalences of the philosophers, has also added to our sense of the complexities of Greek thought. For example, in 1965 Theodor Adorno argued that even though Aristotle preserved the Platonic view of the real and the causal as located in a realm of autonomous and imperishable forms, contrasting their ontological priority to that of formless matter (which is mere possibility), he nevertheless sometimes endowed matter with an independent power to cause chance events. Adorno saw this contradiction both as symptomatic of structural problems inside the idealist concept of causality and as a survival in Aristotle of the older mythical idea of the animating forces of nature.[2] A trio of authors in this volume – Jenny Bryan, James Allen, and Ryan Balot – also modify our perception of the philosophers, presenting strong evidence that the unwavering hostility of the ancient classical philosophers to rhetoric and probability has been exaggerated. Thus, although there is general agreement that "a decisive turning point" introduced a philosophic commitment to the demonstrable certainty of objective truth in fifth-century BCE Athens, reading this volume reinforces the view that the revolution was both limited and incomplete.[3]

In advancing the ongoing efforts to make us more aware of the many strands of Greek thought that have been bequeathed to us, this volume intertwines various discourses with the common thread of *eikos*. The perspective is thus at once unusually sharp and extensive. *Eikos*, as Victoria Wohl explains in her Introduction, has many things to recommend it, including the closeness of its range of meanings to that of its English translation "probable": from "believable, likely, or reasonable" as well as appropriate and seemly (on the positive side of the spectrum) to deceitfully illusionary or counterfeit (at the negative pole). These semantic continuities with a present-day term, combined with stabilities in the array of discourses in which they are found (law, politics, religion, history, and the activities

[1] Arendt 1961: 241. [2] Adorno 2001: 73–76.

[3] There is still, to be sure, a widespread idea that the Greeks sought absolute imperishable truth "to the detriment of every other undertaking": Ogilvie 2004. Thus Ogilvie writes of "a decisive turning point introduced in Greek thought around the fifth century B.C. and to which we, in the West, still subscribe" (2004: 103).

of making, interpreting, and evaluating literary and visual art), imply that much has survived of the ancient idea of probability. Freeing the concept from the role of classical philosophy's rejected other, this volume examines in detail the places where *eikos* required no apology and no defenders, so the collection functions as a nexus where the concept seems to migrate among various disciplinary timelines. *Eikos* transfers at this juncture from the story of philosophy, where it seemed a minor character until quite recently, to the histories of law, politics, historiography, and art, where it played more important roles.

The question this volume provokes is whether or not the evidence of conceptual continuity in these various fields indicates that *eikos* should be seen as a starting place for the history of modern discourses of probability *generally*, including the philosophical and mathematical theoretic discourses. To distinguish the latter version, I will spell it "Probability" with a capital "P." Did *eikos* perhaps eventually contribute to this most prestigious vindication of probability/Probability, the transition point at which it ascended to the very status of "knowledge" and even became a modern variety of the "rationality" that classical philosophy seemed to deny it? Or was it left behind with other pre-modern versions that barely survived the epistemic break of the late seventeenth century and currently have less intellectual clout? Is this early history of probability interesting mainly because it shows how deep the rupture is between the pre-modern and modern concepts? In that case the irony of the history of *eikos* might be that it has gone from serving as a denigrated other of classical philosophy to serving as the unacknowledged and even sometimes explicitly disavowed other of the legitimate progenitors of "scientific" Probability theory. Alternatively, might the ancient history help us to discern a pattern of ruptures and continuities that indicates some significant rifts within the modern term? If so, can we conclude that our current uses of "probability" in the soft humanities disciplines may have quite a bit in common with *eikos* and yet still be distinctly modern?

In answering these questions, I would like to suggest that this volume both opens our view of the variety of ancient Greek thought and also broadens our understanding of the range of meanings of probability today, inviting us to think about both the survival of *eikos* and its partial obsolescence. As Victoria Wohl says in her Introduction, the history of modern probability has been written from two distinct perspectives: that of mathematical and philosophical Probability theory and that of the conceptual probability belonging to the realist novel. In both cases, she rightly contends, many historians have declared the supersession of the classical discourse of probability. The irrelevance of the ancient has seemed most obvious when explaining the rise of Probability. There is general agreement among historians of science and mathematics that the explicit project of justifying claims to non-demonstrative knowledge under conditions of uncertainty relied on the capacity to compute regularities in a series of random events, and that the ambition to make such calculations of degrees of certainty did not appear until the second half of the seventeenth century. When the mathematics, which was developed to solve puzzles associated with games of chance, was applied to real-world problems – such as the laws pertaining to contingent contracts, the values of annuities, and the "political arithmetic" of state administration – statistics was born and Probability slowly became central to many of the very sciences that were once thought to exclude contingent phenomena and probabilistic conjecture.

The historians, however, have many different opinions about both the prehistory and details of this development, as well as about its reach, and they can be seen to differ in their assessments of the absolute discontinuity between the modern idea of Probability and earlier notions of probability. Ian Hacking's seminal book, *The Emergence of Probability*, argues in favor of an abrupt epistemic break, which happened around 1660, in which several preconditions converged to produce the intellectual space in which the concepts comprising Probability theory became thinkable. The modern concept of Probability, he claims, is dual: "It is connected with the degree of belief warranted by evidence, and it is connected with the tendency, displayed by some chance devices, to produce stable relative frequencies."[4] This dual idea required the mutation of the Renaissance notion of the "sign" (developed in the "low sciences" of alchemy and medicine) into the modern idea of evidence as well as the mathematics necessary to assess such evidence. Although Hacking admits that evidence is "primarily a legal notion" and seems aware that classical law was sophisticated enough to distinguish between testimony and circumstances, he would not admit the *eikos* arguments described in these essays into his story of Probability, nor would he take the varying weights given to *eikos*, witness testimony, and documents (described by both Farenga and Gagarin) to be early types of calculating believability.[5] Indeed, he begins by ruling the ancient Greeks and Romans out of his prehistory of the concept on the grounds that they did not have the needed notation for numerals: "a symbolism that makes addition and multiplication easy is a prerequisite for any rich concept of probability" (6). This dismissal of ancient Greek probability is merely a

[4] Hacking 1975: 1. [5] Hacking 1975: 86.

single instance of his general practice; to support the case for a radical and thorough epistemic disruption, Hacking consigns every pre-quantitative use of the term "probable" to the dustbin of "old" usages no longer in circulation. This aspect of his argument has been assailed and demolished from various quarters; Barbara Shapiro, for example, demonstrated in her 1983 *Probability and Certainty in Seventeenth-Century England* that Hacking's claim that the "qualitative" meanings of "Probable" were reduced in the late seventeenth century to "degree of belief warranted by evidence" is obviously wrong.[6] The word has continued to describe verisimilitude, eligibility, appropriateness, the plausible, and so forth. Daniel Garber and Sandy Zabell, to take another example, show that medieval handbooks of law not only distinguished internal from external evidence but also cited classical authors as authorities on the subject.[7]

Lorraine Daston's 1988 *Classical Probability in the Enlightenment* has the benefit of these critiques of Hacking's thesis, and it would seem a more promising work in which to find a place for *eikos* in the prehistory of Probability. Daston separates the questions of the origins of mathematical Probability from the origins of conceptual probability and adds that "some concepts are more readily quantified than others."[8] Moreover, she stresses the intimacy between the emergence and acceptance of the concept and the preceding notion of probability as simply that which we have some good warrant to believe or expect, given our understanding that some things are subject to chance. Daston also traces the mathematization of chance in areas that are close to daily life, for example "proto-quantification" in commercial dealings and the "partial proofs" (or circumstantial evidence) in the law courts. And finally, she demonstrates that even after Probability had been quantified, it continued to be thought of in the Enlightenment as a mode of applying mathematics to normal clear thinking: "nothing more at bottom than good sense reduced to a calculus," as Laplace maintained. The *Encyclopedie* considered Probability the "science of rational belief" and assumed that it described the underlying bases of human behavior. This is close to Hacking's description of the dual nature of the concept, but it stresses that throughout the eighteenth century the two parts were easily reconciled because many Enlightenment thinkers placed their own belief in the natural rationality of Man and his enlightened common sense. The

[6] Shapiro demonstrates that seventeenth-century English intellectuals across a variety of fields (including natural science, religion, law, and history) developed multiple modalities of probabilistic thinking. In commenting on Hacking's argument, she notes that they did so without showing much interest in "calculating chances" (B. J. Shapiro 1983: 277 n. 69).

[7] Garber and Zabell 1979. [8] Daston 1988: 13.

Probability described by Daston thus brought together things that may have later fallen apart (as in the subsequent contest between "objective" Probability, stressing the stochastic calculations, and "subjective" Probability, stressing how we interpret those calculations to arrive at degrees of belief). And it certainly combined things that the ancient philosophers had tended to separate, but which the chapters in this volume indicate had been interleaved in the discursive practices of the larger Greek culture: common sense opinions, awareness of uncertainty, and rationality.

And yet Daston, like Hacking, also does not trace this Enlightenment Probability back to classical Greece, for the third element, the ambition and capacity to use mathematics in order to demonstrate a rationality of common sense, was missing before the seventeenth century. Her few references to Aristotle stress that his "qualitative" idea of probability would have been impossible to quantify and was therefore not a contributor to Laplace's project.[9] Moreover, Laplace may have thought that his mathematics was just a way of describing a universal "good sense" extending back through history, but in fact, according to Daston, it implied a new arrangement of the relevant categories of thought. So Daston, like Hacking, maintains that when mathematics met probability, new semantic emphases in the latter term appeared, which were only slowly recognized as having significantly modified the ancient concept. This point, I believe, might easily be granted without claiming, like Hacking, that all semantic continuity with earlier notions was lost. We might, for example, cite the evidence of the OED, which gives the following as "Definition 1.a.": "The property or fact of being probable, esp. of being uncertain but more likely than not; the extent to which something is likely to happen or be the case; the appearance of truth, or likelihood of being realized, which a statement or event bears in the light of present evidence." This definition covers most of the semantic range of *eikos* without directly specifying quantification, and yet it nevertheless contains subtle signs of the modern revolution in the casual way it takes for granted the opportunity to turn the probable into the knowable through measuring ("the extent to which"), calculating ("more likely than not"), and evaluating evidence ("in the light of present evidence").

These phrases seem almost "proto-quantitative" but so does Aristotle's remark that the probable is what happens "for the most part." The deeper reason why Daston excludes Aristotle is that his idea of probability stands in the way of conceptualizing "a full continuum of degrees" of likelihood.

[9] Daston 1988: 38.

This point is crucial because, as Daston notes, the lack of such a continuum characterized not only Aristotle's view but also that of the entire "qualitative" tradition he inspired: Aristotle sharply distinguished chance events, which are "purposeless and exceptional, from probable ones, which are the rule" (38). The dichotomy between randomness and probability was, moreover, easily reconciled with his dictum that qualitatively different kinds of certainty are appropriate to different objects of knowledge. To be sure, as several of the chapters in this volume remind us, Aristotle did link the universal and the particular through the concept of *eikos* as "what happens for the most part," but the universal had clear ontological (and therefore causal) priority. Daston recognizes the longevity of the Aristotelian tradition that opposed probability to chance, showing its dominance even in John Locke's work and that of the Royal Society. Like Hacking, she views it as finally superseded only by the Port Royal *Logique*. But, although she recognizes the centrality of probability to Aristotle's works and the centrality of those works to the history of the concept up through the heyday of the British Royal Society, she maintains that the essential link between the probable and the random, which makes the mathematics of Probability possible, is excluded by his system. Her reminder that seeing Probability as regularity arising out of contingency, rather than seeing it as a quality from which chance has been nearly expunged, is a key to understanding why the Aristotelian category might have served as a barrier rather than a bridge to the early modern mathematical idea.

Several of the chapters in this volume also reinforce this conceptual gap between the ancient Greek and modern concepts. Daryn Lehoux's account of Aristotle's explanation of why children should be, but never are, replicas of their fathers is a case in point. Here the oddness of the priority of the causal model over the particular instance is strikingly obvious because it not only violates modern scientific understanding but also inverts our common-sense experiential categories. According to Lehoux, Aristotle is forced to recognize that the particulars – the children – always happen to deviate from the ruling model – the father, who carries the typical form. However, instead of concluding that the dissimilarities are the rule, Aristotle calls them monstrosities and makes them conceivable only as adherences to some older ancestor (such as the maternal grandfather) or of a more general type (human). This view may be "probabilistic" in some sense, as Victoria Wohl says in the Introduction, but not in any modern sense; for Aristotle's view is not an anticipation of the modern concept of the chances of genetic descent from both parents. His deviations are not new combinations but merely unfortunate reversions.

Obviously, his theory violates the idea of blended inheritance that was centuries old by the time that "Mendel's laws" became widely accepted, but more importantly it violates our very idea of individual personhood. Far from thinking that particularity is a failure to duplicate a prototype, we celebrate the fact that each person is a unique combination of characteristics inherited from many ancestors, viewing it as a wondrous illustration of nature's law of diversity, the very foundation of the Tree of Life. For us, deviation from progenitors is *The Rule*. Aristotle's idea seems so strange to us because we are unconscious inheritors of both the empiricist view that particulars have ontological priority and the assumption that rules, such as Gregor Mendel's laws of genetic inheritance, are derived from recurring regularities over numerous instances of chance combinations. We presuppose that categories are aggregates of particulars and the consistencies in our biological inheritance depend on chance. In other words, Probability is deep in our basic sense – not just in our rarefied scientific discourse – of how we came into existence as embodied individuals. Modern Probability might, then, be said to inform the thinking even of people who cannot do the math or articulate its leading concepts, and it is reasonable to argue that its most rudimentary assumptions often seem antithetical to some classical Greek uses of the term.

But perhaps not to all. The rise of the realist novel has often been instanced as a sign of the spread of the new common-sense idea of modern Probability-informed probability – focused on "empirical" particulars and open to the idea of meaningful contingency – throughout the culture at large. And yet although the phenomenon of the novel is uniquely modern, the field of literary history nevertheless recognizes the form's complex connections with antique thought. For example, literary historians routinely note that early novelists drew on the distinction between history and "poesie" found in Aristotle's *Poetics* when explaining and defending their fictional practice:

> The poet's job is not to tell what has happened but the kind of things that *can* happen, i.e., the kind of events that are possible according to probability or necessity . . . And in fact that is why the writing of poetry is a more philosophical activity, and one to be taken more seriously, than the writing of history; for poetry tells us rather the universals, history the particulars. (Arist. *Poet.* 1451a36–b7, trans. Else 1957)

This seemed a useful endorsement and status elevation of narrative fiction at a time when many readers expected a "novel" to be a scandalous exposure of some famous individual. Distinguishing himself from scandalmongers,

Henry Fielding paraphrases Aristotle in *Joseph Andrews* when the narrator declares, "I describe not men, but manners; not an individual, but a species."[10]

To be sure, Fielding's "species" is not the same as Aristotle's "kind of person", for Fielding's general category is empirical and social, built out of particular instances of people who occupy various classes and professions (in this case, lawyers), whereas Aristotle's kinds are natural types to which people intrinsically belong (for example nobles, slaves, strong men, weak men). Moreover, the method of the realist novel is not to give us consistent instances of pre-existing kinds but instead to generalize from observed individuals and then to invent fictional instances, who instead of adhering to type are aesthetically "realized" by developing the unique traits and behaviors (often through happenstance) that distinguish them from all others in their category. Ian Watt (1957) was right to note the casual empiricism of the eighteenth-century novel; the form does assume that the most "real" phenomena are observable individuals in the world, and the secondary realities are the types; so creating a "realistic" character (emphatically not to be confused with a real person) means exceeding typicality, falling short of it, learning to aspire to different categories, including the category of uniqueness – in short, deviating back to the seeming reality of particular contingencies. The inversion of ontological priorities certainly makes an unavoidable wrinkle in the smooth transition from Aristotle's poetic "probability" into novelistic probability.

But even as we confront the difference between Aristotle's ontology and that of the neoclassical novelists, we can note the underlying continuities, the most obvious of which is the persistence of the categories themselves – types and individual characters; characteristic action and action out-of-character; probable events and improbable ones. The novel inaugurated a new kind of aesthetic play with these terms, a play that would hardly have existed without the Aristotelian background. The concept of novelistic realism posits two sorts of individuals: those existing in an inferred world and those twice removed from it. The form is structured like a triptych, in which ontologically distinct categories of "the particular" appear on either side of a category of "the general," which mediates them, so that the middle term is encased in contingencies. And yet it remains central to the form even though it does not carry the weight of the real. Because this is fiction, the invented personae can only remain significant and intelligible

10 Fielding 1967: 189. Fielding is alluding not only to Aristotle but also to the ancient Roman debate about the proper objects of satire.

if they are presented as departing from type in ways that exhibit their relations to it. That is one of the differences between actual people and novel personae: the latter are aesthetic constructs, designed to swerve from type in plausibly generalizable and representative ways. This play of the probable involves inventing, even in a single work, ever more refined social, moral, generic, and psychological categories to explain actions, events, and character changes as the narrative proceeds.

Literary scholar Rüdiger Campe sees the novel's aesthetic play with probability as rooted in both the achievement of early modern Probability theory and the legacy of ancient Greek thought about the reversibility of probability and improbability. His 2002 book, *The Game of Probability: Literature and Calculation from Pascal to Kleist* is one of the few studies to treat in detail both the mathematics of Probability and the development of the norm of literary probability in the eighteenth-century novel.[11] Campe argues that the mathematics developed in proximity to many discursive environments but the calculations were explicitly based on the trivial and semantically empty activity of playing games of chance. Only in games of chance could randomness come into focus because only there was it truly a desideratum. Campe explains the difficulty of making randomness an observable phenomenon, of making dice that are not weighted, coins that really are as likely to fall on one side as the other, and of making skill an irrelevant factor. Pure games of chance, "fair games," had to be devised so that all who played had equal odds. The games were thus not just any old example of Probability quantification but the unique arena, the cordoned-off place where the mathematics could develop; they constituted an environment free of meaning and therefore solely devoted to calculation. When other discourses applied those calculations, they took in many of the ideas of the "fair game" out of which they grew, including the valuation of randomness.[12] As the historians of Probability record, Daniel Bernoulli was the first to write of the utility (the good) of contingency, which is that it keeps us actively trying to optimize our good fortune through the calculation of risk.

Campe teases out the significance of this idea for the form of the novel. Tracing it through the thought of Leibniz and Abraham Gotthelf Kästner (an eighteenth-century lecturer in mathematics, philosophy, logic, and law at Leipzig), Campe concludes that by the mid eighteenth century many

[11] Campe 2012, originally published as *Spiel der Wahrscheinlichkeit. Literatur und Berechnung zwischen Pascal und Kleist* (Göttingen, 2002).

[12] For discussions of the preference for "equipossibility" that is related to the desire for a fair game, see Hacking 1971.

thinkers shared a view of ethical behavior that made two incommensurable demands. First, we must admit that Providence selects the best among possible outcomes from a universal perspective, so we should submit patiently to contingent losses and disappointments. But, second, we should pay attention to the desires nature prompts in us and act on them in ways that maximize our chances and minimize our risks, having carefully assessed the particularities of both. Kästner implies that the universal and individual views cannot be fully reconciled, and we must live inside the discrepancy. This double view, Campe argues, maps onto the form of the novel, with its split between the standpoint of an objective spectatorial narration and that of a highly individuated character pursuing individuated aims.

After describing this early modern context in which the particular and contingent became the engines of probability in the novel, Campe turns to the context of classical Greek rhetoric and poetics, looking first at a seeming paradox mentioned by several of the contributors to this volume. This is Aristotle's version in the *Rhetoric*,

> As Agathon says,
>
> One might perchance say that was probable –
> That things improbable oft will hap to men.
>
> For what is improbable does happen, and therefore it is probable that improbable things will happen.
>
> (Arist. *Rh.* 1402a10–14, trans. W. R. Roberts 1954)

Aristotle seems to grant this idea while denying what is sometimes claimed to be its implication, "that 'what is improbable is probable'." The faulty deduction arises from the confusion of general and specific probability, and he instances the problematic reversibility of probability and improbability in Corax's *Art of Rhetoric*:

> If the accused is not open to the charge – for instance if a weakling be tried for violent assault – the defence is that he was not likely to do such a thing. But if he is open to the charge – i.e. if he is a strong man – the defence is still that he was not likely to do such a thing, since he could be sure that people would think he was likely to do it. (Arist. *Rh.* 1402a18–20, trans. W. R. Roberts 1954)

The error of arguing that the very probability of a thing renders it improbable comes, according to Aristotle, "from not adding any clause specifying relationship or reference or manner"; that is, Corax does not properly frame the inversion of the probable and improbable as the result of certain specific conditions that deviate from the general rule.

This may look like the Aristotelian opposition between general and particular, contingent and probable, rearing its head again, but Campe instead sees Aristotle's "old, familiar turn of the argument" as preparing "the way for the framing and modeling of probability in an artificial arrangement" and therefore as a forerunner of the modern novel.[13] Campe's exposition of the various ways in which the novel form trades on framing and modeling devices is immensely complicated, but suffice it to say here that the "artificial arrangements" in which the terms probable/improbable may be inverted link the form both to the early modern mathematics of gaming and to literary and rhetorical figures that were already explored in ancient thought. He contends, for example, that "The modern logic of modeling realities, with their meta- and meta-meta-levels, can . . . be derived from the special effect of a sophism in ancient topics," and that "Ever since Aristotle . . . poetics has tied the paradox of improbable probability to dramatic peripeteia" (275). Campe may, therefore, be just as interested as Hacking and Daston in describing the changes in the concept of Probability/probability after the 1670s, and he does not pay much attention to classical Greek thought when tracing the prehistory of the mathematical concept. But when he turns to the eighteenth-century novel, classical thought becomes an important contributor and numerous continuities emerge.

The eighteenth-century play with Aristotelian categories is even more apparent in another narrative form that began appearing in the eighteenth century: sustained counterfactual history. A sustained counterfactual history is one that integrates hypothetical past-tense speculation ("If Napoleon had marched on St. Petersburg instead of lingering in Moscow, then . . .") into its method rather than just using it occasionally as a rhetorical embellishment. The writers of these counterfactual histories were military historians, and they seem to have been aware of ancient uses of the device, such as the counterfactual passages from Thucydides analyzed by Robert Tordoff in this volume. Nevertheless, they saw themselves as unprecedented because they had the data to pursue the idea systematically. Henry Lloyd's *The History of the Late War in Germany: Between the King of Prussia, and the Empress of Germany and Her Allies* (1766 to 1781), for example, was the first to try to give exhaustive descriptions of the strategic, operational, logistic, and geographical aspects of a war. Lloyd aspired to pure, historical objectivity, hoping to arrive at an orderly guide to the art of warfare. He posed counterfactual questions, informed by numerous statistics

[13] Campe 2012: 275.

drawn from the records of the warring states, which he believed could recreate the "objective possibilities" (to use a phrase from Max Weber).

The crucial difference between this kind of probabilistic narrative practice and the more providential Leibnizian kind out of which Campe sees the novel emerging is that Lloyd and the "critical" military historians who followed him hardly ever thought the historical wars were the best among the possibilities. In war mistakes and errors abound and, by emphasizing them, the up-to-date view of critical military history was, paradoxically, aligned with Aristotle's remarks about the mere accidents of actual history generally. To be sure, even though these historians deplored the errant happenstances, they did not think them meaningless; instead they assiduously catalogued and synthesized the particulars to gain solid information about how to fight future wars more effectively. Only by seeing how wars might have been worse *and* how they might have been better would it be possible to arrive at the general rules of war in order to perfect its art. In short, we should acknowledge the major difference between the poetic generalities Aristotle preferred and the generalities sought by the counterfactual military historians: the former are arrived at by suppressing the merely accidental and the latter by learning from it. Nevertheless, both Aristotle's philosophical poesie and Lloyd's rules of warfare were thought to deliver widely applicable knowledge; in neither case was the particular and contingent sought for its own sake. We might say that this was the persistence of the classical in eighteenth-century neoclassical counterfactualism. And if I had the space to describe our contemporary historical counterfactualism, I could also enumerate many instances in which the classical legacy thrives.

By expanding our view of the multiplicity and diversity of ancient Greek thought and thus undermining our belief in a unified classical worldview, this volume certainly makes us more aware of the range of our own uses of probability. Just as classical writers, including the philosophers, could neither maintain nor enforce a perfectly coherent understanding or consistent valuation of *eikos*, we have been unable to create a unified concept of probability/Probability. In some of our histories of modern Probability, *eikos* is explicitly said to have been superseded; in many of our histories of modern probability, it is said to have been incorporated and heavily modified. But in both, the accounts of the development of the modern concepts need the background of *eikos* for their own historical legibility.

References

Adorno, T. W. (1983) "Valery Proust Museum," in *Prisms*, ed. T. Adorno, trans. S. Weber and S. Weber. Cambridge, Mass.: 175–85.

(2001) *Metaphysics: Concept and Problems*, ed. R. Tiedemann, trans. E. Jephcott. Stanford.

Allan, W. (ed.) (2008) *Euripides: Helen*. Cambridge.

Allen, J. V. (2000) "Galen as (mis)informant about the views of his predecessors," *Archiv für Geschichte der Philosophie* 83: 81–89.

(2001) *Inference from Signs: Ancient Debates about the Nature of Evidence*. New York.

(2007) "Aristotle on the disciplines of argument: rhetoric, dialectic, analytic," *Rhetorica* 25: 87–108.

Almeida, J. A. (2003) *Justice as an Aspect of the Polis Idea in Solon's Political Poems: A Reading of the Fragments in Light of the Researches of New Classical Archaeology*. Leiden.

Alroth, B. (1992) "Changing modes in the representation of cult images," in *The Iconography of Greek Cult in the Archaic and Classical Period*, ed. R. Hägg. Athens: 9–45.

Andorlini, I. and K. Maresch (2006) *Das Archiv des Aurelius Ammon (P. Ammon II)*. Munich.

Andrewes, A. (1961) "Thucydides and the Persians," *Historia* 10: 1–18.

(1970) "Lysias and the Theramenes Papyrus," *Zeitschrift für Papyrologie und Epigraphik* 6: 35–38.

Angier, T. P. S. (2010) *Technē in Aristotle's Ethics: Crafting the Moral Life*. London.

Arendt, H. (1961) *Between Past and Future: Eight Exercises in Political Thought*. New York.

Arrowsmith, W. (1973) "Aristophanes' *Birds*: the fantasy politics of eros," *Arion* 1: 119–67.

Asmis, E. (1986) "*Psychagogia* in Plato's *Phaedrus*," *Illinois Classical Studies* 11: 153–72.

Austin, J. L. (1971) "Performative-Constative," in *The Philosophy of Language*, ed. J. R. Searle. London: 13–22.

Austin, N. (1994) *Helen of Troy and Her Shameless Phantom*. Ithaca.

Balensiefen, L. (1990) *Die Bedeutung des Spiegelbildes als ikonographisches Motiv in der antiken Kunst*. Tübingen.

Balme, D. M. (1987) "The place of biology in Aristotle's philosophy," in *Philosophical Issues in Aristotle's Biology*, ed. A. Gotthelf and J. G. Lennox. Cambridge: 9–20.

Balot, R. K. (2004) "Free speech, courage, and democratic deliberation," in *Free Speech in Classical Antiquity*, ed. I. Sluiter and R. M. Rosen. Leiden: 233–59.

Barnes, J. (1980) "Aristotle and the methods of ethics," *Revue internationale de philosophie* 34: 490–511.

(1991) "Philosophie et dialectique," in *Penser avec Aristote*, ed. M. Sinaceur. Toulouse: 107–15.

Barnouw, J. (2004) *Odysseus, Hero of Practical Intelligence: Deliberation and Signs in Homer's Odyssey*. Lanham, Md.

Barone, C. (1989) "*Eikos* e *thaumaston* nell'*Elena* di Euripide," in *Il meraviglioso e il verosimile: tra antichità e medioevo*, ed. D. Lanza and O. Longo. Florence: 163–71.

Bassi, K. (1993) "Helen and the discourse of denial in Stesichorus' Palinode," *Arethusa* 26: 51–75.

(2005) "Things of the past: objects and time in Greek narrative," *Arethusa* 38: 1–32.

(2007) "Spatial contingencies in Thucydides' *History*," *Classical Antiquity* 26: 171–218.

Bectarte, H. (2006) "'Tenir un miroir' dans l'art funéraire grec antique," in *L'expression des corps. Gestes, attitudes, regards dans l'iconographie antique*, ed. L. Bodiou, D. Frère, and V. Mehl. Rennes: 165–80.

Belfiore, E. S. (1992) *Tragic Pleasures: Aristotle on Plot and Emotion*. Princeton.

Benardete, S. (2000) *Plato's Laws: The Discovery of Being*. Chicago.

Betegh, G. (2010) "What makes a myth *eikōs*? Remarks inspired by Myles Burnyeat's EIKŌS MYTHOS," in *One Book, The Whole Universe: Plato's Timaeus Today*, ed. R. D. Mohr and B. M. Sattler. Las Vegas: 213–24.

Bettinetti, S. (2001) *La statua di culto nella pratica rituale greca*. Bari.

Bilabel, F. (ed.) (1922) *Die kleineren Historikerfragmente auf Papyrus*. Bonn.

Blundell, S. (1995) *Women in Ancient Greece*. London.

Bobonich, C. (1994) "Akrasia and agency in Plato's *Laws* and *Republic*," *Archiv für Geschichte der Philosophie* 76: 3–36.

(1996) "Reading the *Laws*," in *Form and Argument in Late Plato*, ed. C. Gill and M. M. McCabe. Oxford: 249–82.

(2002) *Plato's Utopia Recast: His Later Ethics and Politics*. New York.

Bolton, R. (1997) "The material cause: matter and explanation in Aristotle's natural science," in *Aristotelische Biologie: Intentionen, Methoden, Ergebnisse*, ed. W. Kullmann and S. Föllinger. Stuttgart: 97–126.

Bonitz, H. (1870) *Index Aristotelicus*. Berlin.

Boudon-Millot, V. and A. Pietrobelli (eds.) (2005) "Galien ressuscité: édition princeps du texte grec du *De propriis placitis*," *Revue des études grecques* 118: 168–213.

Bowie, E. L. (1993) "Lies, fiction and slander in early Greek poetry," in *Lies and Fiction in the Ancient World*, ed. C. Gill and T. P. Wiseman. Exeter: 1–37.

Boyer, P. (1994) *The Naturalness of Religious Ideas: A Cognitive Theory of Religion*. Berkeley.

(2001) *Religion Explained: The Evolutionary Origins of Religious Thought*. New York.

Bratman, M. (1999) *Faces of Intention: Selected Essays on Intention and Agency*. Cambridge.

Broadie, S. (2012) *Nature and Divinity in Plato's Timaeus*. New York.

Brock, R. (2013) "Athens' Sicilian expedition: contemporary scenarios of its outcome," in *Hindsight, or the Importance of Unfulfilled Expectations in Greek and Roman History*, ed. A. Powell. Swansea: 49–70.

Brunschwig, J. (ed. and trans.) (1967) *Aristote: Topiques*, vol. 1. Paris.

(1994) "Rhétorique et dialectique, *Rhétorique* et *Topiques*," in *Aristotle's Rhetoric: Philosophical Essays*, ed. D. J. Furley and A. Nehamas. Princeton: 57–96.

Bruss, J. S. (2010) "Ecphrasis in fits and starts? Down to 300 BC," in *Archaic and Classical Greek Epigram*, ed. M. Baumbach, A. Petrovic, and I. Petrovic. Cambridge: 385–402.

Bryan, J. (2012) *Likeness and Likelihood in the Presocratics and Plato*. Cambridge.

Bryson, N. (1983) *Vision and Painting: The Logic of the Gaze*. New Haven.

Bunzl, M. (2004) "Counterfactual history: a user's guide," *American Historical Review* 109: 845–58.

Burgess, S. (2000) "How to build a human body: an idealist's guide," in *Reason and Necessity: Essays on Plato's Timaeus*, ed. M. R. Wright. London: 43–58.

Burnett, A. P. (1971) *Catastrophe Survived: Euripides' Plays of Mixed Reversal*. Oxford.

Burnyeat, M. (1994) "Enthymeme: Aristotle on the logic of persuasion," in *Aristotle's Rhetoric: Philosophical Essays*, ed. D. J. Furley and A. Nehamas. Princeton: 3–55.

(2005) "ΕΙΚΩΣ ΜΥΘΟΣ," *Rhizai* 2: 143–65.

Burton, D. (2003) "Public memorials, private virtues: women on classical Athenian grave monuments," *Mortality* 8: 20–35.

Bury, J. B. (1930) *Selected Essays of J. B. Bury*. Cambridge.

Calderini, A. (1915) "Intorno al P.S.I. 17 (epigrammi funebri)," *Studi della scuola papirologica* 1: 19–44.

Cambiano, G. (2000) "Galeno, Erasistrato e la teleologia di Aristotele," in *Studi su Galeno: scienza, filosofia, retorica e filologia*, ed. D. Manetti. Florence: 45–57.

Camp, J. (2000) "Walls and the *polis*," in *Polis and Politics: Studies in Ancient Greek History*, ed. P. N. Flensted-Jensen, T. H. Rubinstein, and L. Rubinstein. Copenhagen: 41–57.

Campe, R. (2012) *The Game of Probability: Literature and Calculation from Pascal to Kleist*, trans. E. H. Wiggins, Jr. Stanford.

Carawan, E. (2002) "The Athenian amnesty and the 'scrutiny of the laws'," *Journal of Hellenic Studies* 122: 1–23.
(2006) "Amnesty and accountings for the Thirty," *Classical Quarterly* 56: 57–76.
Carr, E. H. (1962) *What Is History?* New York.
Carter, D. M. (2004) "Citizen attribute, negative right: a conceptual difference between ancient and modern ideas of freedom in speech," in *Free Speech in Classical Antiquity*, ed. I. Sluiter and R. M. Rosen. Leiden: 197–220.
Cartledge, P. (1984) "A new lease of life for Lichas, Son of Arkesilas?," *Liverpool Classical Monthly* 9: 98–102.
Cavallo, G. (2000) "Écriture et pratiques intellectuelles dans le monde antique," *Genesis* 15: 97–108.
Childs, W. A. P. (1988) "The classic as realism in Greek art," *Art Journal* 47: 10–14.
Clairmont, C. W. (1970) *Gravestone and Epigram: Greek Memorials from the Archaic and Classical Period.* Mainz.
Cochrane, C. N. (1929) *Thucydides and the Science of History*. London.
Cole, T. (1991a) *The Origins of Rhetoric in Ancient Greece*. Baltimore.
(1991b) "Who was Corax?," *Illinois Classical Studies* 16: 65–84.
Coles, A. (1995) "Biomedical models of reproduction in the fifth century BC and Aristotle's 'Generation of Animals'," *Phronesis* 40: 48–88.
Connell, S. M. (2000) "Aristotle and Galen on sex difference and reproduction," *Studies in History and Philosophy of Science* 31: 405–27.
(2001) "Toward an integrated approach to Aristotle as a biological philosopher," *Review of Metaphysics* 55: 297–322.
Connelly, J. B. (1993) "Narrative and image in Attic vase painting: Ajax and Kassandra at the Trojan Palladion," in *Narrative and Event in Ancient Art*, ed. P. J. Holliday. Cambridge: 88–129.
Cooper, J. M. (1988) "Metaphysics in Aristotle's embryology," *Proceedings of the Cambridge Philological Society* 214: 14–41.
(1994) "Ethical-political theory in Aristotle's *Rhetoric*," in *Aristotle's Rhetoric: Philosophical Essays*, ed. D. J. Furley and A. Nehamas. Princeton: 193–210.
(2004) *Knowledge, Nature, and the Good: Essays on Ancient Philosophy*. Princeton.
Cowley, R. (ed.) (1999) *What If?: The World's Foremost Military Historians Imagine What Might Have Been.* New York.
Crane, G. (1996) *The Blinded Eye: Thucydides and the New Written Word.* Lanham, Md.
Cribiore, R. (1992) "The happy farmer: a student composition from Roman Egypt," *Greek, Roman, and Byzantine Studies* 33: 247–64.
Dale, A. M. (ed.) (1967) *Euripides: Helen*. Oxford.
Dalfen, J. (1985) Review of Synodinou 1981, *Gnomon* 57: 1–4.
Daston, L. (1988) *Classical Probability in the Enlightenment.* Princeton.
Dean-Jones, L. (1994) *Women's Bodies in Classical Greek Science*. Oxford.
De Boer, W. (ed.) (1937) *Galeni De propriorum animi cuiuslibet affectuum dignotione et curatione; De animi cuiuslibet peccatorum dignotione et curatione; De atra bile. Corpus Medicorum Graecorum* V 4,1,1. Leipzig.

Debru, A. (2008) "Physiology," in *The Cambridge Companion to Galen*, ed. R. J. Hankinson. Cambridge: 263–82.

De Cesare, M. (1994) "Una statua di Eracle tra mito ed escatologia: per una lettura unitaria del cratere apulo di New York MMA 50.11.4," *Rend. Linc.* 9th ser., 5: 247–58.

(1997) *Le statue in immagine: studi sulle raffigurazioni di statue nella pittura vascolare greca*. Rome.

Deichgräber, K. (1973) *Pseudhippokrates' über die Nahrung: Text, Kommentar und Würdigung einer stoisch-heraklitisierenden Schrift aus der Zeit um Christi Geburt*. Mainz.

de Jong, I. J. F. (1987) *Narrators and Focalizers: The Presentation of the Story in the Iliad*. Amsterdam.

De Lacy, P. (1979) "Galen's concept of continuity," *Greek, Roman, and Byzantine Studies* 20: 355–69.

(ed.) (1992) *Galeni De semine. Corpus Medicorum Graecorum* V 3,1. Berlin.

(ed.) (1996) *Galeni De elementis ex Hippocratis sententia. Corpus Medicorum Graecorum* V 1,2. Berlin.

(ed.) (2005) *Galeni De Placitis Hippocratis et Platonis. Corpus Medicorum Graecorum* IV 1,2. Berlin.

Demandt, A. (1986) *Ungeschehene Geschichte: Ein Traktat über die Frage, Was Wäre Geschehen, Wenn –?* Göttingen.

Denniston, J. D. (1954) *The Greek Particles*. Oxford.

Diels, H. and W. Kranz (eds.) (1951) *Die Fragmente der Vorsokratiker*. Berlin.

Diller, H. (1936) "Eine stoisch-pneumatische Schrift in Corpus Hippocraticum," *Archiv für Geschichte der Medizin* 29: 178–95.

Dillery, J. (1995) *Xenophon and the History of His Times*. London.

Dillon, S. (2006) *Ancient Greek Portrait Sculpture: Contexts, Subjects, and Styles*. New York.

Donini, G. (1964) "Thucydides 7.42.3: does Thucydides agree with Demosthenes' view?," *Hermes* 92: 116–19.

(2008) "Psychology," in *The Cambridge Companion to Galen*, ed. R. J. Hankinson. Cambridge: 184–209.

Donohue, A. A. (1988) *Xoana and the Origins of Greek Sculpture*. Atlanta.

Dorandi, T. (1991) "Den Autoren über die Schulter geschaut: Arbeitsweise und Autographie bei den antiken Schriftstellern," *Zeitschrift für Papyrologie und Epigraphik* 87: 11–33.

(2000) *Le stylet et la tablette: dans le secret des auteurs antiques*. Paris.

Dover, K. J. (1973) *Thucydides*. Oxford.

(1987) *Greek and the Greeks: Collected Papers*, vol. 1: *Language, Poetry, Drama*. Oxford.

Downing, E. (1990) "*Apate*, *agon*, and literary self-reflexivity in Euripides' *Helen*," in *Cabinet of the Muses: Essays on Classical and Comparative Literature in Honor of Thomas G. Rosenmeyer*, ed. M. Griffith and D. J. Mastronarde. Atlanta: 1–16.

Dueck, D. (2000) *Strabo of Amasia: A Greek Man of Letters in Augustan Rome.* London.

Dunn, F. M. (1996) *Tragedy's End: Closure and Innovation in Euripidean Drama.* New York.

Eden, K. (1986) *Poetic and Legal Fiction in the Aristotelian Tradition.* Princeton.

Edwards, A. T. (1985) *Achilles in the Odyssey: Ideologies of Heroism in the Homeric Epic.* Königstein.

Else, G. F. (1957) *Aristotle's Poetics: The Argument.* Cambridge, Mass.

Elsner, J. (2004) "Seeing and saying: a psychoanalytical account of ekphrasis," *Helios* 31: 157–85.

(2006) "Reflections on the 'Greek revolution' in art: from changes in viewing to the transformation of subjectivity," in *Rethinking Revolutions through Ancient Greece,* ed. S. Goldhill and R. Osborne. Cambridge: 68–95.

Elsner, J. and M. Meyer (eds.) (forthcoming) *Art and Rhetoric in Roman Culture.* Cambridge.

Ernst, W. (1999) "Modular readings (writing the monument): the case of Lapis Satricanus," *Rethinking History* 3: 53–73.

Farenga, V. (2006) *Citizen and Self in Ancient Greece: Individuals Performing Justice and the Law.* Cambridge.

Fearon, J. D. (1991) "Counterfactuals and hypothesis testing in political science," *World Politics* 43: 169–95.

Fejfer, J. (2008) *Roman Portraits in Context.* Berlin.

Fenno, J. (2005) " 'A great wave against the stream': water imagery in Iliadic battle scenes," *American Journal of Philology* 126: 475–504.

Ferguson, N. (ed.) (1997a) *Virtual History: Alternatives and Counterfactuals.* London.

(1997b) "Introduction: virtual history: towards a 'chaotic' theory of the past," in *Virtual History: Alternatives and Counterfactuals,* ed. N. Ferguson. London: 1–90.

Ferrari, G. (2000) "The Iliupersis in Athens," *Harvard Studies in Classical Philology* 100: 119–50.

Ferrari, G. R. F. (1987) *Listening to the Cicadas: A Study of Plato's Phaedrus.* Cambridge.

Fielding, H. (1967) *Joseph Andrews,* ed. M. C. Battestin. Middletown, Conn.

Finkelberg, M. (1986) "Is ΚΛΕΟΣ ΑΦΘΙΤΟΝ a Homeric formula?," *Classical Quarterly* 36: 1–5.

Finley, M. I. (1975) "Myth, memory and history," in *The Use and Abuse of History,* ed. M. I. Finley. New York: 11–33.

Fischhoff, B. (1975) "Hindsight is not equal to foresight: the effect of outcome knowledge on judgement under uncertainty," *Journal of Experimental Psychology: Human Perception and Performance* 1: 288–99.

Flemming, R. (2000) *Medicine and the Making of Roman Women: Gender, Nature, and Authority from Celsus to Galen.* Oxford.

Flory, S. (1988) "Thucydides' hypotheses about the Peloponnesian war," *Transactions of the American Philological Association* 118: 43–56.

Foley, H. P. (2001) *Female Acts in Greek Tragedy*. Princeton.

Ford, A. L. (1992) *Homer: The Poetry of the Past*. Ithaca.

Foster, H. (1988) *Vision and Visuality*. Seattle.

Foucault, M. (2009) *Le courage de la vérité: cours au Collège de France 1983–1984*. Paris.

Fournet, J.-L. (1999) *Hellénisme dans l'Égypte du VIe siècle: la bibliothèque et l'œuvre de Dioscore d'Aphrodité*. Cairo.

(2009) "Homère, *Iliade* I 86–93," *Archiv für Papyrusforschung und verwandte Gebiete* 55: 265–70.

(2010) "Les tribulations d'un pétitionnaire égyptien à Constantinople. Révision de P.Cair.Masp. III 67352," *Proceedings of the XXV International Congress of Papyrology*: 243–52.

Franklin, J. (2001) *The Science of Conjecture: Evidence and Probability before Pascal*. Baltimore.

Frede, D. (1992) "Necessity, chance, and 'what happens for the most part' in Aristotle's *Poetics*," in *Essays on Aristotle's Poetics*, ed. A. O. Rorty. Princeton: 197–219.

(2006) "Stoic determinism," in *The Cambridge Companion to the Stoics*, ed. B. Inwood. Cambridge: 179–205.

(2010) "Puppets on strings: moral psychology in *Laws* Books 1 and 2," in *Plato's Laws: A Critical Guide*, ed. C. Bobonich. Cambridge: 108–26.

Freeland, C. A. (1998) *Feminist Interpretations of Aristotle*. University Park, Pa.

Friedman, R. D. (2007) "Old stories in Euripides' new *Helen*: ΠΑΛΑΙΟΤΗΣ ΓΑΡ ΤΩΙ ΛΟΓΩΙ Γ'ΕΝΕΣΤΙ ΤΙΣ (*Hel.* 1056)," *Phoenix* 61: 195–211.

Frontisi-Ducroux, F. and J. P. Vernant (1997) *Dans l'œil du miroir*. Paris.

Furley, W. D. (2007) "A lesson to all: Lykourgos' fate in the Tbilisi Hymn (P. Ross.Georg. 1.11)," *Zeitschrift für Papyrologie und Epigraphik* 162: 63–84.

Gaertringen, F. H. von (1918) *Aus der Belagerung von Rhodos 304 v. Chr.: Griechischer Papyrus der Kgl. Museen zu Berlin*. Berlin. 752–63.

Gagarin, M. (1994) "Probability and persuasion: Plato and early Greek rhetoric," in *Persuasion: Greek Rhetoric in Action*, ed. I. Worthington. New York: 46–68.

(2002) *Antiphon the Athenian: Oratory, Law, and Justice in the Age of the Sophists*. Austin.

Gaifman, M. (2006) "Statue, cult and reproduction," *Art History* 29: 258–79.

(2012) *Aniconism in Greek Antiquity*. Oxford.

(forthcoming a) "De la figuration de l'invisible à la représentation des rituel," in *Relire Jean-Pierre Vernant*, ed. S. Georgoudi and F. de Polignac. Paris.

(forthcoming b) "Theologies of statues," in *Theologies of Greek Religion*, ed. E. Eidinow, J. Kindt, and R. Osborne. Cambridge.

Gaillard-Seux, P. (2010) "Morsures, piqûres et empoisonnements dans l'*Histoire Naturelle* de Pline l'Ancien," in *Body, Disease and Treatment in a Changing World: Latin Texts and Contexts in Ancient and Medieval Medicine*, ed. D. Langslow and B. Maire. Lausanne: 305–17.

Gallagher, C. (1994) *Nobody's Story: The Vanishing Acts of Women Writers in the Marketplace, 1670–1820*. Berkeley.

(2006) "The rise of fictionality," in *The Novel,* vol. 1: *History, Geography, and Culture*, ed. F. Moretti. Princeton: 336–63.

Gallop, D. (1991) "Can fiction be stranger than truth? An Aristotelian answer," *Philosophy and Literature* 15: 1–18.

Garber, D. and S. Zabell (1979) "On the emergence of probability," *Archive for the History of Exact Sciences* 21: 33–53.

Garland, R. (1995) *The Eye of the Beholder: Deformity and Disability in the Graeco-Roman World.* London.

Gastaldi, S. (1989) "*Eikos* e *thaumaston* nella *Poetica* di Aristotele," in *Il meraviglioso e il verosimile: tra antichità e medioevo*, ed. D. Lanza and O. Longo. Florence: 85–100.

Gernet, L. (1981) *The Anthropology of Ancient Greece*, trans. J. Hamilton and B. Nagy. Baltimore.

Gianotti, G. F. (2005) "Odisseo mendico a Troia (PKöln VI 245)," in *Scrivere, leggere, interpretare: studi di antichità in onore di Sergio Daris*, ed. F. Crevatin and G. Tedeschi. Trieste: 225–32.

Gill, C. (1993) "Plato on falsehood – not fiction," in *Lies and Fiction in the Ancient World*, ed. C. Gill and T. P. Wiseman. Exeter: 38–87.

(2010) *Naturalistic Psychology in Galen and Stoicism.* Oxford.

Gill, C. and T. P. Wiseman (eds.) (1993) *Lies and Fiction in the Ancient World.* Austin.

Gill, M. L. (1989) *Aristotle on Substance: The Paradox of Unity.* Princeton.

Giuliani, L. (1986) *Bildnis und Botschaft: Hermeneutische Untersuchungen zur Bildniskunst der römischen Republik.* Frankfurt.

Glucker, J. (1995) "Probabile, veri simile, and related terms," in *Cicero the Philosopher*, ed. J. G. F. Powell. Oxford: 115–43.

Goebel, G. (1989) "Probability in the earliest rhetorical theory," *Mnemosyne* 42: 41–53.

Goldhill, S. (2007) "What's in a wall?," in *Visualizing the Tragic: Drama, Myth, and Ritual in Greek Art and Literature*, ed. C. Kraus, S. Goldhill, H. P. Foley, and J. Elsner. Oxford: 127–47.

Gombrich, E. H. (1960) *Art and Illusion: A Study in the Psychology of Pictorial Representation.* New York.

Görler, W. (1992) "Eine sprachlicher Zufall und seine Folgen, 'Wahrscheinliches' bei Karneades und Cicero," in *Zum Umgang mit fremden Sprachen in der griechischen-römischen Antike*, ed. C. W. Müller, K. Sier, and J. Werner. Stuttgart: 159–71.

Graham, D. W. (ed.) (2010) *The Texts of Early Greek Philosophy: The Complete Fragments and Selected Testimonies of the Major Presocratics.* Cambridge.

Grethlein, J. (2006) *Das Geschichtsbild der Ilias: Eine Untersuchung aus phänomenologischer und narratologischer Perspektive.* Göttingen.

(2010) "Experientiality and 'narrative reference', with thanks to Thucydides," *History and Theory* 49: 315–35.

Gribble, D. (1998) "Narrator interventions in Thucydides," *Journal of Hellenic Studies* 118: 41–67.

(1999) *Alcibiades and Athens: A Study in Literary Presentation*. Oxford.
Grube, G. M. A. (trans.) (1992) *Plato: Republic*, rev. C. D. C. Reeve. Indianapolis.
Gurd, S. A. (2005) *Iphigenias at Aulis: Textual Multiplicity, Radical Philology*. Ithaca.
(2012) *Work in Progress: Literary Revision as Social Performance in Ancient Rome*. Oxford.
Guthrie, W. K. C. (1971) *The Sophists*. Cambridge.
Gwatkin, W. E. (1957) "The legal arguments in Aischines' *Against Ktesiphon* and Demosthenes' *On the Crown*," *Hesperia* 26: 129–41.
Hackforth, R. (ed.) (1952) *Plato: Phaedrus*. Cambridge.
Hacking, I. (1971) "Equipossibility theories of probability," *British Journal for the Philosophy of Science* 22: 339–55.
(1975) *The Emergence of Probability: A Philosophical Study of Early Ideas about Probability, Induction and Statistical Inference*. London.
(1990) *The Taming of Chance*. Cambridge.
Halliday, M. A. K. and R. Hasan (1976) *Cohesion in English*. London.
Halliwell, S. (ed.) (1987) *The Poetics of Aristotle: Translation and Commentary*. Chapel Hill.
(1992) "Epilogue: the *Poetics* and its interpreters," in *Essays on Aristotle's Poetics*, ed. A. O. Rorty. Princeton: 409–24.
(1998) *Aristotle's Poetics*. Chicago.
(2002) *The Aesthetics of Mimesis: Ancient Texts and Modern Problems*. Princeton.
Halper, E. C. (2003) "Soul, soul's motions, and virtue," in *Plato's Laws: From Theory into Practice*, ed. S. Scolnicov and L. Brisson. Sankt Augustin: 257–67.
Hankinson, R. J. (1989) "Galen and the best of all possible worlds," *Classical Quarterly* 39: 206–27.
(ed.) (1998) *Galen: On Antecedent Causes*. Cambridge.
(ed.) (2008) *The Cambridge Companion to Galen*. Cambridge.
(2009) "Galen on the limitations of knowledge," in *Galen and the World of Knowledge*, ed. C. Gill, T. Whitmarsh, and J. Wilkins. Cambridge: 206–42.
Hanson, A. E. (1982) "Publius Ostorius Scapula: Augustan prefect of Egypt," *Zeitschrift für Papyrologie und Epigraphik* 47: 243–53.
(1984) "The archive of Isodoros of Psophthis and P. Ostorius Scapula, Praefectus Aegypti," *Bulletin of the American Society of Papyrologists* 21: 77–87.
Harris, E. M. (1994) "Law and oratory," in *Persuasion: Greek Rhetoric in Action*, ed. I. Worthington. London and New York: 130–50.
(2000) "Open texture in Athenian law," *Dike* 3: 27–79.
Hartog, F. (1992) "Herodotus and the historiographical operation," *Diacritics* 22: 83–93.
Haselberger, L. (1999) "Old issues, new research, latest discoveries: curvature and other classical refinements," in *Appearance and Essence: Refinements of Classical Architecture – Curvature*, ed. L. Haselberger. Philadelphia: 1–68.

Hauser, S. (2013) "Archaeological methods and theories," *Brill's New Pauly Online*. http://referenceworks.brillonline.com/entries/brill-s-new-pauly/archaeological-methods-and-theories-ct-e1302460. April, 2014.

Hawkins, S. A. and R. Hastie (1990) "Hindsight: biased judgements of past events after the outcomes are known," *Psychological Bulletin* 107: 311–27.

Hawthorn, G. (1991) *Plausible Worlds: Possibility and Understanding in History and the Social Sciences*. Cambridge.

Headland, T. N., K. L. Pike, and M. Harris (eds.) (1990) *Emics and Etics: The Insider/Outsider Debate*. Newbury Park, Calif.

Hellekson, K. (2001) *The Alternate History: Refiguring Historical Time*. Kent, Ohio.

Helmreich, G. (ed.) (1893) *Galeni Pergameni Scripta Minora*, vol. III. Leipzig.

(ed.) (1904) *Claudius Galenus, De Temperamentis Libri III*. Stuttgart.

(ed.) (1907–9) *Galeni De usu partium libri XVII*. Leipzig.

Henderson, J. (1998) "Attic old comedy, frank speech, and democracy," in *Democracy, Empire, and the Arts in Fifth-Century Athens*, ed. D. Boedeker and K. Raaflaub. Cambridge, Mass.: 255–73.

Henderson, M. M. (1975) "Plato's *Menexenus* and the distortion of history," *Acta Classica* 18: 25–46.

Henry, D. (2006) "Aristotle on the mechanism of inheritance," *Journal of the History of Biology* 39: 425–55.

Himmelmann, N. (2000) "Quotations of images of gods and heroes on Attic grave reliefs of the late Classical period," in *Periplous: Papers on Classical Art and Archaeology Presented to Sir John Boardman*, ed. G. R. Tsetskhladze, A. J. N. W. Prag, and A. M. Snodgrass. London: 136–44.

Hoffman, D. C. (2008) "Concerning *eikos*: social expectation and verisimilitude in early Attic rhetoric," *Rhetorica* 26: 1–29.

Holmes, B. (2010) *The Symptom and the Subject: The Emergence of the Physical Body in Ancient Greece*. Princeton.

(2012) "Sympathy between Hippocrates and Galen: the case of Galen's commentary on *Epidemics* II," in *Epidemics in Context: Hippocrates, Galen, and Hunayn between East and West*, ed. P. E. Pormann. Berlin: 49–70.

(2013) "Causality, agency, and the limits of medicine," *Apeiron* 46.3: 302–26.

Hölscher, F. (2010) "Gods and statues: an approach to archaistic images in the fifth century BCE," in *Divine Images and Human Imaginations in Ancient Greece and Rome*, ed. J. Mylonopoulos. Leiden: 105–20.

Hornblower, S. (1987) *Thucydides*. Baltimore.

(1994) "Narratology and narrative techniques in Thucydides," in *Greek Historiography*, ed. S. Hornblower. Oxford: 131–66.

(2011) *Thucydidean Themes*. Oxford.

Hornblower, S. and A. Spawforth (eds.) (1996) *The Oxford Classical Dictionary*, 3rd edn. Oxford.

Hose, M. (2000) "'Wahrscheinlich ist gerade dies, daß Menschen viel Unwahrscheinliches geschieht': über das *Eikos* in der attischen Tragödie," in *Skenika: Beiträge zum antiken Theater und seiner Rezeption*, ed. S. Gödde and T. Heinze. Darmstadt: 17–30.

Howland, J. (2007) "Partisanship and the work of philosophy in Plato's *Timaeus*," *The Review of Politics* 69: 1–27.
Jakobson, R. (1960) "Linguistics and poetics," in *Style in Language*, ed. T. A. Sebeok. Cambridge, Mass.: 350–77.
Johansen, T. K. (2004) *Plato's Natural Philosophy: A Study of the Timaeus-Critas*. Cambridge.
Johnstone, S. (1999) *Disputes and Democracy: The Consequences of Litigation in Ancient Athens*. Austin.
Joly, R. (ed.) (1972) *Hippocrate, Du régime des maladies aiguës: Appendice: De l'aliment; De l'usage des liquides*. Paris.
(1975) "Remarques sur le 'De Alimento' pseudo-hippocratique," in *Le monde grec*, ed. J. Bingen, G. Cambier, and G. Nachtergael. Brussels: 271–76.
Jouanna, J. (1999) *Hippocrates*, trans. M. B. DeBevoise. Baltimore.
(2003) "La notion de la nature chez Galien," in *Galien et la philosophie*, ed. J. Barnes and J. Jouanna. Geneva: 229–68.
Joyce, C. J. (2008) "The Athenian amnesty and scrutiny of 403," *Classical Quarterly* 58: 507–18.
Kahn, C. H. (1963) "Plato's funeral oration: the motive of the *Menexenus*," *Classical Philology* 58: 220–34.
Kalkavage, P. (ed.) (2001) *Plato's Timaeus: Translation, Glossary, Appendices and Introductory Essay*. Newburyport, Mass.
Kaltsas, N. E. (2002) *Sculpture in the National Archaeological Museum, Athens*, trans. D. Hardy. Los Angeles.
Kennedy, G. A. (1963) *The Art of Persuasion in Greece*. Princeton.
Kerferd, G. B. (1981) *The Sophistic Movement*. Cambridge.
Keyser, P. T. (1997) "Science and magic in Galen's recipes (sympathy and efficacy)," in *Galen on Pharmacology: Philosophy, History, and Medicine*, ed. A. Debru. Leiden: 175–98.
King, H. (1998) *Hippocrates' Woman: Reading the Female Body in Ancient Greece*. London.
Kirk, G. S. (ed.) (1990) *The Iliad: A Commentary*, vol. II. Cambridge.
(ed.) (1993) *The Iliad: A Commentary*, vol. VI. Cambridge.
Konstam, N. and H. Hoffman (2004) "Casting the Riace bronzes (2): a sculptor's discovery," *American Journal of Archaeology* 23: 397–402.
Kosman, A. (2010) "Male and female in Aristotle's *Generation of Animals*," in *Being, Nature, and Life in Aristotle*, ed. J. G. Lennox and R. Bolton. Cambridge: 147–67.
Kraus, M. (2006) "Nothing to do with truth? Εἰκός in early Greek rhetoric and philosophy," *Papers on Rhetoric* 7: 129–50.
Kraut, R. (2010) "Ordinary virtue from the *Phaedo* to the *Laws*," in *Plato's Laws: A Critical Guide*, ed. C. Bobonich. Cambridge: 51–70.
Laks, A. (1991) "L'utopie législative de Platon," *Revue philosophique* 4: 417–28.
(2000) "The *Laws*," in *The Cambridge History of Greek and Roman Political Thought*, ed. C. Rowe and M. Schofield. Cambridge: 258–92.
(2001) "In what sense is the city of the *Laws* a second best one?," in *Plato's Laws and Its Historical Significance*, ed. F. L. Lisi. Sankt Augustin: 107–14.

Lampert, L. and C. Planeaux (1998) "Who's who in Plato's *Timaeus-Critias* and why," *The Review of Metaphysics* 52: 87–125.

Lang, M. (1989) "Unreal conditions in Homeric narrative," *Greek, Roman, and Byzantine Studies* 30: 5–26.

Lanza, D. and O. Longo (eds.) (1989) *Il meraviglioso e il verosimile: tra antichità e medioevo*. Florence.

Lapatin, K. D. S. (2001) *Chryselephantine Statuary in the Ancient Mediterranean World*. Oxford.

Laqueur, T. W. (1990) *Making Sex: Body and Gender from the Greeks to Freud*. Cambridge, Mass.

Larivée, A. (2003) "L'incarnation législative du souci de l'âme dans les *Lois*: un héritage socratique," in *Plato's Laws: From Theory into Practice*, ed. S. Scolnicov and L. Brisson. Sankt Augustin: 98–102.

Layton, R. (1977) "Naturalism and cultural relativity in art," in *Form in Indigenous Art: Schematisation in the Art of Aboriginal Australia and Prehistoric Europe*, ed. P. J. Ucko. Canberra: 33–43.

Lazenby, J. F. (2004) *The Peloponnesian War: A Military Study*. London.

Leader, R. (1997) "In death not divided: gender, family, and state on classical Athenian grave stelai," *American Journal of Archaeology* 101: 683–99.

Leaf, W. (trans.) (1971) *The Iliad*. Amsterdam.

Le Blond, J. M. (1938) *Eulogos et l'argument de convenance chez Aristote*. Paris.

Lebow, R. N. (2010) *Forbidden Fruit: Counterfactuals and International Relations*. Princeton.

Lehoux, D. (2012) *What Did the Romans Know? An Inquiry into Science and Worldmaking*. Chicago.

Leith, D. (2009) "The qualitative status of the *onkoi* in Asclepiades' theory of matter," *Oxford Studies in Ancient Philosophy* 36: 283–320.

(2012) "Pores and void in Asclepiades' physical theory," *Phronesis* 57: 164–91.

(2013) "Galen's refutation of atomism," in *Philosophical Themes in Galen. BICS supplement*, ed. P. Adamson and J. Wilberding. London.

Lennox, J. G. (1982) "Teleology, chance, and Aristotle's theory of spontaenous generation," *Journal of the History of Philosophy* 20: 219–38.

(1997) "Material and formal natures in Aristotle's *de Partibus Animalium*," in *Aristotelische Biologie: Intentionen, Methoden, Ergebnisse*, ed. W. Kullmann and S. Föllinger. Stuttgart: 163–82.

Lesher, J. H. (1990) "Early interest in knowledge," in *The Cambridge Companion to Early Greek Philosophy*, ed. A. A. Long. Cambridge: 225–49.

(1991) "Xenophanes on inquiry and discovery: an alternative to the 'Hymn to progress' reading of fr. 18," *Ancient Philosophy* 11: 229–48.

(ed.) (1992) *Xenophanes of Colophon: Fragments*. Toronto.

Lévy, E. (1976) *Athènes devant la défaite de 404: histoire d'une crise idéologique*. Paris.

Lewis, D. K. (1973) *Counterfactuals*. Oxford.

Lewis, J. (2001) "'Dike,' 'moira,' 'bios' and the limits of understanding in Solon 13 (West)," *Dike* 4: 113–35.

(2006) *Solon the Thinker: Political Thought in Archaic Athens*. London.
Lissarrague, F. (2001) *Greek Vases: The Athenians and Their Images*, trans. K. Allen. New York.
(2008) "Image and representation in the pottery of Magna Graecia," in *Performance, Iconography, Reception: Studies in Honour of Oliver Taplin*, ed. M. Revermann and P. Wilson. Oxford: 439–49.
List, C. and P. Pettit (2011) *Group Agency: The Possibility, Design, and Status of Corporate Agents*. Oxford.
Lloyd, G. E. R. (ed.) (1978) *Hippocratic Writings*. Harmondsworth.
(1990) *Demystifying Mentalities*. Cambridge.
Lloyd, M. (ed.) (1994) *Euripides Andromache*. Warminster.
Löbl, R. (1997) *TEXNH-Techne: Untersuchung zur Bedeutung dieses Wortes in der Zeit von Homer bis Aristoteles*. 2 vols. Würzburg.
Lodi, T. (1912) "Epigrammi (P.S.I. 1 17)," *Papiri Greci e Latini* 1: 35–41.
Loening, T. C. (1987) *The Reconciliation Agreement of 403/402 BC in Athens: Its Content and Application*. Stuttgart.
Lonie, I. M. (1964) "Erasistratus, the Erasistrateans, and Aristotle," *Bulletin of the History of Medicine* 38: 426–43.
(ed.) (1981) *The Hippocratic Treatises, "On Generation," "On the Nature of the Child," "Diseases IV": A Commentary*. Berlin.
Loraux, N. (2002) *The Divided City: On Memory and Forgetting in Ancient Athens*, trans. C. Pache. New York.
Lowe, N. J. (2000) "Comic plots and the invention of fiction," in *The Rivals of Aristophanes: Studies in Athenian Old Comedy*, ed. D. Harvey and J. Wilkins. London: 259–72.
Lucas, D. W. (ed.) (1968) *Aristotle: Poetics*. Oxford.
Madero, M. (2010) *Tabula Picta: Painting and Writing in Medieval Law*. Philadelphia.
Maitland, J. (1999) "Poseidon, walls, and narrative complexity in the Homeric *Iliad*," *Classical Quarterly* 49: 1–13.
Marconi, C. (2011) "The birth of an image: the painting of a statue of Herakles and theories of representation in ancient Greek culture," *Res: Journal of Anthropology and Aesthetics* 59/60: 145–67.
Marvin, M. (2008) *The Language of the Muses: The Dialogue between Roman and Greek Sculpture*. Los Angeles.
Mastronarde, D. J. (2010) *The Art of Euripides: Dramatic Technique and Social Context*. Cambridge.
Matheson, S. B. (1986) "Polygnotos: an Iliupersis scene at the Getty Museum," in *Greek Vases in the J. Paul Getty Museum. Occasional Papers on Antiquities 2*, vol. III, ed. J. Frel. Malibu, Calif.: 101–14.
Mattusch, C. (1980) "The Berlin Foundry cup: the casting of Greek bronze statuary in the early fifth century BC," *American Journal of Archaeology* 84: 435–44.
Mayhew, R. (2004) *The Female in Aristotle's Biology: Reason Or Rationalization*. Chicago.
(ed.) (2008) *Plato: Laws 10*. Oxford.

(2010) "The theology of the *Laws*," in *Plato's Laws: A Critical Guide*, ed. C. Bobonich. Cambridge: 197–216.

Mazzoldi, A. (1997) "Cassandra, Aiace e lo ξόανον di Atena," *Quaderni urbinati di cultura classica* 55: 7–21.

McGlew, J. F. (1993) *Tyranny and Political Culture in Ancient Greece*. Ithaca.

Meltzer, G. S. (1994) "'Where is the glory of Troy?' *Kleos* in Euripides' *Helen*," *Classical Antiquity* 13: 234–55.

Métraux, G. P. R. (1995) *Sculptors and Physicians in Fifth-Century Greece: A Preliminary Study*. Montreal.

Meyer, M. (1989) "Alte Männer auf attischen Grabdenkmälern," *Mitteilungen des deutschen archäologischen Instituts athenische Abteilung* 104: 49–82.

Mitchell, W. J. T. (1986) *Iconology: Image, Text, Ideology*. Chicago.

Moles, J. L. (1993) "Truth and untruth in Herodotus and Thucydides," in *Lies and Fiction in the Ancient World*, ed. C. Gill and T. P. Wiseman. Austin: 88–121.

Monoson, S. (1994) "Frank speech, democracy, and philosophy: Plato's debt to a democratic strategy of civic discourse," in *Athenian Political Thought and the Reconstruction of American Democracy*, ed. J. P. Euben, J. R. Wallach, and J. Ober. Ithaca: 172–97.

Morello, R. (2002) "Livy's Alexander digression (9.17–19): counterfactuals and apologetics," *Journal of Roman Studies* 92: 62–85.

Moreno, P. (1974) *Lisippo*. Bari.

(1984) "L'immagine dell'uomo nella scultura e nella pittura: il divino, l'eroico, il quotidiano," in *Civiltà del mezzogiorno*, ed. S. Moscati. Cinisello Balsamo: 84–127.

Moret, J.-M. (1975) *L'Ilioupersis dans la céramique italiote: les mythes et leur expression figurée au IVe siècle*. Rome.

Morgan, J. R. (trans.) (1989) "Heliodorus: an ethiopian story," in *Collected Ancient Greek Novels*, ed. B. P. Reardon. Berkeley: 349–588.

Morris, S. and J. K. Papadopoulos (2005) "Greek towers and slaves: an archaeology of exploitation," *American Journal of Archaeology* 109: 155–225.

Morrow, G. R. (1960) *Plato's Cretan City: A Historical Interpretation of the Laws*. Princeton.

Morson, G. S. (1994) *Narrative and Freedom: The Shadows of Time*. New Haven.

Moss, J. (2013) "Soul-leading: the unity of the *Phaedrus* again," *Oxford Studies in Ancient Philosophy* 43: 1–22.

Mülke, C. (ed.) (2002) *Solons politische Elegien und Iamben (Fr. 1–13, 32–37 W.): Einleitung, Text, Übersetzung, Kommentar*. Munich.

Munn, M. H. (2000) *The School of History: Athens in the Age of Socrates*. Berkeley.

Murray, J. (1988) "Disputation, deception, and dialectic: Plato on the true rhetoric (*Phaedrus* 261–66)," *Philosophy and Rhetoric* 21: 279–89.

Mutschmann, H. (1918) "Die älteste Definition der Rhetorik," *Hermes* 53: 440–43.

Nagy, G. (1990) *Pindar's Homer: The Lyric Possession of an Epic Past*. Baltimore.

Neer, R. T. (1995) "The lion's eye: imitation and uncertainty in Attic red figure," *Representations* 51: 118–53.

(2002) *Style and Politics in Athenian Vase-Painting: The Craft of Democracy, ca. 530–460 BCE*. Cambridge.
(2010) *The Emergence of the Classical Style in Greek Sculpture*. Chicago.
Nehamas, A. and P. Woodruff (eds.) (1995) *Plato: Phaedrus*. Indianapolis.
Neils, J. (2004) "Vases on vases," in *Greek Vase Painting: Form, Figure, and Narrative*, ed. P. G. Warden. Dallas: 28–34.
Nelson, R. S. (2000) "Introduction: Descartes's cow and other domestications of the visual," in *Visuality Before and Beyond the Renaissance: Seeing As Others Saw*, ed. R. S. Nelson. Cambridge: 1–21.
Nesselrath, H.-G. (1992) *Ungeschehenes Geschehen: "Beinahe-Episoden" im griechischen und römischen Epos von Homer bis zur Spätantike*. Stuttgart.
Netz, R. (2009) "Imagination and layered ontology in Greek mathematics," *Configurations* 17: 19–50.
Newsom, R. (1988) *A Likely Story: Probability and Play in Fiction*. New Brunswick.
Nick, G. (2002) *Die Athena Parthenos: Studien zum griechischen Kultbild und seiner Rezeption*. Mainz.
Nickel, D. (ed.) (2001) *Galeni De foetuum formatione. Corpus Medicorum Graecorum* V 3,3. Berlin.
Nightingale, A. W. (1993) "Writing/reading a sacred text: a literary interpretation of Plato's *Laws*," *Classical Philology* 88: 279–300.
(2004) *Spectacles of Truth in Classical Greek Philosophy: Theoria in Its Cultural Context*. Cambridge.
Nightingale, A. W. and D. N. Sedley (eds.) (2010) *Ancient Models of Mind: Studies in Human and Divine Rationality*. Cambridge.
Nutton, V. (1983) "The seeds of disease: an explanation of contagion and infection from the Greeks to the Renaissance," *Medical History* 27: 1–34.
(ed.) (1999) *Galeni de Propriis Placitis*. Berlin.
Oates, J. F., R. S. Bagnall, S. J. Clackson *et al.* (eds.) (2013) *Checklist of Greek, Latin, Demotic and Coptic Papyri, Ostraca and Tablets*, http://scriptorium.lib.duke.edu/papyrustexts/clist.html, April, 2014.
Ober, J. (1989) *Mass and Elite in Democratic Athens: Rhetoric, Ideology, and the Power of the People*. Princeton.
(2008) *Democracy and Knowledge: Innovation and Learning in Classical Athens*. Princeton.
Oenbrink, W. (1997) *Das Bild im Bilde: Zur Darstellung von Göterstatuen und Kultbildern auf griechischen Vasen*. Frankfurt.
Ogilvie, B. (2004) "Truth in France," trans. S. Bhutani, in *Keywords: Truth*, ed. D. Posel. New York: 103–30.
Osborne, R. (1996) "Funerary monuments, the democratic citizen, and the representation of women," in *Démocratie athénienne et culture*, ed. M. B. Sakellariou. Athens: 229–42.
(2011) *The History Written on the Classical Greek Body*. Cambridge.
O'Sullivan, N. (1995) "Aristotle on dramatic probability," *Classical Journal* 91: 47–63.
Ott, S. (1979) "Aristotle among the Basques: the 'cheese analogy' of conception," *Man* 14: 699–711.

Page, D. L. (ed.) (1941) *Select Papyri*, vol. III: *Poetry*. Cambridge, Mass.

Pangle, T. (1976) "The political psychology of religion in Plato's *Laws*," *American Political Science Review* 70: 1059–77.

(ed.) (1980) *The Laws of Plato*. New York.

Parca, M. G. (1991) *Ptocheia, Or Odysseus in Disguise at Troy (P. Köln VI 245)*. Atlanta.

Parens, J. (1995) *Metaphysics as Rhetoric: Alfarabi's Summary of Plato's "Laws."* Albany.

Pasquali, G. (1962) *Storia della tradizione e critica del testo*. Florence.

Patey, D. L. (1984) *Probability and Literary Form: Philosophic Theory and Literary Practice in the Augustan Age*. Cambridge.

Pavel, T. G. (1986) *Fictional Worlds*. Cambridge, Mass.

Pelling, C. B. R. (2000) *Literary Texts and the Greek Historian*. London.

Perry, E. (2005) *The Aesthetics of Emulation in the Visual Arts of Ancient Rome*. Cambridge.

Platt, V. J. (2011) *Facing the Gods: Epiphany and Representation in Graeco-Roman Art, Literature and Religion*. Cambridge.

(forthcoming) "Double vision: epiphanies of the Dioscuri in Greece and Rome," in *Close Encounters: Human–Divine Relations in Greco–Roman Cult and Culture*, ed. G. Petridou, V. Platt, and S. Turner. Farnham.

Pollitt, J. J. (1974) *The Ancient View of Greek Art: Criticism, History, and Terminology*. New Haven.

(1986) *Art in the Hellenistic Age*. Cambridge.

Porter, J. I. (2011) "Making and unmaking: the Achaean wall and the limits of fictionality in Homeric criticism," *Transactions of the American Philological Association* 14: 1–36.

Pouilloux, J. and F. Salviat (1983) "Lichas, Lacédémonien, archonte à Thasos, et le livre VIII de Thucydide," *Comptes rendus de l'Académie des Inscriptions* 376–403.

(1985) "Thucydide après l'exil et la composition de son histoire," *Revue de philologie* 59: 13–20.

Powell, A. (ed.) (2013) *Hindsight, or the Importance of Unfulfilled Expectations in Greek and Roman History*. Swansea.

Powell, J. U. and E. A. Barber (eds.) (1929) *New Chapters in the History of Greek Literature, Second Series; Some Recent Discoveries in Greek Poetry and Prose, Chiefly of the Fourth Century BC, and Later Times*. Oxford.

Pucci, P. (1997) "The *Helen* and Euripides' 'comic' art," *Colby Quarterly* 33: 42–75.

Purves, A. C. (2010) *Space and Time in Ancient Greek Narrative*. New York.

Quillin, J. M. (2002) "Achieving amnesty: the role of events, institutions, and ideas," *Transactions of the American Philological Association* 132: 71–107.

Raaflaub, K. A. (2004) *The Discovery of Freedom in Ancient Greece*, trans. R. Franciscono. Chicago.

Radermacher, L. (ed.) (1951) *Artium scriptores: Reste der voraristotelischen Rhetorik*. Vienna.

Rapp, C. (ed. and trans.) (2002) *Aristoteles Rhetorik*, vol. II. Berlin.

Rawlings, H. R. (1981) *The Structure of Thucydides' History*. Princeton.
Ready, J. (2007) "Toil and trouble: the acquisition of spoils in the *Iliad*," *Transactions of the American Philological Association* 137: 3–43.
Reinhardt, T. (2010) "Plausibility in Plato's *Phaedrus* and the *Rhetorica ad Alexandrum*," *Museum Helveticum* 67: 1–6.
Reiss, H. S. (ed.) (1991) *Kant: Political Writings*. Cambridge.
Rhodes, P. J. and R. Osborne (eds.) (2003) *Greek Historical Inscriptions: 404–323 BC*. Oxford.
Richter, G. M. A. (1955) "The origin of verism in Roman portraits," *Journal of Roman Studies* 45: 39–46.
Ricoeur, P. (1984) *Time and Narrative*, vol. I, trans. K. McLaughlin and D. Pellauer. Chicago.
(1991) "Mimesis and representation," in *A Ricoeur Reader: Reflection and Imagination*, ed. M. J. Valdés. Toronto: 137–55.
Ridgway, B. S. (1981) *Fifth Century Styles in Greek Sculpture*. Princeton.
Roberts, D. (1987) "Parting words: final lines in Sophocles and Euripides," *Classical Quarterly* 37: 51–64.
Roberts, W. R. (trans.) (1954) *Aristotle Rhetoric*. New York.
Roese, N. J. and M. Morrison (2009) "The psychology of counterfactual thinking," *Historical Social Research* 35: 16–26.
Roese, N. J. and J. M. Olson (1995) "Counterfactual thinking: a critical overview," in *What Might Have Been: The Social Psychology of Counterfactual Thinking*, ed. N. J. Roese and J. M. Olson. Mahwah, NJ: 1–55.
Roisman, J. (2004) "Speaker–audience interaction in Athens: a power struggle," in *Free Speech in Classical Antiquity*, ed. I. Sluiter and R. M. Rosen. Leiden: 261–78.
Romilly, J. de (1963) *Thucydides and Athenian Imperialism*, trans. P. Thody. New York.
Rood, T. (1998) *Thucydides: Narrative and Explanation*. Oxford.
Rosenstock, B. (1994) "Socrates as revenant: a reading of the *Menexenus*," *Phoenix* 48: 331–47.
Rösler, W. (1980) "Die Entdeckung der Fiktionalität in der Antike," *Poetica* 12: 283–319.
Rowe, C. J. (ed.) (1986) *Plato: Phaedrus*. Warminster.
(2003a) "Socrates, the laws, and the *Laws*," in *Plato's Laws: From Theory into Practice*, ed. S. Scolnicov and L. Brisson. Sankt Augustin: 87–97.
(2003b) "The status of 'myth' in Plato's *Timaeus*," in *Plato physicus: cosmologia e antropologia nel Timeo*, ed. C. Natali and S. Maso. Amsterdam: 21–31.
(2010) "The relationship of the *Laws* to other dialogues: a proposal," in *Plato's Laws: A Critical Guide*, ed. C. Bobonich. Cambridge: 29–50.
Russo, J., M. Fernandez-Galiano, and A. Heubeck (eds.) (1992) *A Commentary on Homer's Odyssey*, vol. III. Oxford.
Salkever, S. (1993) "Socrates' Aspasian oration: the play of philosophy and politics in Plato's *Menexenus*," *American Political Science Review* 87: 133–43.

Sambursky, S. (1956) "On the possible and the probable in ancient Greece," *Osiris* 12: 35–48.
Sande, S. (1995) "An old hag and her sisters," *Symbolae Osloenses* 70: 30–53.
Saunders, T. J. (1991) *Plato's Penal Code: Tradition, Controversy, and Reform in Greek Penology*. Oxford.
Scarborough, J. (1985) "Erasistratus, student of Theophrastus?," *Bulletin of the History of Medicine* 59: 515–17.
Scheer, T. S. (2000) *Die Gottheit und ihr Bild: Untersuchungen zur Funktion griechischer Kultbilder in Religion und Politik*. Munich.
Schefold, K. (1937) "Statuen auf Vasenbildern," *Jahrbuch des deutschen archäologischen Instituts* 52: 30–75.
Schiappa, E. (1999) *The Beginnings of Rhetorical Theory in Classical Greece*. New Haven.
Schiefsky, M. (2007) "Galen's teleology and functional explanation," *Oxford Studies in Ancient Philosophy* 33: 369–400.
Schmitt, A. (1990) *Selbstständigkeit und Abhängigkeit menschlichen Handelns bei Homer: Hermeneutische Untersuchungen zur Psychologie Homers*. Stuttgart.
Schmitz, T. A. (2000) "Plausibility in the Greek orators," *American Journal of Philology* 121: 47–77.
Schneider-Hermann, G. (1972) "Kultstatue im Tempel auf italischen Vasenbildern," *Bulletin Antieke Beschaving* 47: 31–42.
Schofield, M. (2003) "Religion and philosophy in the *Laws*," in *Plato's Laws: From Theory into Practice*, ed. S. Scolnicov and L. Brisson. Sankt Augustin: 1–13.
(2006) *Plato: Political Philosophy*. Oxford.
(2010) "The *Laws'* two projects," in *Plato's Laws: A Critical Guide*, ed. C. Bobonich. Cambridge: 12–28.
Scodel, R. (1982) "The Achaean wall and the myth of destruction," *Harvard Studies in Classical Philology* 86: 33–50.
(1990) "Euripides and *apatē*," in *Cabinet of the Muses: Essays on Classical and Comparative Literature in Honor of Thomas G. Rosenmeyer*, ed. M. Griffith and D. J. Mastronarde. Atlanta: 75–87.
(1999) *Credible Impossibilities: Conventions and Strategies of Verisimilitude in Homer and Greek Tragedy*. Stuttgart.
Scullion, S. (1999–2000) "Tradition and invention in Euripidean aitiology," in *Euripides and Tragic Theatre in the Late Fifth Century. Illinois Classical Studies* 24–25, ed. M. Cropp, K. Lee, and D. Sansone. 217–33.
Segal, C. (1971) "The two worlds of Euripides' *Helen*," *Transactions of the American Philological Association* 102: 553–614.
Shapiro, B. J. (1983) *Probability and Certainty in Seventeenth-Century England: A Study of the Relationships between Natural Science, Religion, History, Law, and Literature*. Princeton.
Shapiro, H. A. (1989) *Art and Cult under the Tyrants in Athens*. Mainz am Rhein.
Shear, J. L. (2011) *Polis and Revolution: Responding to Oligarchy in Classical Athens*. Cambridge.

Shields, C. (1988) "Soul and body in Aristotle," *Oxford Studies in Ancient Philosophy* 6: 103–37.
Sinos, D. S. (1980) *Achilles, Patroklos, and the Meaning of Philos*. Innsbruck.
Smith, J. E. (1985) "Plato's myths as 'likely accounts,' worthy of belief," *Apeiron* 19: 24–42.
Smith, N. D. (1983) "Plato and Aristotle on the nature of women," *Journal of the History of Philosophy* 21: 467–78.
Smith, R. R. R. (1981) "Greeks, foreigners, and Roman republican portraits," *Journal of Roman Studies* 71: 24–38.
Smith, R. (ed.) (1997) *Aristotle Topics, Books I and VIII*. New York.
Sourvinou-Inwood, C. (1995) *"Reading" Greek Death: To the End of the Classical Period*. Oxford.
Squire, M. (2009) *Image and Text in Graeco-Roman Antiquity*. Cambridge.
Stalley, R. F. (1983) *An Introduction to Plato's Laws*. Oxford.
Stears, K. (1995) "Dead women's society: constructing female gender in classical Athenian funerary sculpture," in *Time, Tradition, and Society in Greek Archaeology: Bridging the "Great Divide,"* ed. N. Spencer. London: 109–31.
(2000) "The times they are a'changing: developments in fifth-century Athenian funerary monuments," in *The Epigraphy of Death: Studies in the History and Society of Greece and Rome*, ed. G. J. Oliver. Liverpool: 25–58.
(2001) "Spinning women: iconography and status in Athenian funerary sculpture," in *Les pierres de l'offrande: autour de l'œuvre de Christoph W. Clairmont*, ed. G. Hoffmann and A. Lezzi-Hafter. Zurich: 107–14.
Ste. Croix, G. E. M. de (1992) "Aristotle on history and poetry (*Poetics*, 9, 1451a36–b11)," in *Essays on Aristotle's Poetics*, ed. A. O. Rorty. Princeton: 23–32.
Steiner, D. (2001) *Images in Mind: Statues in Archaic and Classical Greek Literature and Thought*. Princeton.
Stewart, A. F. (1990) *Greek Sculpture: An Exploration*. New Haven.
(1997) *Art, Desire and the Body in Ancient Greece*. Cambridge.
Strauss, L. (1975) *The Argument and the Action of Plato's Laws*. Chicago.
Suerbaum, W. (1997) "Am Scheideweg zur Zukunft: Alternative Geschehensverläufe bei römischen Historikern," *Gymnasium* 104: 36–54.
Sullivan, S. D. (1989) "A study of the psychic term *NOOS* in the Greek lyric poets (excluding Pindar and Bacchylides)," *Emerita* 57: 129–68.
Sussman, L. A. (ed.) (1994) *The Declamations of Calpurnius Flaccus: Text, Translation, and Commentary*. Leiden.
Sutton, D. F. (1987) *Papyrological Studies in Dionysiac Literature: P. Lit. Lond. 77: A Postclassical Satyr Play and P. Ross. Georg. 1.11: A Hymn to Dionysus*. Oak Park, Ill.
Suzuki, M. (1989) *Metamorphoses of Helen: Authority, Difference, and the Epic*. Ithaca.
Synodinou, K. (1981) *Ἔοικα-Εἰκὸς Καὶ Συγγενικὰ Ἀπὸ Τὸν Ὅμερο Ὡς Τὸν Ἀριστοφάνη. Σημασιολογικὴ Μελέτη*. Ioannina.

Tanner, J. (2000) "Portraits, power, and patronage in the late Roman Republic," *Journal of Roman Studies* 90: 18–50.

(2006) *The Invention of Art History in Ancient Greece: Religion, Society and Artistic Rationalisation*. Cambridge.

Taplin, O. (1992) *Homeric Soundings: The Shaping of the Iliad*. New York.

Taplin, O. and R. Wyles (eds.) (2010) *The Pronomos Vase and Its Context*. Oxford.

Taylor, R. M. (2008) *The Moral Mirror of Roman Art*. New York.

Temkin, O. (trans.) (1991) *Soranus' Gynecology*. Baltimore.

Tetlock, P. E., R. N. Lebow, and G. Parker (eds.) (2006) *Unmaking the West: "What-if" Scenarios That Rewrite World History*. Ann Arbor.

Thompson, E. P. (1978) *The Poverty of Theory and Other Essays*. New York.

Thompson, W. H. (ed.) (1868) *The Phaedrus of Plato*. London.

Tindale, C. W. (2010) *Reason's Dark Champions: Constructive Strategies of Sophistic Argument*. Columbia, SC.

Todd, S. C. (trans.) (2000) *Lysias*. Austin.

Todisco, L. (1990) "Eracle, la statua, l'artefice sul cratere apulo di New York MMA 50.11.4," *Mélanges de l'École française de Rome* 102: 901–57.

Toynbee, A. (1987) *A Study of History*. 2 vols. New York.

Trendall, A. D. (1989) *Red Figure Vases of South Italy and Sicily: A Handbook*. New York.

Trendall, A. D. and A. Cambitoglou (1978) *The Red-Figured Vases of Apulia*. 3 vols. Oxford.

Tsagalis, C. (2008) *Inscribing Sorrow: Fourth-Century Attic Funerary Epigrams*. Berlin.

Turner, S. M. (2009) "Classical Attic grave stelai: gender, death and the viewer." PhD dissertation. Cambridge University.

Vallance, J. T. (1993) "The medical system of Asclepiades of Bithynia," *Aufstieg und Niedergang der römischen Welt* 2.37.1: 693–727.

Van Harten, A. (2003) "Creating happiness: the moral of the myth of Kronos in Plato's *Laws* (*Laws* 4, 713b-714a)," in *Plato's Laws: From Theory into Practice*, ed. S. Scolnicov and L. Brisson. Sankt Augustin: 128–38.

Vegetti, M. (1999) "Historiographical strategies in Galen's physiology (*de Usu Partium, de Naturalibus Facultatibus*)," in *Ancient Histories of Medicine: Essays in Medical Doxography and Historiography in Classical Antiquity*, ed. P. J. van der Eijk. Leiden: 383–95.

Vernant, J.-P. (1990) "Figuration et image," *Mètis* 5: 225–38.

(1991) *Mortals and Immortals: Collected Essays*, ed. F. I. Zeitlin. Princeton.

Veyne, P. (1984) *Writing History: Essay on Epistemology*. Middletown, Conn.

Vickers, B. (1988) *In Defence of Rhetoric*. New York.

Volk, K. (2002) "ΚΛΕΟΣ ΑΦΘΙΤΟΝ revisited," *Classical Philology* 97: 61–68.

Von Fritz, K. (1974) "*Nous, noein*, and their derivatives in pre-Socratic philosophy (excluding Anaxagoras)," in *The Pre-Socratics: A Collection of Critical Essays*, ed. A. P. D. Mourelatos. Garden City, NY: 23–85.

von Staden, H. (1997) "Teleology and mechanism: Aristotelian biology and early Hellenistic medicine," in *Aristotelische Biologie: Intentionen, Methoden, Ergebnisse*, ed. W. Kullmann and S. Föllinger. Stuttgart: 183–208.

(2000) "Body, soul, and nerves: Epicurus, Herophilus, Erasistratus, the Stoics and Galen," in *Psyche and Soma: Physicians and Metaphysicians on the Mind–Body Problem from Antiquity to Enlightenment*, ed. J. P. Wright and P. Potter. Oxford: 79–116.

(2002) "Ὡς ἐπὶ τὸ πολύ: 'Hippocrates' between generalization and individualization," in *Le normal et le pathologique dans la Collection hippocratique: Actes du Xème colloque internationale hippocratique*, ed. A. Thivel and A. Zucker. Nice: 23–43.

Wallace, R. W. (1996) "Law, freedom, and the concept of citizens' rights in democratic Athens," in *Dēmokratia: A Conversation on Democracies, Ancient and Modern*, ed. J. Ober and C. Hedrick. Princeton: 105–19.

(2004) "The power to speak – and not to listen – in ancient Athens," in *Free Speech in Classical Antiquity*, ed. I. Sluiter and R. M. Rosen. Leiden: 221–32.

Watt, I. P. (1957) *The Rise of the Novel: Studies in Defoe, Richardson, and Fielding*. Berkeley.

Weber, H. (2009) "The 'but for' test and other devices – the role of hypothetical events in the law," *Historical Social Research* 34: 118–28.

Werner, D. S. (2012) *Myth and Philosophy in Plato's Phaedrus*. Cambridge.

West, M. L. (1969) "The Achaean wall," *Classical Review* 19: 255–60.

(ed.) (1971) *Iambi et Elegi Graeci Ante Alexandrum Cantati*. Oxford.

Westlake, H. D. (1958) "Ὡς εἰκός in Thucydides," *Hermes* 86: 447–52.

White, H. V. (1978) *Tropics of Discourse: Essays in Cultural Criticism*. Baltimore.

(1987) *The Content of the Form: Narrative Discourse and Historical Representation*. Baltimore.

Will, W. (2000) "Perikles: eine Konjektural-Biographie des Thukydides," in *Virtuelle Antike: Wendepunkte der alten Geschichte*, ed. K. Brodersen. Darmstadt: 27–36.

Wohl, V. (2002) *Love among the Ruins: The Erotics of Democracy in Classical Athens*. Princeton.

(2010) *Law's Cosmos: Juridical Discourse in Athenian Forensic Oratory*. Cambridge.

Wolpert, A. (2002a) *Remembering Defeat: Civil War and Civic Memory in Ancient Athens*. Baltimore.

(2002b) "Lysias 18 and the Athenian memory of the civil war," *Transactions of the American Philological Association* 132: 109–26.

Woodruff, P. (1992) "Aristotle on *mimēsis*," in *Essays on Aristotle's Poetics*, ed. A. O. Rorty. Princeton: 73–95.

(1994) "*Eikos* and bad faith in the paired speeches of Thucydides," *Proceedings of the Boston Area Colloquium in Ancient Philosophy* 10: 115–45.

Woods, J. H. (1974) *The Logic of Fiction: A Philosophical Sounding of Deviant Logic*. The Hague.

Worman, N. (1997) "The body as argument: Helen in four Greek texts," *Classical Antiquity* 16: 151–203.
Wright, M. (2005) *Euripides' Escape-Tragedies: A Study of Helen, Andromeda, and Iphigenia among the Taurians*. Oxford.
Younger, J. (2002) "Women in relief: 'double consciousness' in classical Attic tombstones," in *Among Women: From the Homosocial to the Homoerotic in the Ancient World*, ed. N. S. Rabinowitz and L. Auanger. Austin: 167–210.
Youtie, H. C. and R. Merkelbach (1968) "Ein Michigan-Papyrus über Theramenes," *Zeitschrift für Papyrologie und Epigraphik* 2: 161–69.
Youtie, H. C. (1979) "P. Mich. Inv. 341: a price of wheat," *Zeitschrift für Papyrologie und Epigraphik* 36: 77–81.
Yunis, H. (1996) *Taming Democracy: Models of Political Rhetoric in Classical Athens*. Ithaca.
(2009) "Dialectic and the purpose of rhetoric in Plato's *Phaedrus*," *Proceedings of the Boston Area Colloquium in Ancient Philosophy* 24: 229–48.
(ed.) (2011) *Plato: Phaedrus*. Cambridge.
Zanker, G. (1987) *Realism in Alexandrian Poetry: A Literature and Its Audience*. London.
Zeyl, D. (trans.) (2000). *Plato: Timaeus*. Indianapolis, Ind.
Zuckert, C. H. (2009) *Plato's Philosophers: The Coherence of the Dialogues*. Chicago.

Index locorum

General index